ELECTRICITY, ELECTRONICS, AND WIRING DIAGRAMS FOR HVAC/R

SECOND EDITION

ELECTRICITY, ELECTRONICS, AND WIRING DIAGRAMS FOR HVAC/R

Air Conditioning and Refrigeration Institute

Edward F. Mahoney

PEARSON
Prentice Hall

Upper Saddle River, New Jersey
Columbus, Ohio

Library of Congress Cataloging-in-Publication Data

Mahoney, Edward F.
    Electricity, electronics, and wiring diagrams for HVAC/R / Edward F. Mahoney.—2nd ed.
        p. cm.
    Includes index.
    ISBN 0-13-119085-7
    1. Air conditioning—Electric equipment.   2. Refrigeration and refrigerating machinery—
Electric equipment.   I. Title.
    TK4035.A35M329 2006
    697—dc22

                                                                                    2005007743

Production Editor: Christine Buckendahl
Production Coordination: TechBooks/York, PA Campus
Design Coordinator: Diane Einsberger
Cover Design: Jeff Vanik
Production Manager: Deidra Schwartz
Marketing Manager: Mark Marsden

This book was set in Century Schoolbook by TechBooks/York, PA Campus. It was printed
and bound by Banta Book Group. The cover was printed by Coral Graphic Services, Inc.

Pearson Education LTD.   Pearson Education Australia PTY, Limited
Pearson Education Singapore, Pte. Ltd   Pearson Education North Asia Ltd
Pearson Education, Canada, Ltd   Pearson Educación de Mexico, S.A. de C.V.
Pearson Education–Japan   Pearson Education Malaysia, Pte. Ltd

10 9 8 7 6 5 4 3 2
ISBN 0-13-119085-7

PREFACE

This book is published in two main sections. The first section covers basic electricity and basic electronics related to the needs of air-conditioning refrigeration technicians. The second section of the book covers practical circuits and systems.

In the recent past, there has been an increase in the importance of electricity and electronics in the control and protection of air-conditioning and refrigerator devices.

It is not sufficient to cover electrical concepts lightly. Your career and success might well be based on how well you understand the material presented in the first section of this book.

With a firm foundation in basic concepts, the remainder of the book and the more complex circuits and problems that you will encounter in the field will not prove difficult.

If you master the basics, the rest is easy. This book provides you with the means to do just that.

Thanks to the reviewers of this text for their helpful comments and suggestions: Salvatore Benevegna, Albuquerque Technical Vocational Institute; Arthur T. Miller, Community College of Allegheny County; and Thomas Niesen, Gateway Technical College.

This book is dedicated to my wife, Gloria,
whose patience with me is beyond understanding.
Her knowledge and abilities in grammar, spelling,
and punctuation, as well as her typing skills
made the completion of this project possible.

BRIEF CONTENTS

CONTENTS

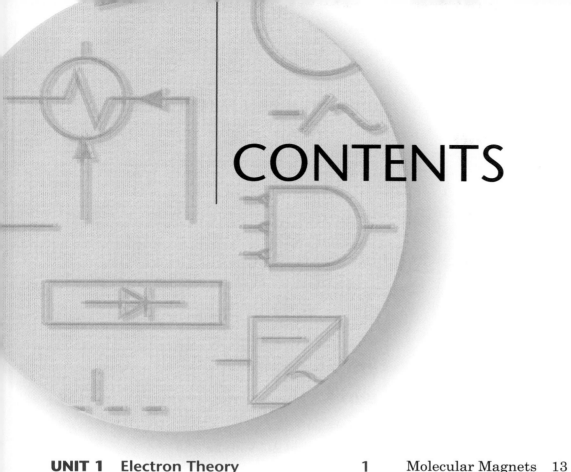

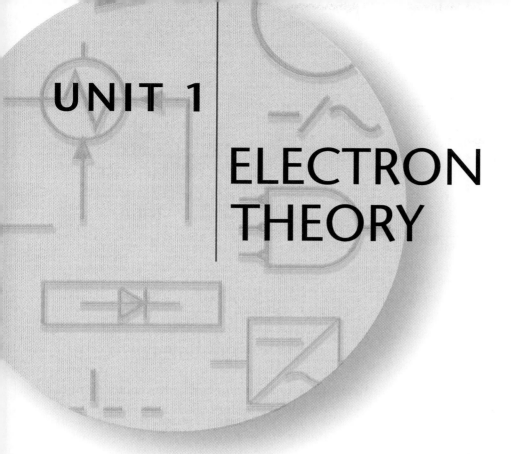

UNIT 1

ELECTRON THEORY

OBJECTIVES

After completion and review of this unit, you should be able to:

- Relate electricity to things common in your environment.
- Picture in your mind electrons and electrons moving.
- Define the terms *voltage* and *current.*
- Recognize the difference between an open and a closed circuit.

The information presented in this unit is theoretical. *Theory* is a line of reasoning that is assumed to be correct. Theory may change from time to time as new information is gathered and new lines of reasoning are formed.

Electrical theory is highly developed at the present time and requires knowledge of higher mathematics for complete understanding. However, mathematical explanations of electricity are not the concern of this text; instead, an elementary approach to the electrical concepts, theories, and formulas that are essential to the air-conditioning and refrigeration technician will be provided.

It is suggested that this unit be read as a story. The material presented is good background information that will assist the reader in understanding the nature of electricity. With a g ood grasp of the nature of electricity, an air-conditioning and refrigeration technician will find troubleshooting an easier task.

THE STRUCTURE OF MATTER

A negatively charged, tiny object called an **electron** must be considered to understand the behavior of electricity. This consideration requires an investigation into the structure of matter—of what things are really made.

Imagine looking at a piece of copper under a microscope as the magnification is increased to an extremely high magnitude. It might be possible to see the actual construction of copper.

With the microscope at ×10 power, the copper looks very much like copper (Figure 1–1). As the magnification is increased to ×100, the rough crystalline

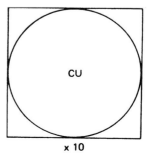

× 10

FIGURE 1–1 Plain copper (Cu).

1

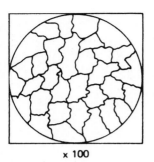

× 100

FIGURE 1–2   Copper (Cu) crystal.

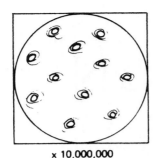

× 10,000,000

FIGURE 1–3   The atomic structure of copper (Cu).

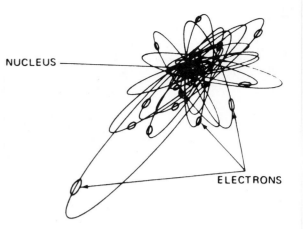

COPPER ATOM

FIGURE 1–5   A copper atom.

structure of copper is seen (Figure 1–2). It takes a large jump in magnification to ×10,000,000 before the beginning of atomic structure becomes evident, as seen in the bumpy surface shown in Figure 1–3. Finally, at a magnification of ×100,000,000, individual **atoms** are seen (Figure 1–4).

What is there to see? The center of a copper atom, the **nucleus**, consists of positively charged and neutral particles—**protons** and **neutrons**. Negatively charged particles, electrons, whirl in four orbits around the center. Electron theory is based on these negatively charged particles. An accepted representation of a copper atom is shown in Figure 1–5.

In the study of **electricity**, the electrons whirling around the nucleus are the center of interest. Moving these electrons on command produces electricity that can be used to do work.

BOHR'S LAW

As scientists experimented with things of an electrical nature, they learned about the characteristics of electricity and about electrical phenomena. Physicist Niels H. D. Bohr (1885–1962) developed a model of electron, proton, and neutron arrangements to represent everything in the universe; this became known as Bohr's quantum theory of atomic structure.

According to Bohr, an atom of hydrogen, the lightest known element, consists of one proton in the nucleus and one electron rotating around it (Figure 1–6). Bohr's theory of the hydrogen atom is regarded as the basis of modern atomic nuclear physics.

As we move up the atomic scale, heavier materials have a greater number of particles in the nucleus and a greater number of electrons whirling around that center. As the theory developed and is now understood, the arrangement of electrons in rings (orbits) around the nucleus is fixed:

1. In the first orbit around the nucleus, there can be no more than 2 electrons.

2. In the second orbit, the maximum is 8 electrons.

3. The third orbit has a maximum of 18 electrons.

4. The fourth orbit has a maximum of 32 electrons.

CONDUCTORS

As indicated, the number of electrons in each orbit around the nucleus of an atom is fixed. The number of electrons in orbit and the number of protons at the center determines the type of material.

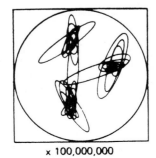

× 100,000,000

FIGURE 1–4   Copper (Cu) atoms.

HYDROGEN ATOM

FIGURE 1–6   A hydrogen atom.

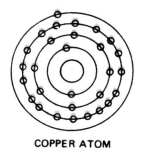

COPPER ATOM

FIGURE 1–7   Electron configuration of a copper atom.

Copper, for example, has a nucleus of 29 protons and 34 neutrons with 29 electrons rotating around it. According to the fixed arrangement of electrons, there are 2 in the first orbit, 8 in the second orbit, and 18 in the third orbit. This accounts for only 28 electrons. The remaining electron orbits alone in the fourth ring (see Figure 1–7). Because this electron is alone in the fourth ring, it is not held as strongly to the nucleus as are the other electrons. This electron is "free" to move whenever an outside force acts on it. Copper is a good conductor of electricity because each of the billions of tiny atoms in a copper conductor has one free electron ready to move when a force is applied to it. A reasonably small copper wire will pass a relatively large amount of electricity.

Another example of a good conductor is aluminum. The aluminum atom has a total of 13 electrons in three orbits. Two are in the first orbit, eight in the second orbit, and three in the third orbit

(Figure 1–8). There are positions for 18 electrons in the third orbit. The three electrons in the third orbit are not tightly bound to the nucleus, and the third orbit can momentarily accommodate extra electrons. Aluminum, therefore, is a good conductor of electricity, but not as good as copper. If the same size of copper and aluminum wire is used, copper will conduct better. If weight and cost are factors to be considered, aluminum—a lightweight metal—is an excellent choice. Using wire of equal length and weight (not size), aluminum will conduct electricity twice as well as copper. With equal weights, the aluminum wire will be much larger, of course.

THE BEHAVIOR OF ELECTRONS

After walking across a rug or sliding out of a car, you sometimes have the strange ability to cause an electric spark when you touch some other object. When you have that ability, you are electrically charged, or you carry an electric charge. This same type of electrical charge can be demonstrated by running a plastic comb through dry hair. The comb will attract lightweight objects, such as small scraps of paper or thread.

Static Electricity

The electricity that is obtained by rubbing certain materials together is called **static electricity**. It is called static because electrons are picked up on an

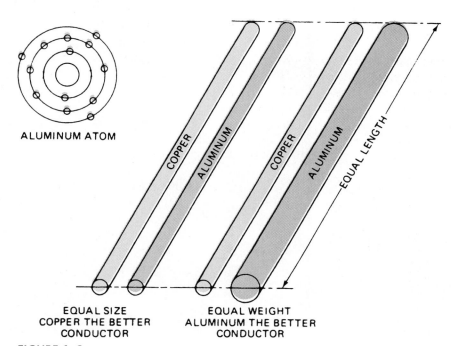

ALUMINUM ATOM

EQUAL SIZE COPPER THE BETTER CONDUCTOR

EQUAL WEIGHT ALUMINUM THE BETTER CONDUCTOR

FIGURE 1–8   Relation of weight and size of aluminum and copper conductors.

item, and the electrical charge remains on the item until it touches some other object that does not have a like charge.

As experiments were performed on static electricity, it was found that two types of charges could be obtained, depending on the nature of the materials rubbed together: If two glass rods are rubbed with a silk cloth, they are charged alike and repel each other; if two hard rubber rods are rubbed with wool, they are also charged alike and also repel each other.

CONCLUSION: Like charges repel each other.

If a charged glass rod is brought near a charged rubber rod, the glass rod and rubber rod will be attracted to each other. The charges on the rods are unlike each other, so they attract each other.

CONCLUSION: Unlike charges attract each other.

In order to classify the charges, they were given names. Benjamin Franklin (1706–1790), an American scientist, suggested **positive** and **negative** charges. The glass rods have a positive charge; the rubber rods have a negative charge.

Static electricity is electricity at rest.

Dynamic Electricity

Static electricity is not a practical form of electricity. The more practical form of electricity that is used to provide power and energy all over the world is **dynamic electricity**.

Dynamic electricity is electricity in motion. In an automobile, the battery provides dynamic electricity, as does the **alternator (generator)**. A power company uses water at high pressure behind dams to move large alternators that provide dynamic electricity. In some areas where there are no rivers for dams, large steam plants are used to turn alternators that generate dynamic electricity.

A new kind of power plant that uses atomic energy to provide the heat needed for steam turbine operation has been constructed in many areas of the country. One example is located at Turkey Point, Florida; it produces much of the electricity used in south Florida (Figure 1–9).

FIGURE 1–9   Turkey Point, Florida, power plant. (Photo courtesy of Florida Power & Light Company)

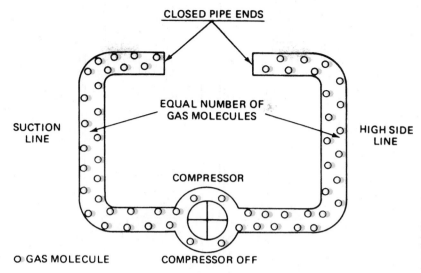

FIGURE 1–10 A balanced system.

Another energy source that may be of greater importance in the future, geothermal energy, is presently being investigated. To tap this energy source, a deep hole is drilled into the earth to an area of intense heat. Water is introduced into the hole, where it is converted to superheated steam at the lower level. The steam returns to the earth's surface to operate turbines that rotate the electric alternators that provide electricity.

Regardless of the energy source, dynamic electricity is produced by generators or alternators converting mechanical energy into electrical energy.

VOLTAGE AND CURRENT

A well-developed understanding of the basic terms relating to dynamic electricity is necessary to troubleshoot air-conditioning and refrigeration electric systems effectively.

Voltage and *current* are terms associated with dynamic electricity. Voltage and current will be covered together, because it is not possible to obtain one without some of the other.

Voltage is the pressure that causes electrons to move.

Current is the movement of electrons.

The Compressor Analogy of Electricity

As is often the case, it is easier to understand a new subject when it is related to some past experience. The electrical phenomena will be related to an air-conditioning compressor system.

In this analogy, it is not necessary to be concerned with temperature changes in the refrigeration system; only the gas movement need be considered. Figure 1–10 shows a compressor with inlet (suction) and outlet (discharge) lines attached. The lines are sealed at the ends so that no gas can enter or exit the system. When the compressor is not running, the system will be in balance with an equal number of gas molecules on either side. Balance is the state in nature that everything is trying to reach.

Figure 1–11 shows the compressor in operation. Some of the gas molecules are moved by the compressor from the suction line (on the left) to the discharge side (on the right). There is a relative excess of gas molecules in the discharge line (compression), causing a pressure difference between the two lines.

If a small pipe is connected from the discharge line to the suction line, as shown in Figure 1–12, some of the gas molecules will move through the pipe. There will be continuous movement of gas molecules through the system as long as the compressor is operating. The suction side of the system will continue to be under a vacuum, and the discharge side will continue at high pressure as long as the small tubing offers some restriction (resistance) to the movement (flow) of gas molecules through the system.

This system should seem reasonable to the student of refrigeration and air-conditioning technology, because it is similar to the basic system most commonly associated with the refrigeration cycle. It is equally reasonable to use a similar system to understand how electricity works.

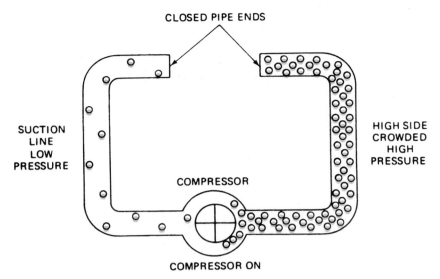

CLOSED PIPE ENDS

SUCTION
LINE
LOW
PRESSURE

HIGH SIDE
CROWDED
HIGH
PRESSURE

COMPRESSOR

COMPRESSOR ON

FIGURE 1–11   An unbalanced system.

In Figure 1–13, an electric generator has been substituted for the compressor and the air lines have been replaced with copper wire. The circuit is open because there is no connection between the wire from the left side of the generator and the wire from the right side. Within the copper wire, there are a large number of copper atoms. Each atom has one free electron (easily moved) in its outer orbit.

The electrical generator, when running, is capable of moving electrons. This is similar to the compressor's ability to move gas molecules. How the generator moves electrons is covered in later units.

When the generator is not running, there are an equal number of free electrons in the wires on either side of the generator: The system is in balance. When turned on, the generator draws some of the free electrons from the left side and pushes them to

the right side, as shown in Figure 1–14. The movement of electrons from one side to the other creates an unbalance. As long as there is an unbalance, there will be a pressure called voltage. The pressure is measured in **volts** and is exerted in an effort to try to balance the system. Another term for voltage is **electromotive force**, abbreviated **EMF**.

Because there is an excess of electrons on the right side, this side of the generator is negative (or we could say there is an absence of electrons on the left side). Because the nucleus of the copper atoms has not changed, the protons remain the same on both sides of the generator. The situation now is that there are more protons on the left side than electrons. Thus, the left side of the generator is positive. As long as the generator is operating, the system will remain unbalanced; voltage will exist between any point on the right side and any point on the left side.

If a voltage-measuring device (called a **voltmeter**) is connected across the open wire ends, the meter needle will move (Figure 1–15). This indicates the unbalance in electrical pressure that exists between the two points. Voltage is always measured between two points in a circuit. (See Unit 3.)

> **WARNING:** Never connect a wire across the output terminals of a generator. The wire will short out the generator. It is used here only as a means of explanation.

If a small wire is connected between the left side of the generator and the right side, some of the excess electrons will move from the right side through the wire to the left side (Figure 1–16). The generator will

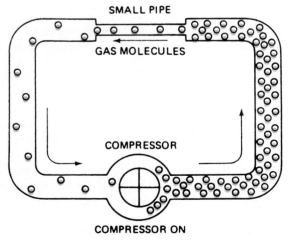

SMALL PIPE

GAS MOLECULES

COMPRESSOR

COMPRESSOR ON

FIGURE 1–12   Molecule movement.

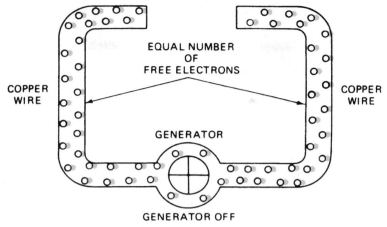

FIGURE 1–13 Electrical circuit in balance.

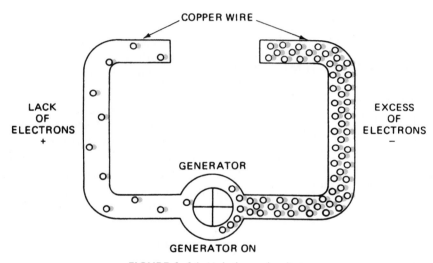

FIGURE 1–14 Unbalanced voltage.

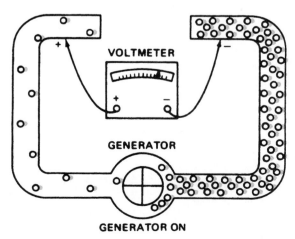

FIGURE 1–15 Measurement of voltage.

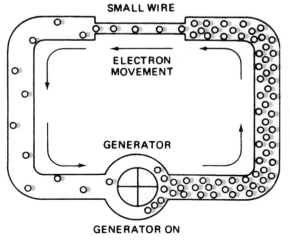

FIGURE 1–16 Electron flow, complete circuit.

continue to move electrons, just as the compressor moves gas. An electric current will flow in the circuit as long as the circuit is complete and the generator is running. As long as there is an unbalance and a circuit, current will flow. Current, or electrons in motion, is measured in **amperes**, commonly referred to as **amps**. It takes 6,250,000,000,000,000,000 (6.25×10^{18}) electrons passing a point in one second to make 1 ampere. This quantity is called a **coulomb**.

Voltage is caused by an imbalance in electrons. It is not the electrons but the electrons that have moved that create the unbalance. Similarly, the current in the circuit is the electrons in motion.

SUMMARY

- Whenever there is an unbalance in an electrical system, a pressure will exist trying to balance the system.

- In an electrical system the unbalance, or pressure between the two points in the system, is the electromotive force (EMF)—most often called voltage. Remember that voltage is measured between two points.

- If a circuit is provided between the unbalanced points, current (a movement of electrons) will flow through the circuit.

- If a circuit is not connected between the unbalanced points (an open circuit), no current will flow.

PRACTICAL EXPERIENCE

Required equipment   A plastic or rubber pocket comb and paper

Exercise 1

1. Tear a small piece of paper into smaller pieces, approximately a half-inch (12–13 mm) square.
2. Pass the comb through your hair several times.
3. Place the comb near the pieces of paper.

Quiz

1. Did the comb attract the pieces of paper?
   a. If no, before proceeding, determine why not.
   b. If yes, proceed with question 2.
2. What do you think caused this attraction?
3. Is this exercise an example of static or dynamic electricity?
4. Explain your answer to question 3.

FIGURE 1–17   A 120 V power plug.

Exercise 2

1. On a wall-type air-conditioning unit, locate the power plug.
2. Observe the shape of the power plug.
3. Draw the shape of the power plug on a separate sheet of paper.
4. Locate the nameplate on the air-conditioning unit.
5. Find the voltage rating of the unit on that nameplate.
6. If the voltage rating of the unit is 120 volts (V), the power plug should be shaped like a large, standard home appliance plug such as the one shown in Figure 1–17.
7. Is that the case?
8. If the voltage rating of the unit as indicated on the nameplate were 220 V at less than 15 amperes, the plug would be similar to the ones shown in Figures 1–18 and 1–19.
9. Examine other air-conditioning unit plugs.
10. Determine the voltage and current limitations of the plugs by their shape.
11. Check the air-conditioner nameplate for voltage and current levels.

FIGURE 1–18   A 220 V power plug.

FIGURE 1–19   Another example of a 220 V power plug.

12. Visit a hardware or home maintenance store. Check in the electrical department for different types of outlets and plugs and find their voltage and current rating.

REVIEW QUESTIONS

1. Static electricity is electricity at rest.
   T_____   F_____

2. Dynamic electricity is electricity in motion.
   T_____   F_____

3. Like charges repel each other.
   T_____   F_____

4. Unlike charges repel each other.
   T_____   F_____

5. Voltage is a pressure that causes electrons to move. T_____   F_____

6. Current is the movement of electrons.
   T_____   F_____

7. The natural state is for everything to be in a balanced condition.
   T_____   F_____

8. Current is measured in amperes.
   T_____   F_____

9. Electromotive force, EMF, is measured in volts.
   T_____   F_____

10. With equal size and length of conductors, copper is a better conductor than aluminum.
    T_____   F_____

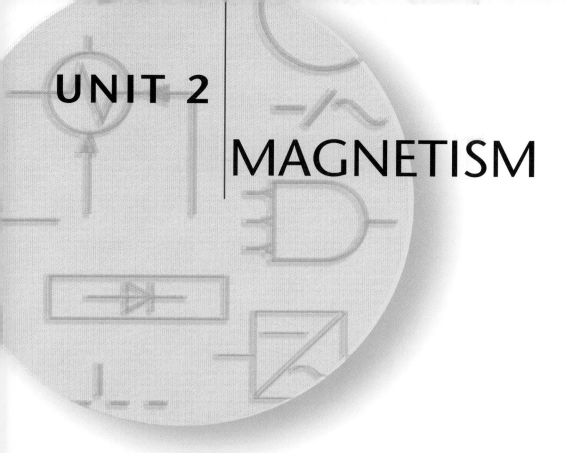

UNIT 2 | MAGNETISM

OBJECTIVES

After completion and review of this unit, you should be able to:

- Understand the relationship of natural magnets and artificial magnets.
- Understand the relationship of magnetic polarity and magnetic fields.
- Understand the relationship of electromagnets and magnetic strength.

Magnets and magnetism are involved in many components of air-conditioning and refrigeration electrical circuits. Compressors are driven by electric motors that function through magnetism. Many electrical controls operate on magnetic principles. A basic understanding of magnetic principles and magnetic circuits will be helpful to all air-conditioning and refrigeration technicians.

NATURAL MAGNETS

Ancient peoples knew of the special properties of certain stones, notably those found in Magnesia, Asia Minor, that attracted small bits of iron. The stones are called **magnets**, after the location in which they were found. Because most ancient peoples had little

scientific knowledge, the action of the magnets was attributed to magic. Natural magnets are stones with a high iron-ore content. Because of its special characteristics, this type of stone was given the name *magnetite*.

Around the year AD 1000, the property of natural magnets to point in a fixed direction was discovered. Because these stones were now useful in navigation, they were given a new name, *loadstone* (also spelled **lodestone**). The name loadstone means "leading stone" and is still used to indicate special iron ore with **magnetic** characteristics.

ARTIFICIAL MAGNETS

A piece of iron or steel can be magnetized by placing it in contact with a natural magnet or loadstone. Another method of producing an artificial magnet is to wrap a coil of wire around an iron or steel core and then pass an electrical current through the wire.

If the material being magnetized is hardened steel, it will retain its magnetism for a considerable length of time. If soft iron is used, a strong magnetism will appear during the magnetizing process, but little magnetism will remain after the magnetizing force is removed. **Permanent magnets** are made from hardened steel and other special alloys because of their useful characteristics. Soft iron or steel is used as the core of relays and solenoids, in which strong magnetism must be present only when current is flowing in the coils.

MAGNETIC POLARITY

If a piece of metal such as a steel bar is magnetized, the magnetic effects concentrate at the ends of the bar. These points of magnetic concentration are called the **poles** of the magnet. Away from the ends and around the bar, an invisible force is present, the effects of which can be seen if small bits of iron or steel are placed near the magnet. The invisible force around the magnet is called a **magnetic field**.

The planet Earth is a permanent magnet, with one end near the North Pole and the other near the South Pole (Figure 2–1). If a bar magnet is balanced near the center by a small thread, it tends to line up with Earth's magnetic field. One end of the bar magnet points north. It is therefore labeled the north-seeking pole and usually is marked with the letter *N*. The other end of the bar points south and is marked *S*, for south-seeking pole (Figure 2–2).

By definition, the magnetic field of a bar magnet is said to emerge from its north pole and to reenter the magnet at its south pole. Some lines will emerge from the sides of the magnet, but the major concentration is at the poles (Figure 2–3).

If two bar magnets are hung at their balance points and then brought together, the repelling and attracting forces can be demonstrated. In Figure 2–4, two magnets are hung from strings, and the repulsion of two like poles is shown. As the magnetic bars approach each other, the two north poles swing away from each other. In Figure 2–5, the attraction

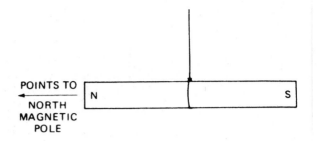

A BAR MAGNET BALANCED AND HUNG BY A STRING WILL ALIGN ITSELF WITH THE EARTH'S MAGNETIC FIELD

FIGURE 2–2   Bar magnet compass.

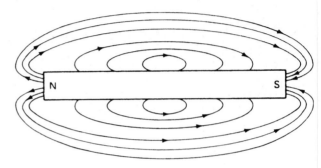

FIGURE 2–3   Magnetic field.

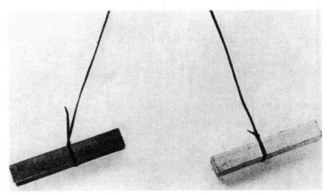

FIGURE 2–4   Like poles repel. (Courtesy of BET Inc.)

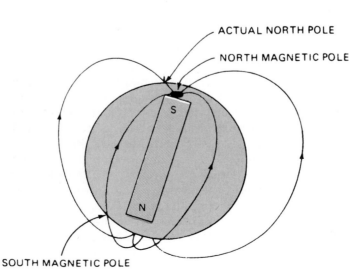

FIGURE 2–1   Earth as a magnet.

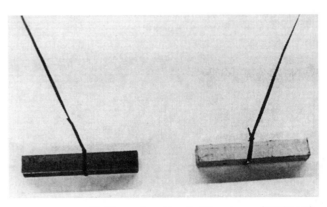

FIGURE 2–5   Unlike poles attract. (Courtesy of BET Inc.)

UNMAGNETIZED

HAPHAZARD MOLECULAR MAGNETS

FIGURE 2–6   Molecular magnets.

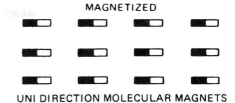

MAGNETIZED

UNI DIRECTION MOLECULAR MAGNETS

FIGURE 2–7   Aligned molecular magnets.

of two unlike poles is shown. As the two unlike poles approach, they are attracted to each other, as the pull on the strings indicates.

MAGNETIC MATERIALS

Most materials are not magnetic. Copper, aluminum, brass, wood, paper, and glass are examples of nonmagnetic materials. Magnetic fields will pass through these materials, but the material will not be magnetized. Some materials are magnetic; iron, steel, cobalt, and nickel are examples of magnetic materials. Many alloys of iron, cobalt, and nickel also provide excellent magnetic properties. Some of these alloys contain nonmagnetic materials, such as aluminum, but the final alloy is magnetic and, in most cases, very highly so.

MOLECULAR MAGNETS

In ordinary steel, each molecule of steel is a tiny, permanent magnet. In unmagnetized steel, the molecular magnets are arranged haphazardly throughout the metal (Figure 2–6). Because each molecular

magnet is pointing in a different direction, the total magnetic effect is zero. The steel is not a magnet and will not attract magnetic materials.

When steel becomes magnetized, the magnetizing process rotates the molecular structure (magnets) so that most molecules are pointing in the same direction. This arrangement is shown in Figure 2–7.

MAGNETIC FIELD

Although the magnetic field is invisible, its effect can be demonstrated. If a piece of paper is placed over a bar magnet and the paper is sprinkled with iron filings, a pattern will form. The magnetic force, or field, will arrange the filings in lines running from one end of the bar magnet to the other, as shown in Figure 2–8.

There are no insulators from magnetic fields. Magnetic lines of force can pass through all materials. They will be deflected or bent by other magnetic fields but not stopped or blocked.

The term *antimagnetic* is in most cases a play on words. A wristwatch, for example, may be marked antimagnetic. Because the delicate mechanisms within

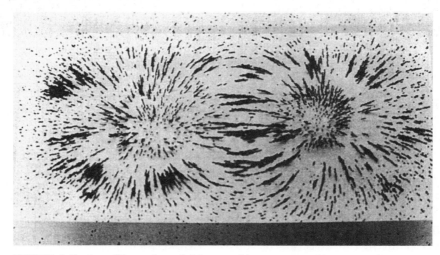

FIGURE 2–8   Iron filings show field around bar magnet. (Courtesy of BET Inc.)

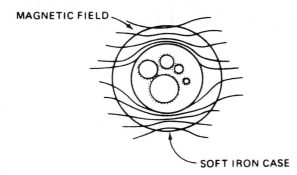

FIGURE 2–9   Magnetic shielding.

FIGURE 2–10(a)   No current.

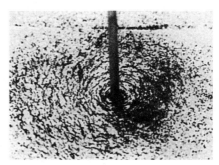

FIGURE 2–10(b)   Current flow. (Courtesy of BET Inc.)

a watch may be very susceptible to magnetic fields, these small, delicate parts are protected by the frame of the watch, which is made of soft iron (Figure 2–9). Magnetic fields will pass through the iron frame rather than the hard steel parts. The small parts are protected, not because they are antimagnetic, but because the magnetic fields take the easiest path through the soft iron.

ELECTROMAGNETISM

As indicated previously, a magnet can be created by passing an electric current through a coil of wire. This can be explained by one of the basic tenets of electricity: A magnetic field exists around each electron in motion. When electrons move through a wire, the magnetic field set up around the wire is proportional to the amount of current in the wire. This is a basic relation between electricity and magnetism. The magnetic field around a wire carrying current is in the form of concentric circles (Figure 2–10).

The method for demonstrating the magnetic field around a wire is shown in Figure 2–11. A wire carrying current is passed through a sheet of paper. Compasses placed on the paper demonstrate the existence of a magnetic field, as well as the direction of the field. The compass needles align themselves with the magnetic field.

Left-Hand Rule

If the direction of electron flow in a wire is known, the direction of the magnetic field around the wire can be determined by following the **left-hand rule**: Grasp the current-carrying wire in the left hand with the thumb pointing in the direction of electron flow. The fingers will point in the direction of the **magnetic lines of force** (Figure 2–12).

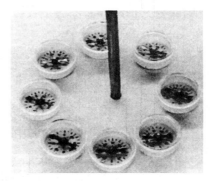

FIGURE 2–11(a)   No current, compasses all point north. (Courtesy of BET Inc.)

FIGURE 2–11(b)   Electrons traveling down, compasses line up with field. (Courtesy of BET Inc.)

Electromagnets

If a current-carrying wire is formed into a loop or coil, the concentric lines of force will all be in the same direction through the center of the loop (Figure 2–13).

If loops are placed close together, there will be a further concentration of lines of force. Some lines will merge and go around the combined loops, as shown in Figure 2–14.

If several turns of insulated wire are formed into a coil, lines of force will enter one end of the coil, pass through it, and emerge at the other end. The lines of force will be completed outside of the coil, as shown in Figure 2–15.

Magnetic Polarity of Electromagnets

The polarity of an electromagnet may also be determined by using the left-hand rule. The direction of electron current must be known in order to use the rule. Grasp the coil in the left hand with the fingers pointing in the direction of electron flow. The thumb will point in the direction of the lines of force through the center of the coil and to the north pole of the electromagnet (Figure 2–16).

Magnetic Strength of Electromagnets

The strength of an electromagnet depends on the size, length, material of the core, number of turns in the coil, and amount of current flowing through it. Other conditions being the same, an electromagnet with a soft iron core will be stronger than one with a core of steel.

If the direction of the field is known but the direction of electron flow is not, the rule can be used to determine the current direction. Grasp the wire

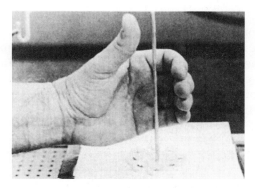
FIGURE 2–12   Demonstrating left-hand rule. (Courtesy of BET Inc.)

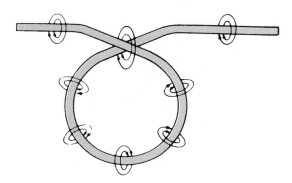
FIGURE 2–13   Field around wire. (Courtesy of BET Inc.)

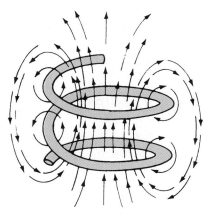
FIGURE 2–14   Field around coil.

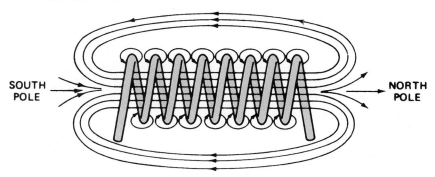
SOUTH POLE   NORTH POLE
FIGURE 2–15   Concentration of magnetic lines.

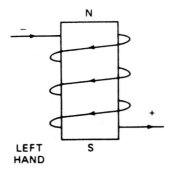

FIGURE 2–16   Left-hand rule.

with the left hand with fingers pointing in the direction of the magnetic lines. The thumb will point in the direction of electron flow.

Ampere-Turns

The magnetizing force of a coil is based on the ampere-turns of the coil. The ampere-turns of a coil are equal to the amount of current flowing through the coil multiplied by the number of turns. In Figure 2–17, the ampere-turn rating of different coil–current combinations is given.

The magnetizing force of a one-turn coil with 1 ampere of current flowing through it is very weak: 1 ampere-turn. A two-turn coil with 5 amperes flowing through it has a magnetizing force 10 times stronger, or 10 ampere-turns. A five-turn coil with 2 amperes of current has the same magnetizing force, 10 ampere-turns. A three-turn coil with 10 amperes of current has the strongest magnetizing force of the group, 30 ampere-turns.

Iron Core Coils

When a magnetic material is used as the core of an electromagnet, the strength of the magnet is greatly increased. This is very important in the design of components such as electric motors, relays, and solenoids. With an iron core, a stronger magnet is obtained in a coil with the same current flow. This produces a stronger magnet at a lower operating cost. High-energy, efficient motors are those with more iron in the core and more windings in the

coils. The higher initial cost is usually offset in a year or so by savings in operating costs.

A large number of devices that operate on the magnetic principle are found in air-conditioning and refrigeration electrical systems. Specific information on how these devices function will be found in later units of this text.

SUMMARY

- Permanent magnets are made with hardened steel.
- Soft iron is used in temporary magnets.
- Like poles of magnets repel each other.
- Unlike poles of magnets attract each other.
- There are no insulators for magnetism or magnetic lines of force.
- Soft iron is often used to shield items from magnetism by providing a better path for the magnetic lines of force.
- When an electric current flows in a wire, a magnetic field exists around the wire.
- When a current-carrying wire is formed into a loop or coil, the magnetic field is concentrated in the center of the loop.
- When an iron core is used as the core of a coil carrying an electric current, the magnetic field is greatly increased.

PRACTICAL EXPERIENCE

Equipment required   A permanent magnet; copper, aluminum, and iron material; a screwdriver

Procedure

1. In sequence, place the magnet close to the copper, aluminum, and iron. Describe the action that took place as the magnet was placed close to the suggested materials.
2. Place the magnet at the screwdriver point. Did the magnet and screwdriver attract each other?

1 TURN 1 AMP	2 TURNS 5 AMPS	5 TURNS 2 AMPS	3 TURNS 10 AMPS
TURN x 1 AMP = 1 AT	2 TURNS x 5 AMPS = 10 AT	5 TURNS x 2 AMPS = 10 AT	3 TURNS x 10 AMPS = 30 AT

FIGURE 2–17   Ampere turns.

3. Stroke the screwdriver with a pole of the magnet from about 2 inches from the tip to the tip itself.

4. Place the tip of the screwdriver at the piece of iron. Is there an attraction exhibited? The action in steps 3 and 4 demonstrate the ease with which magnetic materials iron and steel may be magnetized.

Conclusion   Certain magnetic materials, iron and steel, are easily magnetized when they come into contact with a magnet.

REVIEW QUESTIONS

1. With magnets, like poles (repel, attract) each other.

2. The concentration of a magnetic field in a bar magnet is at the (side, ends) of the bar.

3. Two adjacent wires carrying current in the same direction will (attract, repel) each other.

4. Two adjacent wires carrying current in the opposite direction will (attract, repel) each other.

5. If a wire carrying current is formed into a loop or coil, the magnetic field will (increase, decrease).

6. If a magnetic material, such as soft iron, is used as the core of an electromagnet, the strength of the magnet will (increase, decrease).

7. Soft iron is used as the core of (permanent, temporary) magnets.

8. Hard steel is used as the core of (permanent, temporary) magnets.

9. (Nothing, glass) will isolate a material from magnetic lines of force.

10. Electric motors containing more iron and windings are usually (cheaper, more expensive) to operate.

UNIT 3

OHM'S LAW AND THE ELECTRIC CIRCUIT

OBJECTIVES

After completion and review of this unit, you should be able to:

- Recognize resistors as components in electrical/electronic circuits.
- Recognize the schematic symbol for fixed resistors.
- Recognize the schematic symbol for variable resistors.
- Use Ohm's law to determine one of the three values—*voltage, current,* or *resistance*—when two of the values are known.

In Units 1 and 2, voltage and current were covered. Another factor is of equal importance in the study of electricity and electrical circuits: **resistance**. In the discussion of voltage and current, the following observation was made: When a generator is operating, electrons are pulled from the wire on one side and forced out on the wire on the other side. This creates an unbalance called voltage, shown in Figure 3–1. As long as the generator operates, it will maintain the unbalance, and a voltage will exist between the two wires.

Note in the circuit of Figure 3–1 that no wire connects the output wires of the generator. The

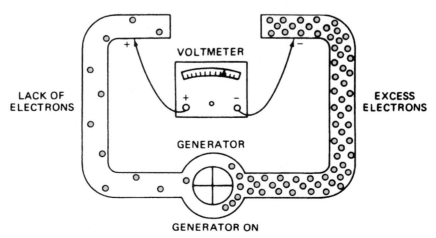

FIGURE 3–1   Generator producing voltage.

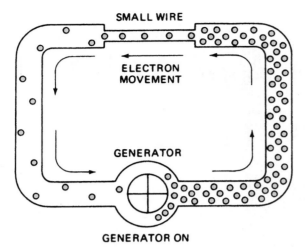

FIGURE 3–2   Current flow.

resistance of air is extremely high and may be considered infinite. With such a high resistance, no current will flow.

If a small wire is connected between the ends of the original wires, a current will flow in the circuit, as shown in Figure 3–2. The amount of current that flows in the circuit depends on the type of material from which the wire is made, the size (diameter) of the wire, and its length. In the circuit in Figure 3–2, the small wire is said to offer resistance to the current flow. Although resistance is offered, current will flow.

The connecting wire in the circuit in Figure 3–3 is the same length and of the same material. However, the cross-sectional area is twice the size of the wire used in Figure 3–2. More current flows in the circuit in Figure 3–3 than in the circuit in Figure 3–2 because resistance is reduced. Actually, twice as much current flows when the larger wire is used.

The resistance of a wire depends on four things:

1. Material used to make the wire (available free electrons)
2. Cross-sectional area of the wire (size)
3. Length of the wire
4. Temperature (normally as temperature increases, resistance increases)

Resistors come in many shapes and sizes. Figure 3–4 shows four resistors: three are different types of wire-wound power resistors, and one is a carbon resistor.

The physical size of a resistor relates to how much power the resistor can handle (how much heat the resistor can radiate, or get rid of). The physical size of a resistor has nothing to do with its resistance value in ohms.

Another resistive element that may be familiar is the heat strip from an air-conditioning system (Figure 3–5). The component is made of nichrome wire that has been coiled to decrease the space needed to accommodate the total wire length. The heat strip is a resistive element.

COMPLETE ELECTRIC CIRCUITS

For current to flow in an electric circuit, the circuit must be complete. Another way of stating this is that current must be able to flow from the source through

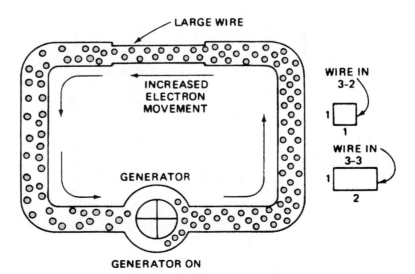

FIGURE 3–3   Increased current flow.

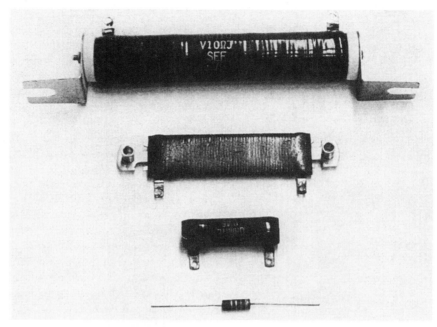

FIGURE 3–4   Resistors. (Courtesy of BET Inc.)

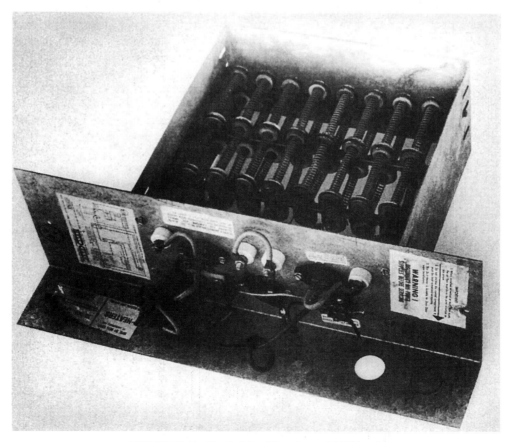

FIGURE 3–5   Heat strip. (Courtesy of BET Inc.)

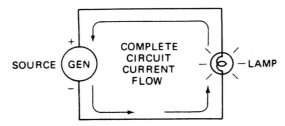

FIGURE 3–6 Complete circuit allows current flow. Lamp is lit.

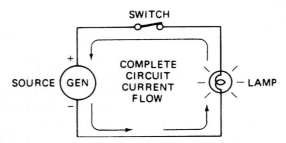

FIGURE 3–7 With a closed switch, the circuit is still complete. Lamp is lit.

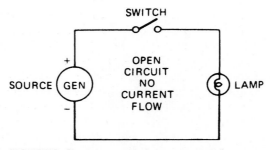

FIGURE 3–8 Open switch, no current flow. Lamp is not lit.

the load, back to the source, through the source back to the load, and so on.

In Figure 3–6, an electric generator is the energy source, and a lamp is the load. As long as the circuit is complete, current will flow from the generator (source) through the lamp (load) and back through the generator. There is a continuous path for current to flow in. It is a complete electric circuit.

If a switch is included in the circuit, the same current that flows through the lamp flows through the switch. Two conditions can exist. When the switch is **closed**, the circuit is complete and current can flow (Figure 3–7). When the switch is in the **open** position, the circuit is not complete (Figure 3–8). Current cannot flow through the open switch; therefore, current cannot flow through the lamp.

Figure 3–9 shows two circuits. Note the position of the switches. In both circuits, the switches and the lamp are connected in series. In the circuit in Figure 3–9a, if either switch A-1 or A-2 is open, the circuit is not complete: The lamp will not light. Similarly, in the circuit in Figure 3–9b, if either switch B-1 or B-2 is open, the circuit is not complete. The lamp will not light. A complete circuit may be made up of the generator, the load, and the connecting wires.

The connecting wires in a circuit are usually considered not to have resistance. Actually, wires do have resistance but for practical application, the resistance of wires (copper) is so small that it need not be considered in circuit calculation.

Electrically, both circuits in Figure 3–9 are the same. They are both series circuits. The same current must flow through each component of the circuit.

Figure 3–10 shows a simple control circuit for an air-conditioning compressor contactor. It is a series circuit. The contactor will be energized as long as 24 volts are available from the source and both the high-pressure cutout and the thermostat select switches are closed. If either switch should open, the circuit to the contactor coil will open, no current can

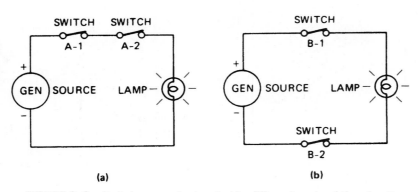

FIGURE 3–9 Switches may be located in different parts of the circuit.

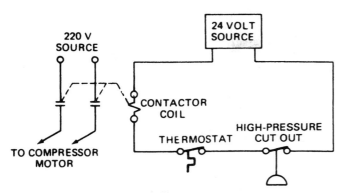

FIGURE 3–10   Low-voltage circuit controlling a high-voltage circuit.

flow in the circuit, and the contactor will de-energize and remove power from the compressor motor.

Consider Figure 3–11, a power source connected to a group of lamps. In this circuit, switch S1 is closed. There is a complete circuit from the A terminal of the generator through S1, through the lamp L1, to generator B and through the generator to A. The circuit is complete. Lamp L1 will be lit.

If switch S2 is closed, a complete circuit exists from generator A through S2, through lamp L2, and back through the generator. Lamp L2 will be lit.

When switch S3 is open, a complete path does not exist for current flow in the L3 lamp circuit. Lamp L3 will not be lit.

If switch S4 is closed, a complete circuit exists through S4, the lamp L4, and the generator. Lamp L4 will be lit.

If current is to flow in any electric circuit, a complete path must exist from the power source, through the load, and back to the power source.

Common or "Ground" Connection

In a standard electrical system, two wires are used to connect components to the power source. A wire connected to the A terminal of the generator is

FIGURE 3–12   Negative terminal of the generator connected to the engine and chassis of the automobile to provide a *common* or *ground* connection.

shown in Figure 3–11. Another wire is shown connected to the B terminal of the generator. These two wires are then connected to lamp and switches.

Another method of completing a circuit is to use a **common** or **ground** for one of the wires. A common or ground connection is usually used where there is a metal chassis or frame involved in the overall product.

An automobile is a good example of the use of a common or ground. In an automobile the chassis, frame undercarriage, and engine are all mechanically connected together. This forms one large electrical common connection. One side of the alternator or generator is directly connected to the engine block, which is common with the rest of the system. This can be seen on the schematic shown in Figure 3–12.

The completion of a circuit similar to the one in Figure 3–11 is shown in Figure 3–13. Current leaving the A terminal of the generator continues on through the closed switches, the lamp, and finally reaches the ground. The ground, being metal, completes the path back to the B terminal of the generator.

In most situations, components in air-conditioning and refrigeration circuits use wires for both the supply and return sides of the power system, as shown in Figure 3–11. Keep in mind, however, that some systems may use a common electrical connection. This is apparent in many electronic systems; for example, a printed circuit board. The common is around the edge of the circuit board.

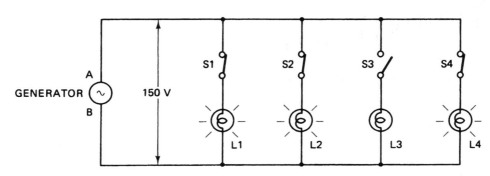

FIGURE 3–11   Parallel lamp circuit with individual switches.

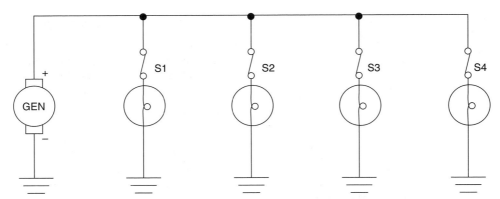

FIGURE 3–13   Lamps and switches connected in a circuit using common or ground as the return to the generator.

OHM'S LAW

The relationship of voltage, current, and resistance may be determined by a relatively simple formula known as **Ohm's law**. It was devised by a German physicist, Georg Simon Ohm, in 1848. He discovered that the current in any circuit is directly proportional to the voltage applied and is inversely proportional to the resistance.

$$I = \frac{E}{R}$$

or

$$R = \frac{E}{I}$$

or

$$E = I \times R$$

The following rules can be derived from Ohm's law:

1. If the resistance is kept the same and the voltage is increased, the current will increase.
2. If the voltage is kept the same and the resistance is increased, the current will decrease.

In honor of the discoverer of the law, the name of the unit used to measure resistance is the **ohm**; it is indicated by the Greek letter *omega* (Ω). The symbol for resistance is the letter R. The symbol for voltage is the letter E (from EMF). The symbol for current is the letter I (from current intensity).

By using any one of the three forms of Ohm's law, when two factors are known in a circuit, the third factor can always be found.

EXAMPLE 1   (*I* is the unknown)

A 12-ohm resistor has 24 volts across it. How much current is flowing through it? (See Figure 3–14.)

Solution

$$I = \frac{E}{R}$$

$$I = \frac{24}{12}$$

$$I = 2 \text{ amps}$$

Mathematically, the answer is 2 amperes. If an **ammeter** is connected in series with the 12-ohm resistor, as shown in Figure 3–15, the meter will indicate 2 amperes. ∎

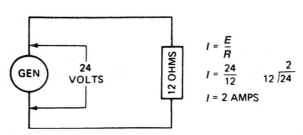

FIGURE 3–14   Using Ohm's law to determine current.

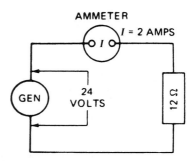

FIGURE 3–15   Connecting an ammeter to indicate current.

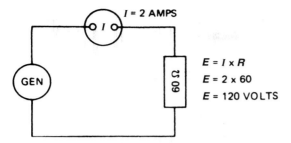

FIGURE 3–16   Using Ohm's law to determine voltage.

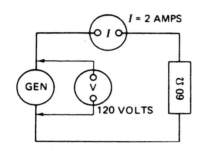

FIGURE 3–17   Measuring supply voltage.

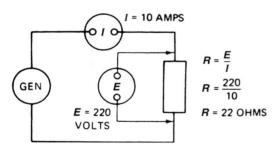

FIGURE 3–18   Measuring current flow.

EXAMPLE 2   (*E* is the unknown)

A resistor of 60 ohms is connected across an electrical power source. An ammeter connected in series with the resistor indicates 2 amperes (Figure 3–16). What is the voltage of the power source?

Solution

$$E = I \times R$$
$$E = 2 \times 60$$
$$E = 120 \text{ volts}$$

Mathematically, the answer is 120 volts. If a voltmeter is connected across the power source, it will indicate 120 volts (Figure 3–17).   ∎

EXAMPLE 3   (*R* is the unknown)

A resistor is connected across a 220-volt power source. An ammeter connected in series with the resistor indicates 10 amperes (Figure 3–18). What is the value of the resistor?

Solution

$$R = \frac{E}{I}$$
$$R = \frac{220}{10}$$
$$R = 22 \text{ ohms}$$

Mathematically, the answer is 22 Ω. If an **ohmmeter** is connected to the resistive device, it will indicate 22 Ω (Figure 3–19).

In normal, everyday operations, the air-conditioning and refrigeration technician is not required to solve Ohm's law problems. An understanding of Ohm's law does, however, provide the means to a better understanding of electricity. A good understanding

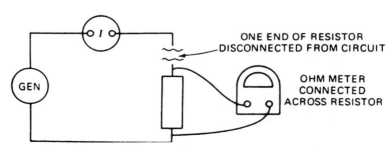

FIGURE 3–19   Using an ohmmeter to measure resistance.

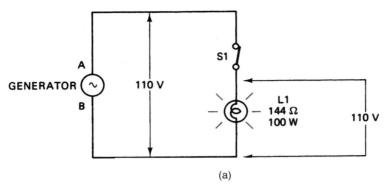

FIGURE 3–20(a)   Closed switch has low resistance; voltage appears across the lamp.

of electricity is important to the technician when troubleshooting equipment.   ■

Open Circuit—Ohm's Law

Ohm's law is true for the whole circuit or any part of a circuit. Consider the circuit in Figure 3–20a. Switch S1 is closed. The resistance of the closed switch could be 0.001 ohm or even lower. The switch resistance is so low compared to the lamp resistance that the switch resistance does not effectively contribute to the total resistance of the circuit (144 ohms versus 144.001 ohms).

By Ohm's law, the current in the circuit is

$$I = \frac{E}{R}$$

$$I = \frac{120}{144}$$

$$I = 0.833 \text{ amperes}$$

In Figure 3–20b, switch S1 is open. The resistance of the open switch could be 100,000,000 ohms. The resistance of the lamp is insignificant when

compared to the open switch resistance. The supply voltage, 120 volts, will appear across the open switch. By Ohm's law, the current is

$$I = \frac{E}{R}$$

$$I = \frac{120}{100,000,000}$$

$$I = 0.0000000012 \text{ amperes}$$

Lamp L1 will not be lit, as there is too little current flowing through it. For all practical purposes, there is no current. (The actual resistance of the open switch would depend on a number of factors. One is air resistance. The resistance of air varies according to moisture content, among other factors.)

DIAGRAMS

Diagrams are used to provide circuit information. Four types of diagrams are particularly useful in showing component connection. They are (1) pictorial, (2) schematic, (3) ladder, and (4) wiring. Figure 3–21

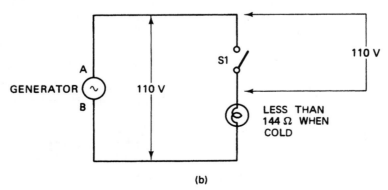

FIGURE 3–20(b)   Open switch has extremely high resistance; voltage appears across the switch.

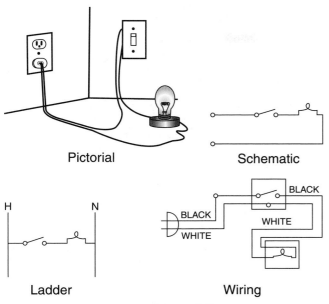

FIGURE 3–21   Types of diagrams.

shows examples of the same circuit in the four diagram types. Some of the diagrams may look similar because the circuit is so simple. The difference will become more obvious as the circuits become more complex.

RESISTOR COLOR CODE

Normally, we are only interested in the resistance of loads such as motors, coils, and heaters. However, in electronic circuits and some other special applications, a device called a resistor (Figure 3–22) is used for the sole purpose of providing a resistance and providing no other function. You will find resistors on circuit boards, but you will usually not troubleshoot these boards down to the component level. But sometimes you will. For example, if you use your ohmmeter incorrectly, you can ruin it by blowing out one of the resistors. If you can find the failed resistor (usually by the burn mark around it), it can be re-

placed with a resistor of the same resistance value. The resistance value of a resistor can be determined from the colors of the four bands. Each different color stands for a different number, as follows:

Black	= 0
Brown	= 1
Red	= 2
Orange	= 3
Yellow	= 4
Green	= 5
Blue	= 6
Violet	= 7
Gray	= 8
White	= 9

The first colored band (closest to the end of the resistor) is read as the first number. The second band is read as the second number. The third band is read as the number of zeros that must be added to the first two numbers. The fourth band tells you the manufacturing tolerance of the resistor, with the following values:

Gold	= 5%
Silver	= 10%
None	= 20%

The manufacturing tolerance tells you how close the actual resistance of the resistor will be to the value that you read by the colors.

EXAMPLE 4

A burned-out resistor has the colors gray, red, orange, and gold. What value resistor will you buy to replace it?

Solution

The first band, gray, stands for 8. The second band, red, stands for 2. The third band, orange, stands for 3, which means three zeros. The value of the resistor is 82,000 ohms. The fourth band, gold, means that the actual resistance is within 5% of the 82,000 ohms that you just determined.   ∎

TROUBLESHOOTING

A practical troubleshooting situation could exist in the circuit shown in Figure 3–20a. Here the switch is closed, and the lamp should be lit.

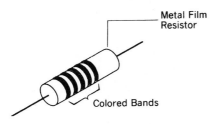

FIGURE 3–22   Color-coded resistor.

But, if the lamp is not lit, there is a problem. The first procedure normally would be to replace the lamp with one that is known to be good. If the new lamp still does not light, troubleshooting to find the problem in the circuit is required.

1. Measure the supply voltage at the generator. The voltage between the output A and B terminals should be 110 volts.
2. Measure the voltage across the lamp socket.
   a. If there are 110 volts across the socket and the good lamp is not lit, the lamp socket is bad.
   b. If 110 volts are not present at the socket, measure the voltage at the switch.
   c. The voltage across the closed switch should be zero volts. If the voltage across the closed switch is 110 volts, the switch is bad.

SUMMARY

- A complete circuit is necessary for current to flow.
- Ohm's law is the relationship between current voltage and resistance in an electric circuit.
- If two of the factors in Ohm's law are known, the third can always be found.
- Ohm's law is true for the entire circuit or for any part of the circuit.

PRACTICAL EXPERIENCE

Determine the value of the unknown in the following problems.

1. A 22-ohm resistive heater has 220 volts across it. How much current is flowing through it?
2. A resistor of 15 ohms has 11 amperes flowing through it. How much voltage is across the resistor?
3. The voltage measured across a heat strip is 220 volts. A clamp-on ammeter used on a line feeding this circuit indicates 9 amperes. What is the resistance of the element?

Conclusion The relationship between current, voltage, and resistance in a component is fixed by a law called Ohm's law. The formula may be stated in three ways:

1. The current flow through a component is equal to the voltage across the component divided by the resistance of the component.

$$I = E/R$$

2. The resistance of the component is equal to the voltage across the component divided by the current flow through the component.

$$R = E/I$$

3. The voltage across the component is equal to the current flow through the component multiplied by the resistance of the component.

$$E = I \times R$$

REVIEW QUESTIONS

1. The resistance of a piece of wire depends on four things. What are they?
2. In an electric circuit, if the resistance is kept the same and the voltage is increased, the current will (increase, decrease).
3. In an electric circuit, if the voltage is kept the same and the resistance is increased, the current will (increase, decrease).
4. If a 6-ohm resistor has 20 amperes flowing through it, the voltage across the resistor will be _____ volts.
5. If 24 volts is measured across an 8-ohm resistor, how much current must be flowing through it?
6. An ammeter connected in series with a resistor indicates 11 amperes. A voltmeter connected across the resistor indicates 220 volts. What is the value of the resistor?
7. What voltage is needed to cause 4 amperes to flow through a 110-ohm resistor?
8. If the voltage across a 15-ohm resistor is measured at 90 volts, how much current is flowing through the resistor?
9. If the voltage applied to a resistor is doubled, will the current double or decrease to one-half the original value?
10. What size resistor will cause a current of 3 amperes to flow when 120 volts is applied to it?

UNIT 4

SERIES CIRCUITS

OBJECTIVES

After completion and review of this unit, you should be able to:

- Recognize a complete circuit.
- Recognize a series circuit by component connections.
- Solve series circuit problems.

In Unit 3, Ohm's law and its application to simple electric circuits were discussed. In this unit, the discussion will be expanded to cover Ohm's law as it relates to series circuits.

An example of a simple series circuit is the **relay** contacts of an air-conditioning compressor and the related compressor motor. (See Figure 4–1.) The contacts are in series with the motor. All the current that flows through the compressor must flow through the contacts. When the contacts are open, the circuit is open; no current flows through the compressor motor.

A more complete presentation of series circuits follows; the relationship between voltage, current, and resistance in a series circuit is covered in this unit.

WIRE RESISTORS

A common material used in the construction of wire-wound resistors is nichrome. Nichrome wire has a relatively high resistance for short lengths of wire. For example, 100 feet of nichrome wire could have a resistance of about 100 ohms. A practical, usable resistor can be made out of nichrome wire by winding it on a core of ceramic with terminals at either end (Figure 4–2).

The circuit in Figure 4–3 consists of a source (a generator), connecting wires, and a load (a 100-ohm wire-wound resistor). According to Ohm's law, if the

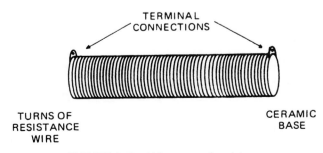

FIGURE 4–2   Wire-wound resistor.

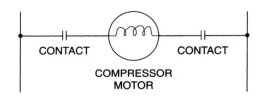

FIGURE 4–1   Contacts in series with compressor motor.

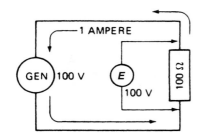

FIGURE 4–3   Single-resistor circuit.

generator is producing 100 volts, then 1 ampere of current will flow in the resistor. A voltmeter connected across the resistor will read 100 volts.

$$I = \frac{E}{R}$$

$$I = \frac{100}{100}$$

$$I = 1 \, \text{ampere}$$

If the resistor is broken at the exact center (Figure 4–4), there will be two resistors of 50 ohms each (two 50-foot lengths of the nichrome wire). The two broken ends of wire could be clamped together to remake the 100-ohm resistor (Figure 4–5). The two 50-ohm resistors, tied in series, make the 100-ohm resistor: 50 + 50 = 100.

The two reconnected 50-ohm resistor sections can be connected into the original circuit (Figure 4–6). This circuit is a **series circuit**. The same current, 1 ampere, flows through each of the 50-ohm resistor sections. Ohm's law may be used to calculate the voltage across R_1, the first resistor section.

$$E = I \times R_1$$
$$E = 1 \times 50$$
$$E = 50 \, \text{volts}$$

Ohm's law also is used to calculate the voltage across R_2, the second resistor section.

$$E = I \times R_2$$

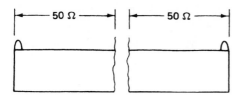

FIGURE 4–4   100 Ω resistor broken at the exact center.

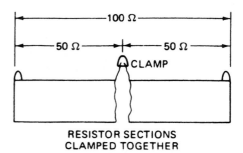

FIGURE 4–5   Two 50-ohm resistors joined together.

$$E = 1 \times 50$$
$$E = 50 \, \text{volts}$$

Note that the sum of the voltages across R_1 and R_2 equals the supplied voltage, 50 V + 50 V = 100 V. Also note that the total resistance, 100 ohms, is the sum of the individual resistors, 50 Ω + 50 Ω = 100 Ω.

LAWS OF SERIES CIRCUITS

We can now state the laws of series circuits.

1. The same current flows through each component of a series circuit.
2. The total voltage in a series circuit is equal to the sum of the voltages across the individual components.
3. The total resistance of a series circuit is equal to the sum of the resistances of the individual components.

EXAMPLE 1

A circuit consists of a 25-ohm resistor and a 50-ohm resistor connected in series across a 150-volt generator (Figure 4–7). Determine the current flow in the circuit and the voltage across each resistor.

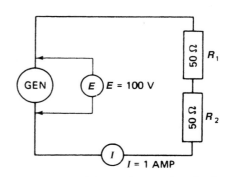

FIGURE 4–6   Two-resistor series circuit.

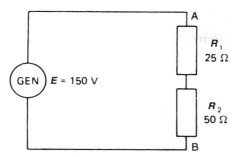

FIGURE 4-7   Circuits of 25 ohms and 50 ohms in series.

Solution

The total resistance is equal to the sum of the individual resistances.

$$R_T = R_1 + R_2$$
$$R_T = 25 + 50$$
$$R_T = 75 \text{ ohms}$$

Use Ohm's law on the whole circuit.

$$I = \frac{E}{R}$$
$$I = \frac{150 \text{ V}}{75}$$
$$I = 2 \text{ amperes}$$

Because the same current flows through each component, the voltage across each component may be determined by using Ohm's law.

$$E_{R_1} = I \times R \qquad E_{R_2} = I \times R$$
$$E_{R_1} = 2 \times 25 \qquad E_{R_2} = 2 \times 50$$
$$E_{R_1} = 50 \text{ volts} \qquad E_{R_2} = 100 \text{ volts}$$

The supply voltage is equal to the sum of the voltages across the individual components.

$$E_T = E_{R_1} + E_{R_2}$$
$$E_T = 50 + 100$$
$$E_T = 150 \text{ volts} \qquad \blacksquare$$

EXAMPLE 2

Determine the value of each resistor, the voltage across each resistor, and the total resistance of the circuit shown in Figure 4–8. The generator supplies 240 volts; the ammeter indicates 4 amperes.

Solution

Solve for R_T, using Ohm's law for the whole circuit.

$$R_T = \frac{E_T}{I}$$
$$R_T = \frac{240}{4}$$
$$R_T = 60 \text{ ohms}$$

Solve for R_2.

$$R_2 = \frac{E_{R_2}}{I}$$
$$R_2 = \frac{80}{4}$$
$$R_2 = 20 \text{ ohms}$$

Solve for ER_3.

$$E_{R_3} = I \times R_3$$
$$E_{R_3} = 4 \times 5$$
$$E_{R_3} = 20 \text{ volts}$$

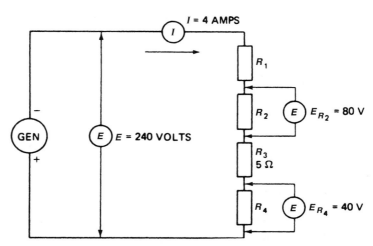

FIGURE 4-8   Using Ohm's law in series circuits.

Solve for R_4.

$$R_4 = \frac{E_{R_4}}{I}$$

$$R_4 = \frac{40}{4}$$

$$R_4 = 10 \text{ ohms}$$

Solve for R_1, using the total resistance law.

$$R_T = R_1 + R_2 + R_3 + R_4$$

$$60 = R_1 + 20 + 5 + 10$$

$$R_1 = 60 - 35$$

$$R_1 = 25 \text{ ohms}$$

Check the solution, using total voltage.

$$R_1 = 25 \text{ ohms}$$

$$E_{R_1} = I \times R_1$$

$$E_{R_1} = 4 \times 25$$

$$E_{R_1} = 100 \text{ V}$$

$$E_T = E_{R_1} + E_{R_2} + E_{R_3} + E_{R_4}$$

$$240 = 100 + 80 + 20 + 40$$

$$240 = 240$$ ∎

The solution checks.

The refrigeration and air-conditioning technician will not often have to solve electrical problems mathematically. But when the technician is required to solve these problems, it is important to know the laws of series circuits. The laws of series circuits and other laws of circuits that you will learn later also are useful when troubleshooting.

TROUBLESHOOTING IN A SIMPLE SERIES CIRCUIT

Troubleshooting is the process a technician uses to locate problem areas and problem components in a system.

In Figure 4–9, three resistors are shown connected in series with 120 volts supplied to the circuit. By using Ohm's law and the laws of series circuits, we can determine that the total resistance (R_T) of the circuit is 60 ohms, $R_T = R_1 + R_2 + R_3$. Further, the current through the circuit can be found using Ohm's law.

$$I = E/R_T = 120\,\text{V}/60 = 2\,\text{amp}$$

THE VOLTAGE ACROSS

$$E_{R_1} = I \times R_1 = 2 \times 30 = 60\,\text{V}$$

$$E_{R_2} = I \times R_2 = 2 \times 10 = 20\,\text{V}$$

$$E_{R_3} = I \times R_3 = 2 \times 20 = 40\,\text{V}$$

$$E_{R_3} = I \times R_3 = 2 \times 20 = 40\,\text{V}$$

Voltmeter measurements taken as shown in Figure 4–9 should indicate V_1 is 60 volts, V_2 is 20 volts, and V_3 is 40 volts. Because 120 volts is the supplied voltage, everything seems to check out. In a troubleshooting situation, the technician is usually looking for something that does not fit the normal situation, something that does not check out. In the circuit in Figure 4–9, the technician's voltage measurements are

$$V_{R_1} = 0\,\text{V};\ V_{R_2} = 0\,\text{V};\ V_{R_3} = 120\,\text{V}.$$

FIGURE 4–9   Three-resistor series circuit.

The technician would know immediately there was trouble in the circuit. In this case, because all the voltage is across R_3, the resistor is probably open or rather than being 20 ohms, its resistance has risen to some value well above 20 ohms.

Another possibility is that resistors R_1 and R_2 are shorted out or bypassed; that is, there is a direct connection between the top of R_1 and the bottom of R_2. In such a situation, the full 120 V would appear across R_3.

SUMMARY

- In a series circuit, the sum of the voltage across the components is equal to the applied voltage.
- The same current flows through each component in a series circuit.
- The total resistance of a series circuit is equal to the sum of the resistances of the individual components.

PRACTICAL EXPERIENCE

Required equipment Volt-ohm milliammeter (VOM) and resistors with values within the same range (1 to 100, 100 to 1000, 1000 to 10,000, and so on). At least three resistors are required.

Procedure

1. Select ohmmeter service on the VOM.
2. Check the zeroing of the meter by shorting the leads together and making any necessary adjustment.
3. Measure and record the value of each of the resistors. (Change the meter scale and rezero the meter as necessary.)

   Resistor 1 _____

   Resistor 2 _____

   Resistor 3 _____

   Resistor 4 _____

   Resistor 5 _____

4. Connect one end of resistor 1 (R_1) to one end of resistor 2 (R_2), as shown in Figure 4–7.
5. Connect the ohmmeter to measure the resistance of the two resistors in series. From A to B is _____ ohms.

6. Add the values of resistor 1 and resistor 2 as recorded in step 3.
7. Is the value determined in step 6 equal to the value measured in step 5?
8. Repeat steps 5, 6, and 7 with different resistors.
9. Three or four resistors can be connected in series. The total resistance should be the sum of the individual resistances.

Conclusions

1. The total resistance of a series circuit is equal to the sum of the individual resistors.
2. The sum of the voltages across the components connected in series is equal to the applied voltage.
3. The same current flows through each component in a series circuit.

REVIEW QUESTIONS

Draw a sketch of each circuit before attempting to solve these problems.

1. A circuit consists of four 20-ohm resistors connected in series. The total resistance of the circuit is _____ ohms.
2. A circuit consists of an 8-ohm resistor and a 4-ohm resistor connected in series. The supply voltage to the circuit is 24 volts. How much current will flow in the circuit? How much voltage will be measured across each resistor?
3. A circuit consists of a 10-ohm resistor and a 2-ohm resistor connected to a 24-volt power source. The 10-ohm resistor has 20 volts across it. How much current is flowing in the 2-ohm resistor?
4. A circuit consists of three resistors connected in series to a power source. Each resistor has 6 volts across it. What is the voltage of the power source?
5. A series circuit consists of two resistors connected to a power source of 24 volts. The first resistor has 16 volts across it; the second is a 4-ohm resistor. What is the total resistance of the circuit?
6. The (same, different) current flows through each resistor in a series circuit.
7. Two 10-ohm resistors are connected in a series circuit. The voltage across the circuit is

40 volts. The current through each resistor is
_____ amps.

8. Four 8-ohm resistors are connected in a series circuit. The current through one of the resistors is 2 amperes. The voltage supplied to the circuit is _____ volts.

9. A current flow of 3 amperes is flowing in a two-resistor series circuit. The supply voltage is

24 volts. One of the resistors has a value of 6 ohms. The other resistor is _____ ohms.

10. A three-resistor series circuit is made up of 6-ohm resistors. A voltmeter placed across one of the resistors reads 9 volts. The source voltage is _____ volts.

UNIT 5

PARALLEL AND SERIES–PARALLEL CIRCUITS

OBJECTIVES

After completion and review of this unit, you should be able to understand that:

- Parallel circuits are circuits in which the total line current splits into two or more paths before returning to the current source.
- Series–parallel circuits are circuits in which some of the components of the circuit are connected in series and other components of the circuit are connected in parallel.
- Ohm's law is applicable to the whole circuit or the individual components of the circuit in both the parallel or series–parallel circuit.

An example of a parallel circuit is in the connection of the compressor and fan in many air-conditioning condenser units. The fan motor is connected directly across the compressor motor. Whenever the compressor motor is operating, the fan will be operating. (See Figure 5–1.) Both motors receive the same voltage. This is one of the conditions describing a parallel circuit. The same voltage appears across components connected in parallel.

RESISTIVE PARALLEL COMPONENTS

Consider a piece of paper 2 inches (50.8 mm) wide and 4 inches (101.6 mm) long. A very thin layer of carbon material is deposited on this paper to form a resistor, as shown in Figure 5–2. The path for current, then, is 2 inches (50.8 mm) wide by 0.01 inch (0.254 mm) high.

A voltage of 4 volts is impressed across the resistor and a current of 2 amperes is seen to be flowing through it, as shown in Figure 5–3. By applying Ohm's law, the resistance of the homemade resistor can be calculated.

$$R = \frac{E}{I}$$

$$R = \frac{4}{2}$$

$$R = 2\ \Omega$$

Next, a very sharp instrument is used to cut our homemade resistor in half lengthwise without

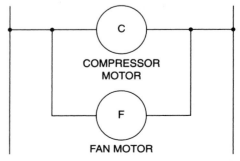

FIGURE 5–1 Compressor motor and fan motor in parallel.

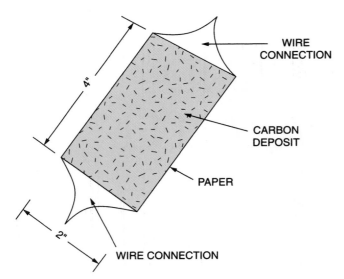

FIGURE 5–2   Homemade resistor.

removing any of the carbon deposit (Figure 5–4). Since none of the carbon deposit was removed, the total circuit was not changed electrically. With 4 volts impressed, a current of 2 amperes flows. The current path is still a total of 2.0 inches (50.8 mm) wide by 0.01 inches (0.254 mm) high. Since there are now two paths for current to flow in, the current will split. Since the two paths are of the same material and of equal dimensions, the current will split equally. One ampere will flow through the resistor on the left and one ampere will flow through the resistor on the right.

The value of each resistor may be found using Ohm's law. The battery is connected directly across the resistors. Four volts is impressed on each resistor. The current flow through each resistor is 1 ampere. According to Ohm's law, each resistor is

$$R = \frac{E}{I}$$

$$R = \frac{4}{1}$$

$$R = 4 \ \Omega$$

Note that each resistor is 4 Ω whereas the total resistance of the circuit is only 2 Ω.

The two resistors form what is called a **parallel circuit**. A parallel circuit is a circuit in which there is more than one path for current to flow.

Some other interesting facts may be determined using that same parallel circuit:

1. The same voltage is impressed across each component in a parallel circuit. Many components in circuits may have equal amounts of voltage impressed across them, although they may not be in parallel. If components are in parallel, the voltage must be the *same* voltage. In the circuit in Figure 5–5, the supply voltage is 20 volts. The measured voltages are 10 volts across R_1, 10 volts across R_2, and 10 volts across R_3. The same 10 volts appear across R_2 and R_3. They are in parallel. R_1 is not in parallel, even though it has 10 volts across it.

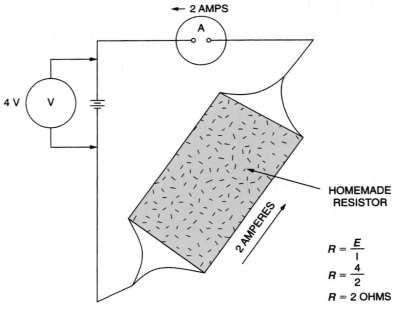

FIGURE 5–3   Application of Ohm's law to homemade resistor.

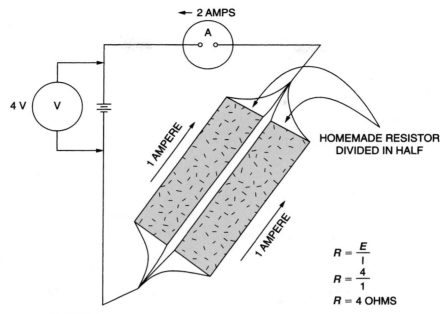

FIGURE 5–4 Divided resistor provides two paths for current flow.

2. The total current is the sum of the currents in the individual branches of the parallel circuit.
3. The effective resistance is smaller than the smallest resistor in a parallel circuit.

PARALLEL CIRCUITS

Parallel circuits are found when the same source voltage is required across two or more components. A typical parallel circuit is shown in Figure 5–6. If the generator is producing 120 volts at its output terminals, there will be 120 volts across each lamp.

Current will flow through each lamp. If each lamp has a resistance of 60 ohms, the current will be 2 amperes in each, according to Ohm's law.

$$I = \frac{E}{R}$$

$$I = \frac{120}{60}$$

$$I = 2 \, \text{amps}$$

The current for both lamps comes through the generator, the power source. There will be 4 amperes flowing through the generator (2 + 2 = 4).

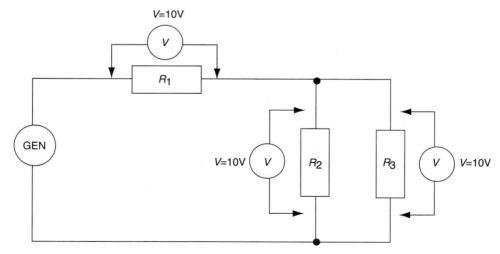

FIGURE 5–5 Three-resistor series–parallel circuit.

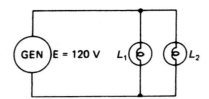

FIGURE 5–6   Parallel circuit.

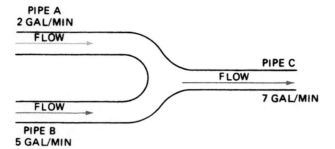

FIGURE 5–7   Parallel water circuit.

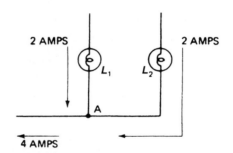

FIGURE 5–8   Sum of currents.

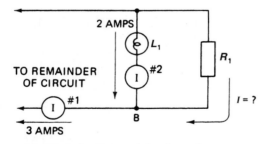

FIGURE 5–9   Determining branch current.

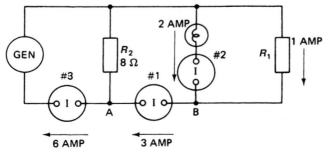

FIGURE 5–10   Branch and line currents.

Current flow in a parallel circuit is similar to water flow in a pipe system. In the water–pipe junction shown in Figure 5–7, 2 gallons of water per minute are flowing in pipe A, entering from the left. In pipe B, 5 gallons of water per minute are flowing, entering from the left. There must be 7 gallons of water per minute flowing out of pipe C. The water cannot disappear. The sum of the water flows into the junction must equal the water flow out of the junction.

In an electric circuit, the sum of the currents entering a junction must equal the current leaving the junction. In Figure 5–8, the lower junction of Figure 5–6 is shown. If 2 amperes are flowing to junction A from lamp L1, and 2 amperes are flowing to it from lamp L2, there must be 4 amperes leaving junction A.

Sometimes it is necessary to use Ohm's law to find a branch current value. (A **branch** is one of the paths for current flow in a parallel circuit.) The circuit in Figure 5–9 is a portion of a total circuit. Ammeter 1 indicates 3 amperes and ammeter 2, in the lamp circuit, indicates 2 amperes. Since 3 amperes leave junction B and only 2 amperes come to junction B through the lamp, there must be 1 ampere flowing through resistor R_1 to the junction B.

The total circuit of Figure 5–9 could contain another branch, as in Figure 5–10. The current entering junction A, coming from junction B, is 3 amperes. The current leaving junction A, going to the generator, is 6 amperes. Where did the other 3 amperes come from? There must be 3 amperes flowing into junction A through R_2. Given that R_2 is an 8-ohm resistor, other unknown values of the circuit may be determined. To determine the voltage across R_2, use Ohm's law to solve for E.

$$E = I \times R$$
$$E = 3 \times 8$$
$$E = 24\,\text{V}$$

The voltage across each component in parallel is the same. The voltage across R_2 is 24 volts, the voltage across L1 is 24 volts, and the voltage across R_1 is 24 volts. To determine the resistance of lamp L1, use Ohm's law for R.

$$R_{L1} = \frac{E}{R}$$
$$R_{L1} = \frac{24}{2}$$
$$R_{L1} = 12\,\text{ohms}$$

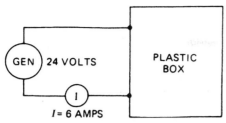

FIGURE 5–11   Black-box analogy.

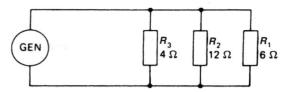

FIGURE 5–13   Three parallel resistors.

$$R_T = \frac{E}{I}$$

$$R_T = \frac{24}{6}$$

$$R_T = 4 \text{ ohms}$$

The Product-over-the-Sum Method

When the box in Figure 5–11 was opened, a 12-ohm resistor was found connected in parallel with a 6-ohm resistor (Figure 5–12). The resistance of the parallel combination could also be found using the product-over-the-sum method, as in the formula

$$R_T = \frac{R_1 \times R_2 \, (\text{product of } R_1 \text{ and } R_2)}{R_1 + R_2 \, (\text{sum of } R_1 \text{ and } R_2)}$$

Use the product-over-the-sum method to solve for the parallel combination in Figure 5–12.

$$R_T = \frac{12 \times 6}{12 + 6}$$

$$R_T = \frac{72}{18}$$

$$R_T = 4 \text{ ohms}$$

This is the same total resistance value that was calculated using Ohm's law.

If more than two resistors are connected in parallel, the product-over-the-sum method may be used on two resistors at a time. An example of this procedure is given using the circuit in Figure 5–13.

To determine the resistance of R_1

$$R_1 = \frac{E}{I}$$

$$R_1 = \frac{24}{1}$$

$$R_1 = 24 \text{ ohms}$$

To determine the resistance of the circuit, the total current and voltage must be used. The total current as indicated by ammeter 3 (Figure 5–10) is 6 amperes. The total voltage is 24 volts.

$$R_T = \frac{E}{I}$$

$$R_T = \frac{24}{6}$$

$$R_T = 4 \text{ ohms}$$

The total resistance of the circuit is 4 ohms, which is less than the resistance of any of the three components: R_1 (24 ohms), L1 (12 ohms), or R_2 (8 ohms).

The total resistance of a parallel circuit is less than the smallest branch resistance.

There are other interesting relationships in parallel circuits. The circuit in Figure 5–11 contains a generator producing 24 volts, wires from the generator to a plastic box, and an ammeter indicating 6 amperes. There is no indication as to what is in the box. Solve for the total resistance (R_T) of the circuit in the box.

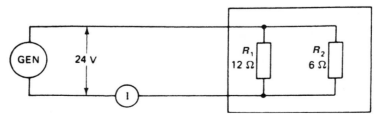

FIGURE 5–12   Inside the black box.

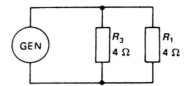

FIGURE 5–14   Equivalent circuit.

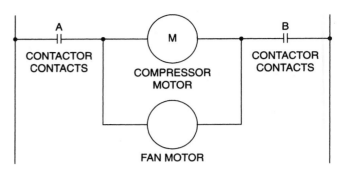

FIGURE 5–16   Series–parallel circuit: compressor contacts in series, motors in parallel.

What is the total resistance of the circuit? It has been shown that the combination of R_1 and R_2 is equal to a 4-ohm resistor. Figure 5–14 shows the equivalent circuit after resistors R_1 and R_2 are combined. The product-over-the-sum method may now be used to determine the resistance of the total circuit (R_T).

$$R_T = \frac{R_1 \times R_2}{R_1 + R_2}$$

$$R_T = \frac{4 \times 4}{4 + 4}$$

$$R_T = 2 \, \text{ohms}$$

Equal Parallel Resistors

The total resistance of two resistors of equal value connected in parallel is thus equal to the value of one resistor divided by the number of equal resistors. If three 9-ohm resistors are connected in parallel, R_T will be equal to 9 divided by 3 (Figure 5–15).

There are other methods of solving for the total resistance of parallel circuits that will not be covered in this text. The other methods require higher mathematics. At this time, however, you should understand the following:

1. The same voltage appears across each component of a parallel circuit.
2. The total current in a parallel circuit is equal to the sum of the currents in the branches.

3. The total resistance (R_T) of a parallel circuit is always smaller than the smallest resistor connected in the parallel combination.

SERIES–PARALLEL CIRCUITS

An example of a series–parallel circuit can be made using contactor contacts, a compressor motor, and a condenser fan motor. Look at Figure 5–16. The contactor contacts A and B are in series with the remainder of the circuit, the compressor motor, and the condenser fan motor; the motors are in parallel.

All of the circuit's current must flow through the contactor contacts. The current then splits, with the major portion going to the compressor and a small portion going to the condenser fan.

Ohm's law may be applied to resistive components connected in series–parallel circuits; however, the relationship of the components must be considered when doing so.

A series–parallel circuit is made up of a combination of components. Some are connected in series, and others are connected in parallel. When working with series–parallel circuits, it is necessary to use the laws of series circuits on the series elements and the laws of parallel circuits on the parallel elements.

The circuit in Figure 5–17 is a series–parallel circuit. The 10-ohm resistor R_3 is connected in series with the parallel combination R_1 and R_2. To solve for

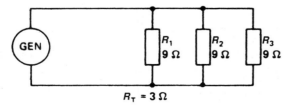

FIGURE 5–15   Equal resistors in parallel.

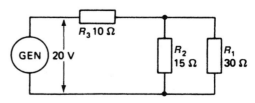

FIGURE 5–17   Series–parallel circuit.

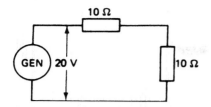

FIGURE 5–18   Equivalent series circuit.

the current flow in the circuit, the resistors must be combined. Resistors R_1 and R_2 are in parallel.

$$R_T = \frac{R_1 \times R_2}{R_1 + R_2}$$

$$R_T = \frac{15 \times 30}{15 + 30}$$

$$R_T = \frac{450}{45}$$

$$R_T = 10 \text{ ohms}$$

The equivalent circuit is a series circuit (Figure 5–18).

$$R_T = R_1 + R_2$$

$$R_T = 10 + 10$$

$$R_T = 20 \text{ ohms}$$

The generator is producing 20 volts. According to Ohm's law, the current through the generator is

$$I = \frac{E}{R}$$

$$I = \frac{20}{20}$$

$$I = 1 \text{ ampere}$$

Look back at the original circuit shown in Figure 5–17. The current through the series 10-ohm resistor R_3 is 1 ampere and there are 10 volts across it. With 10 volts across R_3, there are 10 volts left (from the 20 volts supplied) to appear across R_1 and R_2. Since the same voltage appears across resistors in parallel, the current flow through R_1 is

$$I = \frac{E}{R}$$

$$I = \frac{10}{30}$$

$$I = \frac{1}{3} \text{ or } 0.333 + \text{ ampere}$$

The current flow through R_2 is

$$I = \frac{E}{R}$$

$$I = \frac{10}{15}$$

$$I = \frac{2}{3} \text{ or } 0.666 + \text{ ampere}$$

The total current entering the junction (1 ampere) is equal to the sum of the currents in the branches.

$$\frac{1}{3} \text{ ampere } + \frac{2}{3} \text{ ampere } = 1 \text{ ampere}$$

or

0.3333 ampere plus 0.6666 ampere = 0.9999 or 1 ampere

SOLVING FOR AN UNKNOWN RESISTOR

There are times when the information known about a circuit is other than the resistance values. For example, using the circuit shown in Figure 5–17 again, the known information might be shown in Figure 5–19: The generator is producing 20 volts, the value of resistor R_1 is 30 ohms, the value of resistor R_2 is 20 ohms, and a voltmeter across R_1 indicates 10 volts. The unknown information can be found by using the laws of series circuits, parallel circuits, and Ohm's law.

The current flow through R_1 can be found using Ohm's law. It has already been shown that with 10 volts across 30 ohms the current is ⅓ ampere. According to the laws of parallel circuits, the voltage across R_2 is the same as across R_1, 10 volts. The current through R_2 can be determined: 10 volts across 15 ohms provides ⅔ ampere. The current entering junction point A is equal to the sum of the currents leaving it. Therefore, 1 ampere (⅓ ampere plus ⅔ ampere) enters junction point A.

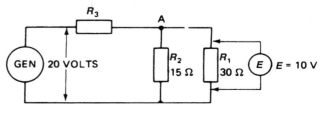

FIGURE 5–19   Using Ohm's law in series–parallel circuits.

This 1 ampere is the current flowing through the unknown series resistor R_3. Since the supply voltage is 20 volts and there are 10 volts across the parallel resistors R_1 and R_2, there must be 10 volts across R_3 (the supply of 20 volts minus the 10 volts across the parallel branch leaves 10 volts).

According to Ohm's law

$$R_3 = \frac{E}{I}$$

$$R_3 = \frac{10}{1}$$

$$R_3 = 10 \, \text{ohms}$$

SWITCHES IN SERIES–PARALLEL CIRCUITS

In air-conditioning and refrigeration electrical systems, switches are used to control the operation of devices such as compressor motors, fan motors, heaters, relays, and indicator lights. These switches are connected in series, parallel, or series parallel as required to accomplish the desired circuit action. Some examples of switch circuit combinations are shown in Figures 5–20a through 5–20f.

The lamps shown in earlier examples of circuit diagrams could easily have been compressor motors or fan motors. The switches could have been circuit breakers, temperature controls, or pressure controls. The important point to remember is how switches control various electric **alternating current (AC)** circuits.

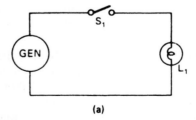

(a)

FIGURE 5–20(a) Switch S1 must be closed if lamp L1 is to light.

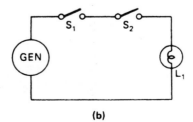

(b)

FIGURE 5–20(b) Switches S1 and S2 must be closed if lamp L1 is to light.

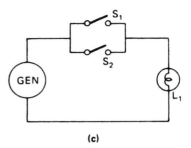

(c)

FIGURE 5–20(c) Either switch S1 or S2 must be closed if lamp L1 is to light.

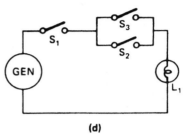

(d)

FIGURE 5–20(d) Switches S1 and S2 or S3 must be closed if lamp L1 is to light.

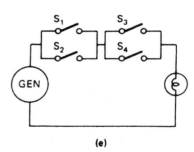

(e)

FIGURE 5–20(e) Switches S1 and S2 and S3 or switch S4 must be closed if lamp L1 is to light.

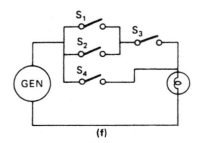

(f)

FIGURE 5–20(f) Switch S1 or S2 with switch S3 will cause lamp L1 to light, or switch S4 alone will light lamp L1.

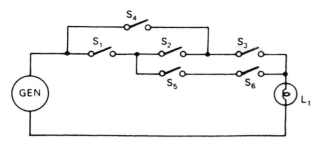

FIGURE 5–21 Switch circuit.

SNEAK CIRCUITS

Whenever switches are connected in series parallel, the possibility of sneak circuits exists. **A sneak circuit** is one that provides for the operation of a component at a time when operation is not wanted. Sneak circuits are not always obvious at the time a circuit is designed. Some sneak circuits only cause operation under odd conditions and are therefore difficult to discover.

Consider the circuit shown in Figure 5–21. The path through the switch is normally considered to start at the left top and proceed through the switches to the lamp. Different paths and switches might be considered, but most people would follow a path from left to right. The circuit designer no doubt drew the circuit while considering paths from left to right.

A sneak circuit might exist in a path through S2 from right to left. If switches S4, S2, S5, and S6 are closed, there is a complete path for current flow, as shown in Figure 5–22. The circuit path may be an unwanted path causing lamp L1 to be on during a period when it should not be on. When a sneak circuit is discovered, a redesign is required. Most sneak circuits would not be as obvious as the one shown in this example. In some cases, a considerable amount of circuit investigation is necessary before the prob-

lem is discovered. When a sneak circuit is causing a problem, the technician investigating the problem normally assumes that something is malfunctioning, but in fact, every component is operating correctly.

Sneak circuits are normally in-plant manufacturing problems, not field problems. However, they do sometimes occur in the field, and technicians must be aware of them.

SUMMARY

- In parallel resistive circuits, the total current is the sum of the currents in the branches.
- The same voltage appears across each component in parallel.
- The total resistance may be found by using the product-over-the-sum method.
- In series–parallel circuits, the laws of series circuits apply to the portions of the circuit connected in series.
- In series–parallel circuits, the laws of parallel circuits apply to the portion of the circuit connected in parallel.
- Ohm's law may be applied to the whole circuit or any part of the circuit.

PRACTICAL EXPERIENCE

Required equipment An ohmmeter, a 120-V air-conditioner (cover removed), and a wiring diagram of the system.

Procedure

1. Examine the plug attached to the line cord of the air-conditioning (A/C) unit.

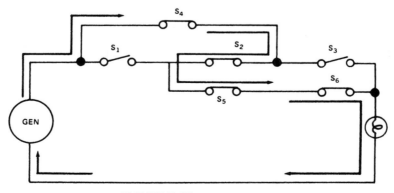

FIGURE 5–22 Sneak circuit.

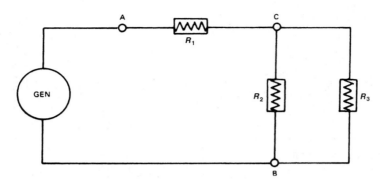

FIGURE 5–23   Test circuit.

2. Determine which plug terminal is the hot terminal and which is the neutral terminal.

3. Follow the neutral terminal to a connection within the A/C unit. (Use the ohmmeter to make this check.)

4. Follow the hot line from the plug to the first connection within the A/C unit.

5. Is there a switch in the hot line controlling the complete A/C unit?

6. Would you consider this switch in series with the rest of the unit?

7. Refer to the wiring diagram of the system.

8. Determine how the thermostat contacts are connected with reference to the compressor motor.

9. Are the contacts in series or parallel?

10. How would you consider the overall connection of the compressor circuit with the fan circuit?

Conclusions

1. Ohm's law is true for the whole circuit or any part of a circuit.

2. In a series–parallel circuit, the laws of series circuits are applied to the component connected in series and the laws of parallel circuits are applied to components connected in parallel.

REVIEW QUESTIONS

Draw a sketch of each circuit before attempting to solve these problems.

1. Four 40-ohm resistors are connected in parallel. What is the total resistance of the combination?

2. A 40-ohm resistor and a 20-ohm resistor are connected in parallel. The voltage across the 40-ohm resistor is 80 volts. What is the total current in the circuit?

3. A 10-ohm resistance is needed in a control circuit. Three 30-ohm resistors are available. How could these resistors be connected to obtain the required resistance?

4. A resistance is needed in a control circuit. The resistance must be more than 2 ohms but less than 4 ohms. Three 2-ohm resistors are available. How would you connect the 2-ohm resistors to obtain the required resistance?

5. Figure 5–21 is a lamp-and-switch circuit. Which switches must be closed to light the lamp, using the least number of switches?

6. In Figure 5–23, if the voltage across R_1 is 8 volts and the voltage across R_3 is 6 volts, the voltage across R_2 must be _____ volts.

7. Using the same voltages given in question 6, the generator voltage must be _____ volts.

8. In Figure 5–23, if the voltage from point A to point C is 18 volts and the voltage from point A to point B is 24 volts, then the voltage from point C to point B must be _____ volts.

9. In Figure 5–23, if the current through R_1 is 4 amperes and the current through R_2 is 3 amperes, then the current through R_3 must be _____ amperes.

10. In Figure 5–23, if the current through R_2 is 2 amperes and the current through R_3 is 3 amperes, and R_1 is a 10-ohm resistor, then the voltage across R_1 is _____ volts.

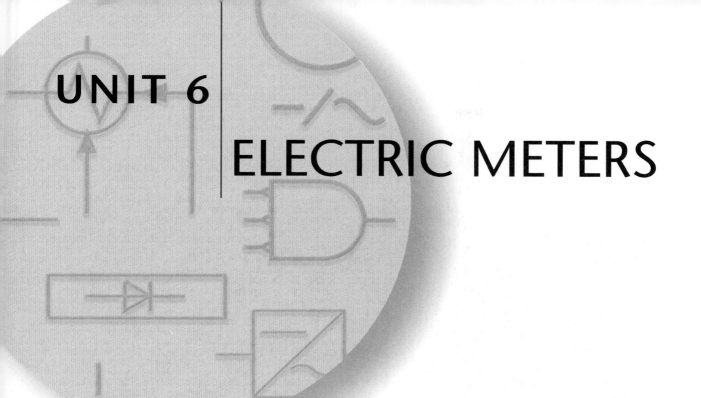

UNIT 6 | ELECTRIC METERS

OBJECTIVES

After completion and review of this unit, you should be able to understand:

- The construction of analog meter movements.
- The internal meter connections that provide for meter movement applications such as an ammeter, voltmeter, or ohmmeter.
- In ammeter service, a shunt is used in parallel with the meter movement providing proportional amounts of current through the shunt and meter movements.
- In voltmeter service, a series resistor provides for current flow through the meter movement proportional to the voltage applied.
- When the meter movement is used to indicate resistance, an internal battery provides power while the resistance to be measured determines the current flow through the meter movement.
- The clamp-on ammeter provides a means of measuring current without the need to disconnect wires.
- A digital meter is an inexpensive, rugged meter that is easy to read.

An electric meter is used to determine the characteristics of an electric circuit or component. The common types of meters used by the air-conditioning and refrigeration technician include the ammeter, voltmeter, and ohmmeter. Occasionally, the technician will have the opportunity to use a wattmeter. Meters used in the trade today include both analog and digital types.

ANALOG METERS

The internal operation of analog meters is provided as a review of circuit operations. Series, parallel, and series–parallel circuits provide for proper operation of analog meters.

An analog electric meter uses a meter movement that operates on the electric motor principle. To obtain meter pointer deflection, a current must pass through the meter-movement coil. This current sets up a magnetic field in the meter-movement coil that is repelled by the fixed magnetic field provided by the permanent magnets. The stronger the current flow through the meter-movement coil, the greater the deflection of the coil and pointer, as shown in Figure 6–1.

In the meter movement shown in Figure 6–1, the permanent magnets are indicated by the capital N for north pole and capital S for south pole. The movement deflection coil is pivoted at the center and has a pointer attached to it. A coil spring attached to the meter movement holds the movement and pointer at zero deflection when no current flows through the movement coil.

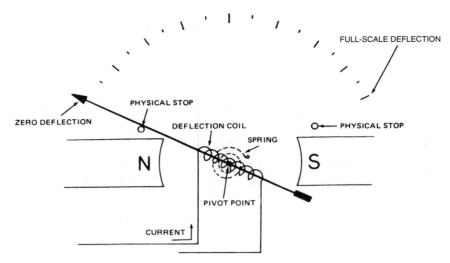

FIGURE 6–1 Meter movement.

Physical stops in the form of pins are located just beyond the zero and full-scale deflection points to limit the movement of the pointer. When current flows through the meter-movement deflection coil, a magnetic field is set up around this coil. The magnetic field of the deflection coil will be repelled by the field of the permanent magnets. The meter movement will rotate around the pivot point, causing the pointer to move up the scale.

As the meter movement rotates up the scale, pressure will increase on the coil spring attached to the movement. The rotation will stop when the pressure of the repelling magnetic fields is balanced by the reverse pressure offered by the coil spring.

A different amount of current flow through the meter movement coil will cause a different amount of meter movement and pointer deflection.

AMMETERS

An ammeter is used to measure current flow. Current flow is the movement of electrons through the circuit. To measure current flow, the circuit must be opened and the ammeter inserted so that the current to be measured flows through the ammeter. The circuit connections are shown in Figure 6–2.

Normally, the current to be measured is much higher than the amount needed to deflect the meter movement to full scale. To make meters useful in line-current measuring systems, a **shunt** is installed within the meter. The shunt is a low-resistance current path that allows a proportional amount of current to flow through the meter movement. The remaining current by-passes the meter movement

by going through the shunt. This is shown in Figure 6–3.

Suppose that a 1-ampere full-scale meter is designed using a meter movement that requires 0.001 amperes (1 milliampere, mA) full-scale deflection. The meter movement resistance is 100 ohms. According to Ohm's law, there will be $0.001\,\text{A} \times 100\,\Omega = 0.1\,\text{V}$ across the meter movement at full-scale deflection. The scale of the meter would be changed to indicate 1 ampere instead of 1 mA, but it would still take only 1 mA through the meter movement to cause full-scale deflection.

This arrangement is shown in Figure 6–4. One ampere is coming down the line. Only 1 mA goes through the meter movement, whereas 999 mA goes through the shunt. The shunt is in parallel with the meter movement. The resistance of the shunt is

$$R_{shunt} = \frac{0.1\,\text{V}}{999\,\text{mA}} = \frac{0.1\,\text{V}}{0.999\,\text{A}} = 0.1001$$

The shunt in the ammeter is a low-resistance path for current flow. An ammeter is therefore a

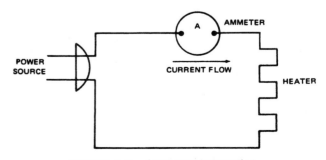

FIGURE 6–2 Ammeter connection.

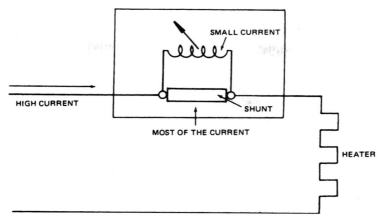

FIGURE 6–3   Shunt circuit.

low-resistance device. An ammeter is never connected across the circuit. The low resistance of the ammeter will short the circuit, usually causing the line fuse to open. Ammeters are never connected in parallel with any component. Ammeters are always connected in series.

Many meters contain a switch system that provides more than one full-scale deflection sensitivity. A single meter could have a 1-, 10-, and 100-ampere full-scale reading. Figure 6–5 shows an example of an analog volt-ohm-milliammeter.

Clamp-on Ammeters

One problem in using a standard ammeter is that the circuit must be opened to insert the ammeter. In large, high-current situations, opening the circuit

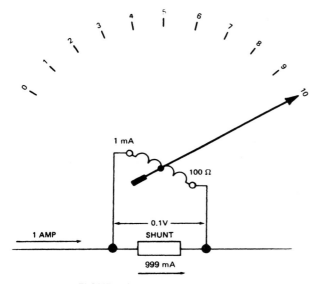

FIGURE 6–4   Shunt operation.

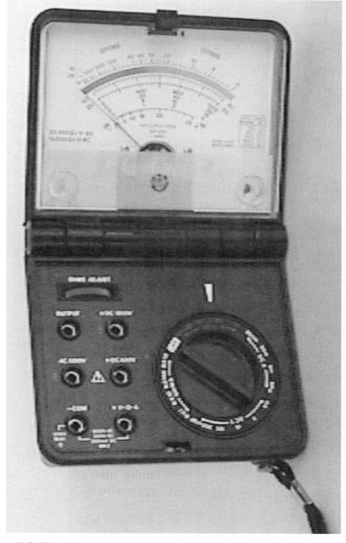

FIGURE 6–5   Volt-ohm-milliammeter. (Courtesy of BET Inc.)

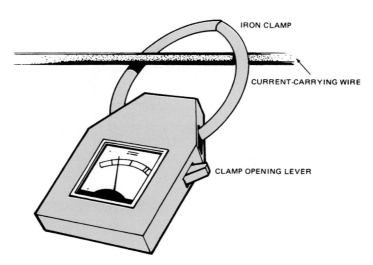

FIGURE 6–6   Clamp-on ammeter.

can be a time-consuming project. This is especially true if current must be measured in a number of different current-carrying lines. The clamp-on ammeter was developed to simplify the current measurement procedure.

Unit 2 explained that any wire carrying current has a magnetic field around it. If the current is an alternating current (AC), the current will vary in the form of a sine wave. The varying current will produce a varying magnetic field around the wire. The strength of the magnetic field is directly proportional to the strength or magnitude of the current flow.

The clamp-on ammeter uses a voltage step-up, current step-down transformer in the current-measuring process. An iron-clamping mechanism is placed around the wire carrying AC current, as shown in Figure 6–6. Inside the clamp-on ammeter, the iron clamp completes a magnetic circuit through a current transformer. The transformer produces an output that is connected to a standard meter movement. The meter movement has a current scale that indicates the current flow in the AC current-carrying wire. After the current is measured, the clamp-opening lever is depressed, which opens the clamp, allowing the meter to be removed from the current-carrying wire.

A number of manufacturers provide clamp-on ammeters. The meters have different specifications, particularly with regard to current ranges. Always read equipment specifications before using the equipment.

Measuring Small Currents on a High-Scale Meter

The clamp-on ammeter reacts to the magnetic field developed around the wire passing through the clamp. (See Figure 6–6.) If the wire is formed into a coil, the clamp-on ammeter will react to the total strength of the magnetic field in the clamp area.

In Figure 6–7, the clamp is placed around five turns of the current-carrying wire. The meter will indicate five times the current in the wire. In this case, there is 1 ampere of current flow in the wire. The ammeter indicates 5 amperes. The meter reader then divides by 5 to get an accurate measurement of 1 ampere of current flow in the wire. Keep in mind that it might be difficult to accurately read an indication of 1 ampere on a 20-ampere scale.

Voltmeters

A voltmeter can be developed with the same meter movement used in the ammeter. That meter movement required 1 milliampere for full-scale deflection and had 100 ohms of resistance.

A 300-volt meter is shown in Figure 6–8. The scale has been drawn using 300 V as the full-scale deflection. Since voltage is measured across a circuit, the voltmeter will have to be placed across the circuit in order to measure voltage. There will be 300 volts at the terminals of the voltmeter (Fig. 6–8). According to Ohm's law, the total resistance will equal the voltage divided by the current.

$$R = \frac{300}{0.001} = 300,000 \text{ ohms}$$

The scaling resistor would be 299,900 ohms; the meter-movement resistance of 100 ohms added to it equals the 300,000-ohm total. The procedure used in making a voltmeter is to determine the required

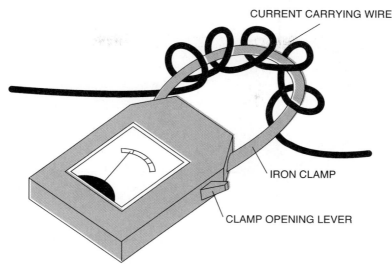

CURRENT CARRYING WIRE

IRON CLAMP

CLAMP OPENING LEVER

FIGURE 6–7 Clamp-on ammeter, indicating five times the current flow in the wire.

resistance by dividing the desired full-scale voltage by the meter-movement full-scale current. The scaling resistor is then the total resistance minus the resistance of the meter movement.

OHMMETERS

An ohmmeter can be developed using the same 1-milliampere full-scale meter movement. An ohmmeter requires the use of an internal power source. The standard source is a 1.5-volt cell. Observe the circuit shown in Figure 6–9. If the terminals of the meter are shorted together, a complete series circuit will exist. The power source is 1.5 volts. The series resistances are 100 ohms of meter resistance, the

1300-ohm resistor, and the 300-ohm potentiometer. The potentiometer is adjusted to provide a total resistance of 1500 ohms for the series circuit. Using Ohm's law, we then find

$$I = \frac{E}{R} = \frac{1.50}{1500} = 0.001 \, \text{A}$$

The current flow through the circuit is 1 milliampere, the full-scale current of the meter movement. The meter movement rotates to full scale, where the scale is marked 0 ohms. The meter leads are shorted together. The resistance is 0 ohms.

If the meter leads are connected across a 1500-ohm resistor, the series circuit will contain the 1500-ohm resistor, the 100-ohm meter movement, the

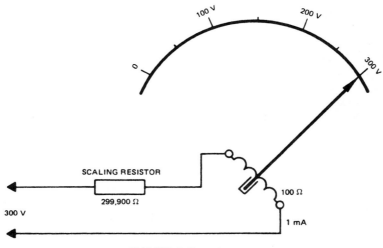

100 V   200 V

0   300 V

SCALING RESISTOR

299,900 Ω

300 V

100 Ω

1 mA

FIGURE 6–8 Voltmeter.

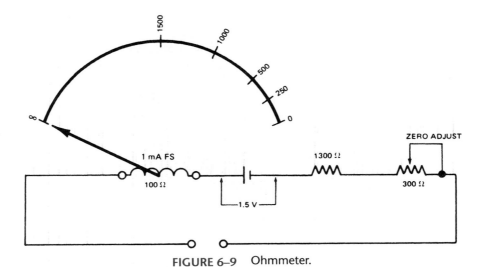

FIGURE 6–9   Ohmmeter.

1300-ohm resistor, and the potentiometer set at 100 ohms. The total resistance is 3000 ohms. This can be checked using Ohm's law.

$$I = \frac{E}{R} = \frac{1.5}{3000} = 0.005 \text{ A}$$

Thus 0.5 milliampere flows through the circuit. The meter movement will move to approximately half-scale, where the marking is 1500 ohms.

It is obvious that the value of the resistor connected between the terminals of the meter will determine the total current flow through the series circuit and therefore the amount of current through the meter movement. The current determines how far up the scale the meter will rotate. A high resistance will keep the current low and provide little

meter movement. A low resistance will provide a higher current and the meter may rotate close to 0 ohms. Many ohmmeters provide three or four multipliers for resistance measurement. Common multipliers include $R \times 1, R \times 10, R \times 100, R \times 1000,$ and $R \times 10,000$.

DIGITAL METERS

Many new types of meters are becoming available as new digital electronic circuitry is developed. Digital meters are easier to read than the standard meters since the meter reader has only to read the digits. With the older analog meters, the meter reader must interpolate, that is, decide what the meter is indicating. It takes practice to obtain correct answers. Figure 6–10 is a photograph of a digital multimeter.

SUMMARY

- Analog meters contain a meter movement (restrained electric motor).
- The same analog meter movement is often used to indicate current, voltage, and resistance.
- The connection of the meter movement to other electrical components and the method of connection determine whether current, voltage, or resistance will be measured.
- It requires experience to properly read an analog meter.
- A digital meter is not necessarily more accurate than an analog meter, although it is easier to read.

FIGURE 6–10   Digital multimeter. (Courtesy of BET Inc.)

PRACTICAL EXPERIENCE

Required equipment   Operable A/C system with access to leads to the compressor motor and condenser fan, clamp-on ammeter, and analog and digital voltmeters

Procedure

1. In an operating system, measure the line voltage with an analog voltmeter. _____ volts

2. Measure the same voltage with a digital voltmeter. _____ volts

3. Which *meter* is more accurate? Discuss your answer to step 3 with colleagues.

4. Using the clamp-on ammeter, measure the current flow to the compressor motor. _____ amps

5. Using the clamp-on ammeter, measure the current flow to the fan motor. _____ amps

6. For the purpose of steps 4 and 5, is the clamp-on ammeter easier to use than a standard ammeter?

Conclusions

1. The analog–meter movement is a spring-restricted small motor.

2. The analog meter responds to current flow through the meter movement coil.

3. In ammeter service, a shunt is used in parallel with the meter movement, providing proportional amounts of current through the shunt and meter movement.

4. In voltmeter service, a series resistor provides for current flow through the meter movement proportional to the voltage applied.

5. When the meter movement is used to indicate resistance, an internal battery provides power, whereas the resistance to be measured determines the current flow through the meter movement.

6. The clamp on the ammeter provides a means of measuring current without the need for disconnecting wires.

7. Digital meters serve as inexpensive rugged meters that are easy to read.

REVIEW QUESTIONS

The meter shown in Figure 6–11 is a volt-ohm-ammeter. The switch in the meter provides for the selection of volts, current, or resistance. The position of the switch selects the range of the meter.

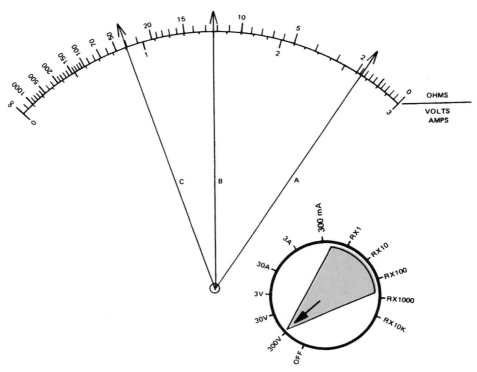

FIGURE 6–11   Volt-ohm-ammeter.

The meter is shown with a pointer placed at different positions. Read the following questions and indicate what value the meter is indicating.

1. The selector switch is in the 30 V position. Pointer B position indicates _____ volts.

2. The selector switch is in the 3 A position. Pointer A indicates _____ amperes.

3. The selector switch is in the 3 V position. Pointer C indicates _____ volts.

4. The selector switch is in the 300 V position. Pointer A indicates _____ volts.

5. The selector switch is in the R × 10 position. Pointer A indicates _____ ohms.

6. The selector switch is in the R × 1000 position. Pointer B indicates _____ ohms.

7. The selector switch is in the 30 A position. Pointer B indicates _____ amperes.

8. The selector switch is in the R × 100 position. Pointer C indicates _____ ohms.

9. The selector switch is in the R × 1 position. Pointer B indicates _____ ohms.

10. The selector switch is in the 300 mA position. Pointer A indicates _____ mA.

UNIT 7

BATTERIES AND ELECTROMOTIVE FORCE

OBJECTIVES

After completion and review of this unit, you should be able to:

- Understand batteries as a source of direct current.
- Demonstrate the relationship of balanced and unbalanced circuits.

BATTERIES

The purpose of this unit is not to explain how chemical energy is stored, converted, and then released as electrical energy when needed, but to cover the connection of cells, internal resistance, and the terminal voltage of batteries.

An Italian chemist-physicist, Count Alessandro Volta (1745–1827), invented the **battery** in 1780. His battery, the voltaic pile, was the first source of constant-current electricity. Because of his discovery, the *volt,* a unit of electromotive force, was named in his honor in 1881.

Whenever two different metals are placed in an acid or salts solution in an insulated case, a voltage is produced between the metals by chemical action. This combination of the two metals and the solution is known as a **cell** (Figure 7–1). One plate is positive and the other is negative. All cells and batteries provide a source of direct current. The term *cell* refers to a single unit, the two metals in a solution.

The term *battery* refers to cell combinations in series or in parallel that provide for higher current or voltage capabilities. A battery is used for storing and converting chemical energy into electrical energy. Commonly, the term *battery* is now used instead of cell; for example, a flashlight battery is actually a flashlight cell.

There are two main types of cells: primary and secondary. Primary cells are temporary (non-rechargeable) and use up the materials of which they are composed while providing electrical energy. Figure 7–2 shows some examples of primary cells. Secondary cells change in chemical composition while providing electrical energy but may be recharged.

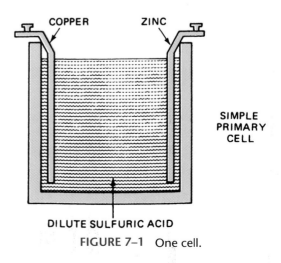

COPPER ZINC

SIMPLE PRIMARY CELL

DILUTE SULFURIC ACID

FIGURE 7–1   One cell.

FIGURE 7–2   Primary cells. (Courtesy of BET Inc.)

Recharging converts the cell back to its original condition by passing current through it in the opposite direction. A flashlight battery is a good example of a primary cell, and a car battery is a good example of a secondary cell.

Chargers have become available that will recharge primary cells, including some flashlight batteries. Most batteries are not returned to their original condition, however. By design, some batteries are intended to be recharged, whereas others are intended to be discarded when discharged.

The symbols used for cells and batteries are shown in Figure 7–3. The longer line is the positive terminal of the cell; the shorter line is the negative terminal.

The voltage output of a cell is determined by the materials used to make up the cell. The standard carbon–zinc (flashlight) cell comes in many sizes. The small penlight cell has an output of 1.5 volts. The common D size flashlight cell also has an output of 1.5 volts. The D size cell can, however, deliver higher current for a longer time than the penlight cell.

INTERNAL RESISTANCE

Everything through which current flows has resistance. Batteries have an internal resistance. If the open-circuit voltage of a cell is measured and current

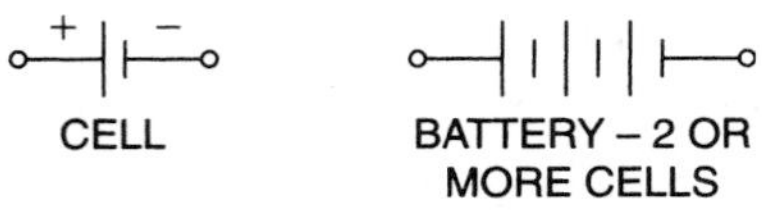

FIGURE 7–3   Symbols for cells and batteries.

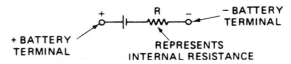

FIGURE 7–4   Battery or cell with internal resistance.

is then supplied (by that cell) to an external circuit, its voltage will decrease. This decrease in voltage is equal to the voltage developed across the internal resistance of the cell. Figure 7–4 illustrates a battery with the internal resistance shown before the output terminals.

A cell with a voltmeter in parallel, an ammeter in series, and a load resistor of 10 ohms in series controlled by a switch is shown in Figure 7–5. When the switch is open, the cell voltage is 1.5 volts. When the switch is closed, the ammeter indicates 0.14 ampere of current. The voltmeter indicates 1.4 volts. This means that 0.1 volt is being developed across the internal resistance of the battery (1.5 − 1.4 = 0.1). The internal resistance of the cell may be determined by using Ohm's law.

$$R_i = \frac{E}{I}$$

$$R_i = \frac{0.1}{0.14}$$

$$R_i = 0.714 \, \text{ohm}$$

All batteries do not have the same internal resistance. As a cell becomes discharged, its internal resistance increases. As the internal resistance of the cell increases, its ability to supply current decreases.

BATTERIES IN SERIES

Cells are connected in series whenever a voltage higher than that which can be supplied by a single cell is needed. Two flashlight cells are shown in

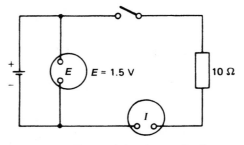

FIGURE 7–5   Determining internal resistance.

series in Figure 7–6. The total voltage available is 1.5 plus 1.5 or 3 volts. This is standard for a two-cell flashlight. If three cells are connected in series, as in Figure 7–7, the available voltage is 4.5 volts (1.5 + 1.5 + 1.5 = 4.5).

CELLS IN PARALLEL

Whenever higher current capability is needed, cells are connected in parallel. Certain precautions must be taken when connecting cells in parallel. Most importantly, the cells must have the same terminal voltage. For example, a 1.2-volt cell should not be connected in parallel with a 1.5-volt cell.

A load device drawing a constant current of 3 amperes requires a voltage of at least 1.3 volts (Figure 7–8). Cells of 1.5 volts are available. Each cell has an internal resistance of 0.1 ohm. With one cell connected to the load device, there is 0.3 volt developed across the internal resistance of the cell, leaving 1.2 volts to appear across the load. If two cells are connected in parallel, the situation improves. In Figure 7–9, the load device is still drawing 3 amperes. Each cell, however, is supplying only 1.5 amperes. The voltage developed across the internal resistance is

$$E = I \times R$$
$$E = 1.5 \times 0.1$$
$$E = 0.15 \text{ volt}$$

The voltage at the terminals of the cells and across the load is

$$E \text{ out} = 1.5 - 0.15$$
$$E \text{ out} = 1.35 \text{ volt}$$

This voltage meets the original requirement of at least 1.3 volts across the load.

Whether or not more cells should be connected in parallel depends on how long current is to be drawn. If three cells are connected in parallel, each has to supply only 1 ampere. The voltage developed across the internal resistance is only 0.1 volt. The terminal voltage of each cell and the voltage to the load is 1.4 volts (Figure 7–10). The three-cell combination will last longer than the two-cell combination.

Internal resistance is present in every device that produces electricity, whether it is a battery, generator, bimetal strip, or solar cell. The effect of internal resistance is the same on these other devices as on batteries.

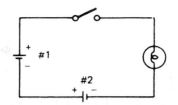

FIGURE 7–6   Two cells in series.

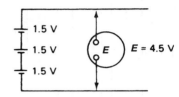

FIGURE 7–7   Three cells in series.

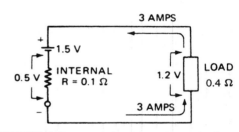

FIGURE 7–8   Internal voltage drop, single cell.

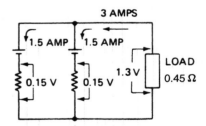

FIGURE 7–9   Lower internal voltage drop with parallel cells.

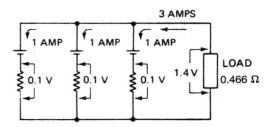

FIGURE 7–10   Further decrease in voltage drop with a three-cell supply.

UNBALANCED AND BALANCED CIRCUITS

It has been shown that when two batteries are connected in series, the total voltage available is the sum of the two battery voltages. The individual battery voltages are also available (Figure 7–11). This is similar to regular household electric power, in which 240 volts AC may be available from the two hot lines, and 120 volts are available from either hot line to the neutral or ground wire (Figure 7–12).

Unbalanced Circuits

An example of an unbalanced circuit is shown in Figure 7–13. One resistor of 6 ohms is connected across a 12-volt section of the power source. A current of 2 amperes will flow in the circuit of battery A and in the 6-ohm resistor when there is no current flow in battery B. If a 12-ohm resistor is added to the system, the circuit will still be unbalanced, as shown in Figure 7–14. Note that at the junction below the 6-ohm resistor, the current splits: 1 ampere returns to battery A while the other 1 ampere flows down through the 12-ohm resistor and through battery B. This 1 ampere joins with the 1 ampere returning on the neutral wire to provide the 2 amperes flowing through battery A.

Balanced Circuits

If both resistors are of the same value, no current flows in the neutral wire (Figure 7–15). Although balanced circuits are possible under laboratory conditions, they are almost never obtained in actual practice in household situations. The unbalanced circuit is the most common.

The circuit in Figure 7–16 shows the lower voltage—as well as the higher voltage—components in

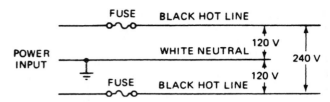

FIGURE 7–12   Standard house AC supply.

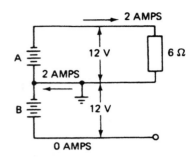

FIGURE 7–13   Unbalanced system.

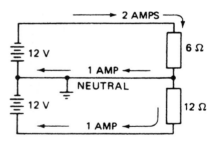

FIGURE 7–14   Unbalanced system, both lines loaded.

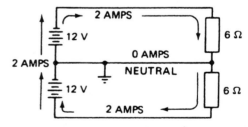

FIGURE 7–15   Balanced system.

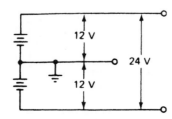

FIGURE 7–11   Two batteries with a ground.

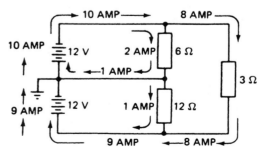

FIGURE 7–16   Unbalanced system using high and low voltage.

an unbalanced system. In each junction, the current entering the junction is equal to the current leaving the junction. Ohm's law is true for each component of the system.

The relationship shown in Figure 7–16 parallels that of the 240/120 systems used in homes. The voltage levels in homes are different from those shown with batteries, but the circuit relationships are the same.

SUMMARY

- A cell is a combination of two dissimilar metals in an acid or alkaline solution.
- A battery is the combination of cells in series or in parallel.
- If cells are connected in series, the available voltage is the sum of the individual cell voltages.
- If cells are connected in parallel, the available current is the sum of the currents available from the individual cells.
- Batteries or cells should be connected in parallel only when they are of the same voltage.
- As a battery becomes discharged, its internal resistance increases.

PRACTICAL EXPERIENCE

Required equipment Flashlight assembly, clip leads, and voltmeter

Procedure

1. Remove the lamp portion of the flashlight.
2. Using clip leads, make connections completing the flashlight circuit.
3. With the light lit, measure the voltage across the lamp. _____ volts
4. Disconnect the lamp assembly.
5. Measure the battery voltage. _____ volts

6. Is the light on voltage lower than the light off voltage? (NOTE: If an ammeter is available, the current flow [lamp on] may be measured. The internal resistance of the battery is equal to the no-load voltage minus the full-load voltage divided by the load current.)

Conclusions

1. A balanced system is present when the current in both hot lines of a system is equal and there is no current in the neutral line.
2. In the home or in industry, the systems are more likely to be unbalanced than balanced.

REVIEW QUESTIONS

1. Whenever a higher current capability is needed, batteries are connected in _____.
2. When a higher voltage is needed, batteries are connected in _____.
3. Current flow through the internal resistance of a battery causes the output voltage to _____.
4. A battery has a no-load terminal voltage of 13 volts. When 1 ampere of current is drawn from the battery, the terminal voltage is 12 volts. What is the internal resistance of the battery?
5. As a battery becomes discharged, its internal resistance _____.
6. A 12-volt battery has an internal resistance of 0.5 ohm. How much current can be drawn from the battery before the terminal voltage decreases to 10 volts?
7. How many 1.5-volt batteries must be connected in series to obtain 13.5 volts?
8. In a balanced circuit, _____ current flows in the neutral wire.
9. Is the balanced or unbalanced circuit more common in actual practice?
10. Refer to Figure 7–16. The voltage across the 3-ohm resistor is _____ volts.

UNIT 8

ALTERNATING CURRENT

OBJECTIVES

After completion and review of this unit, you should be able to understand:

- ■ The development of current and voltage by mechanical means (generator–alternator).
- ■ The law regarding the movement of electrons when passing through a magnetic field.
- ■ The generation of voltage and current in rotating coils and rotating magnetic fields.
- ■ The efficiency of polyphase generators.
- ■ The effects of wye and delta connections.

In the previous discussion of electrical current, the implication was that electrons always move in one direction. Actually, the most common form of electrical energy generated today is alternating current (AC), which changes direction 60 times each second.

The standard frequency in the United States is 60 cycles per second or 60 hertz (Hz). Another common frequency that is used in some countries is 50 hertz.

Alternating current is the most common form of electrical energy because it is the natural form of a generated current. Alternating current may be delivered over great distances with much less loss in power than direct current (DC). For these reasons,

alternating current is generally less expensive than direct current for most applications.

GENERATION OF AC VOLTAGE

In Unit 2, we discussed the fact that an electric current flowing in a wire produces a magnetic field around the wire. Another relationship between magnetic fields and electricity is the following: Whenever a wire passes through a magnetic field, a voltage is induced in the wire.

The amount of voltage induced in the wire depends on two factors: the strength of the magnetic field and the speed at which the wire cuts through the field.

In generating electric voltages

Stronger magnetic field = higher voltage

Higher speed = higher voltage

Generated Electromotive Force (Voltage)

A *generator* is a device that converts mechanical energy into electrical energy by mechanically rotating coils of wire in a magnetic field. An *alternator* is a device that converts mechanical energy into electrical energy by rotating magnetic fields past coils of wire.

In Figure 8–1, a piece of wire is shown moving down through a uniform magnetic field. In position 1,

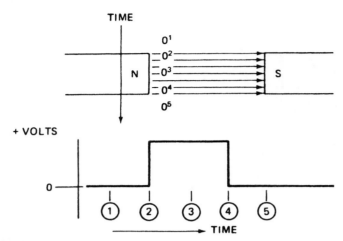

FIGURE 8–1   Voltage output when a wire moves through a uniform magnetic field.

the wire is not cutting through the field, and no voltage will be generated. As the wire enters the magnetic field at position 2, a level of voltage is established. As the wire moves through the uniform field at position 3, at a constant speed, the level of voltage remains constant. At position 4, the wire leaves the magnetic field, and the voltage drops to zero. In position 5, the wire does not cut through the magnetic field, and the voltage remains at zero.

Direction of Generated Voltage

When a piece of wire passes through a magnetic field, some of the free electrons in the wire are forced to move. The direction in which the electrons move, and therefore the polarity of the voltage, depends on two things:

1. The direction in which the wire moves through the magnetic field
2. The direction of the magnetic field

Left-Hand Rule

There is a rule associated with generator action, the left-hand rule. This rule states that if the lines of force are directed into the palm of the left hand with the thumb pointed in the direction of wire motion, the extended fingers will point in the direction of electron movement. (Remember, lines of force travel from the north (N) pole to the south (S) pole.)

Figure 8–2 shows an example of the rule. The direction of electron movement is indicated in Figure 8–2 with a (+) at the tail of an arrow. The electrons move from the (+) end of the wire to the other end, creating a voltage. The (+) end is positive, since electrons move away from this end of the wire. The other end of the wire is negative because the electrons crowd in there. If the other end of the wire were pictured, a dot would be shown to indicate the negative end of the wire. The dot suggests that point of an arrow and the electrons coming out.

Rotating Coils

For practical mechanical reasons, generators are made up of coils rotating in a magnetic field. When a coil rotates in a magnetic field, an alternating

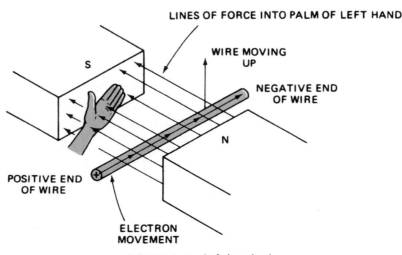

FIGURE 8–2   Left-hand rule.

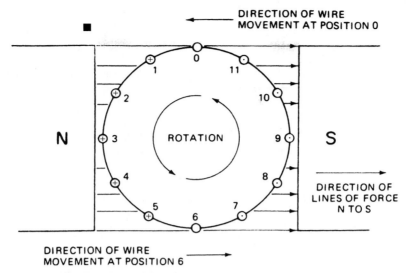

FIGURE 8–3   Rotating wire in a magnetic field.

voltage is produced. In Figure 8–3, an example of coil section, one wire is shown. As the wire rotates through a complete revolution, a complete cycle of alternating voltage is produced. The complete revolution is shown, divided into 12 equal parts. At position 0, the wire is moving parallel to the magnetic lines. The wire does not cut through them. No voltage is produced. As the wire moves from position 0 to position 1, it cuts through a few magnetic lines. A (+) indicates the direction of electron movement and produced voltage. The graph in Figure 8–4 indicates the level of voltage. As the wire moves from position 1 to position 2, it cuts through an increasing number of magnetic lines. A corresponding increase in voltage is indicated on the graph. At position 3, the wire is cutting directly down through the magnetic

lines. The maximum voltage is generated, as indicated on the graph. As the wire moves to positions 4 and 5, it cuts through fewer lines, until at position 6 it is again moving parallel to the magnetic lines. In the graph, this is shown as decreasing voltage to a zero value at position 6. As the wire moves from position 6 to position 7, the direction of movement through the magnetic lines changes.

The level of voltage starting at 7 is increasingly negative, reaching a maximum at position 9. The voltage decreases at 10 and 11, falling to zero volts at position 0. This completes one revolution and one complete cycle of voltage. As the wire continues to rotate, another cycle starts.

According to the left-hand rule for positions 1 through 5, as the wire moves down, electrons

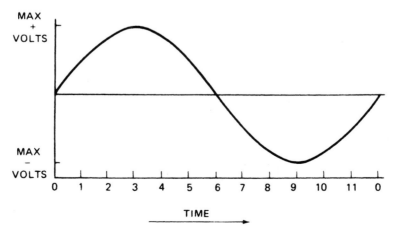

FIGURE 8–4   Sine-wave generation from rotating a wire in a magnetic field.

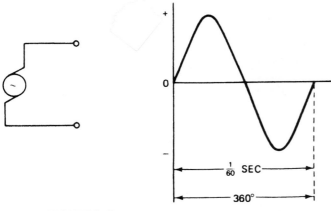

FIGURE 8–5 Standard U.S. 60-cycle voltage.

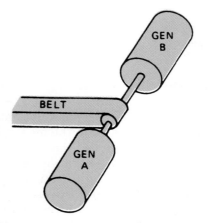

FIGURE 8–6 Two generators operating at the same speed.

would be moving away from you as indicated by the tail of an arrow (+) sign in each wire. In positions 7 through 11, the wire is moving up. The left-hand rule provides for electron motion coming at you, as indicated by a dot (.) in the center of the wire symbolizing the point of an arrow in Figure 8–3.

The voltage is called alternating voltage because the output of the generator will first be positive and then negative at each output terminal.

Most power companies in the United States generate electricity with a frequency of 60 **hertz (Hz)** or 60 cycles per second (cps). There are 60 complete cycles of positive and negative alternations during each second. Equipment designed to operate with voltage at this frequency is marked on the nameplate at 60 cps or 60 Hz. Many foreign countries produce voltages and operate equipment at 50 cycles per second (50 cps or 50 Hz).

Polyphase Generators and Alternators

Generators that produce a single output, alternating voltage are called *single-phase generators* (Figure 8–5). The voltage completes one cycle (360°) in one second, for standard U.S. supply.

A second generator could be connected to the same drive source and produce an output voltage (Figure 8–6). If the armatures of the two generators are exactly aligned, as in Figure 8–7, the output voltages will be in phase with each other.

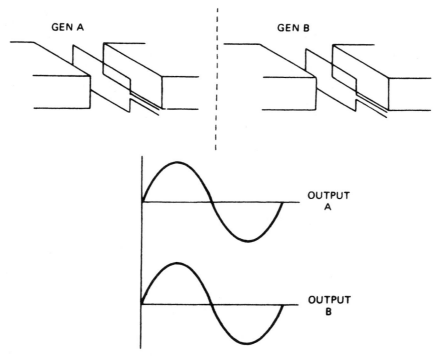

FIGURE 8–7 Two in-phase sine waves.

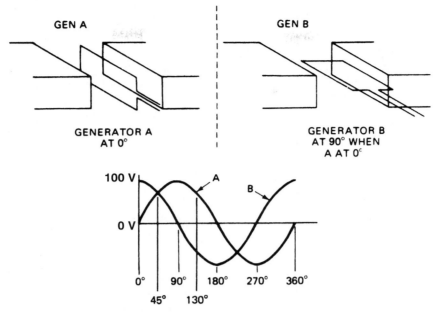

GEN A

GENERATOR A
AT 0°

GEN B

GENERATOR B
AT 90° WHEN
A AT 0°

FIGURE 8–8 Two sine waves of voltage 90° out of phase (two phase).

Another alignment of the armatures could be made that would not provide in-phase voltages at the outputs. In Figure 8–8, the armature of generator B has been rotated clockwise 90°. The output of generator B is 90° out of phase with the output of generator A. The two outputs of the generators could be connected (Figure 8–9). The voltages available would then be as shown in Figure 8–9.

The output between the two generators is not the sum of the two voltages because the voltages are not in phase and cannot be added directly. Since the voltages are 90° out of phase with each other, the output between A and B will be 100 times the square root of 2 ($\sqrt{2} = 1.414$).

$$100 \times 1.414 = 141.4 \text{ volts}$$

SUM OF THE INSTANTANEOUS VALUES

Consider the addition of instantaneous values of voltages at intervals throughout the cycle. In Figure 8–8, phase A is 0 volts at 0 degrees, whereas phase B is at maximum, 100 volts at 0 degrees. The sum of the two voltages at this instant is 100 volts and is shown in Figure 8–10 as the 0° voltage.

At 45°, both phases A and B are 70.7 volts. The sum of the two voltages is 141.4 volts. This is shown as the peak positive voltage at 45° in Figure 8–10. If

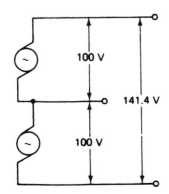

FIGURE 8–9 In the two-phase system the line voltage is not the direct sum of the individual voltages.

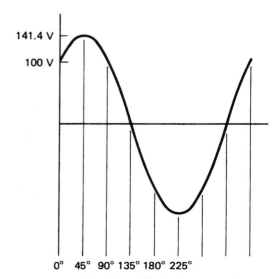

FIGURE 8–10 Sum of instantaneous values.

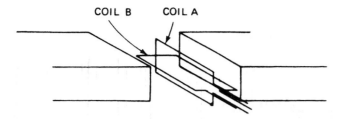

FIGURE 8–11 Two-phase generator.

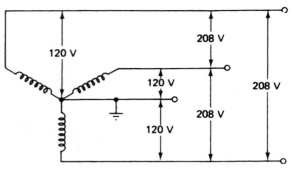

FIGURE 8–13 Wye connection.

the procedure through 360° is continued, the waveform shown as Figure 8–10 will be completed. Note that the phase angle of the waveform in Figure 8–10 is halfway between phases A and B of Figure 8–8.

The sum of the two voltages in Figure 8–8 is a voltage, 141.4 volts peak. This voltage lags phase B by 45° and leads phase A by 45°.

If two-phase power is required, a single generator can be constructed with armature coils placed 90° out of phase with each other (Figure 8–11). The output of the generator with windings 90° out of phase with each other would be the same as that shown in Figure 8–8: The voltages would be 90° out of phase with each other. Two-phase generators or alternators are not very common.

In actual practice, it is more practical to have three rather than two windings in an alternator. Most AC power is produced in alternators since they are easier to construct and do not require brushes for the high-power output. The space in the alternator is put to almost complete use. The load that is presented to the prime mover, whether it is a steam turbine or water pressure from a dam system, is more constant.

In **three-phase** alternators, the windings are placed 120° out of phase with each other. The alternator output voltages produced are shown in Figure 8–12.

Wye and Delta Connections

There are two methods of connecting the windings of alternators; each method produces different output voltages. They are called the **wye** and **delta** connections. The wye connection provides for higher voltage output, whereas the delta connection provides for higher current capabilities.

The output windings of the alternator in Figure 8–12 could be connected as shown in Figure 8–13. The common connection could be connected to a ground and could provide a neutral wire. If each winding is producing 120 volts, 208 volts is available between each pair of phase wires. In the wye connection, the output voltage is equal to the winding voltage times the square root of 3 ($\sqrt{3} = 1.732$).

$$120 \times 1.732 = 208$$

The voltage between any output phase wire and neutral would be the winding voltage of 120 volts.

In the delta connection of windings, the output voltage is equal to the winding voltage (Figure 8–14). There are 120 volts between each of the three-phase output lines. Current flow in a delta system splits. If load resistors of 6.93 ohms (Figure 8–14)

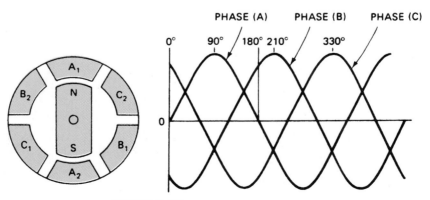

FIGURE 8–12 Three-phase alternator.

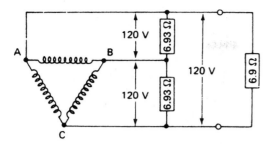

FIGURE 8–14   Delta connection.

are connected between the output lines, the current flow is

$$I = \frac{E}{R}$$

$$I = \frac{120}{6.93}$$

$$I = 17.32 \text{ amps}$$

This is the current in each resistor.

When the current in a line reaches the alternator terminals, it splits to two windings. Each winding has only 10 amperes flowing in it. In the delta connection, the winding current equals the line current divided by the square root of 3.

$$I \text{ winding} = \frac{I \text{ line}}{\sqrt{3}}$$

$$I_w = \frac{17.32}{1.732}$$

$$I_W = 10 \text{ amps}$$

The three-phase system of producing electric power is the most common in the world today.

EFFECTIVE VALUE OF VOLTAGE AND CURRENT

When a **direct current (DC)** voltage is applied to a resistor, a level of current is established, and it flows in the circuit. In a DC circuit, the effects are constant (Figure 8–15).

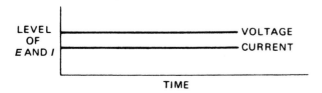

FIGURE 8–15   Representation of DC voltage and current.

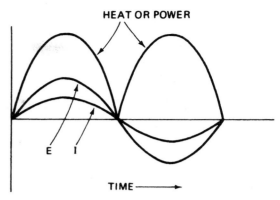

FIGURE 8–16   Power developed with AC system.

In an AC circuit, the voltage is varying continuously. When an AC voltage is applied to a resistor, varying amounts of heat are produced as the voltage varies (Figure 8–16).

Although the applied voltage varies in both the positive and negative directions, the heat is shown as being all positive. It does not make any difference in the resistor which way current flows through it; the result is heat. In Figure 8–16, it is obvious that the same amount of heat is not produced at all times: there are peaks and valleys. At the 0°, 180°, and 360° marks, no heat is produced. The overall result is that if an alternating voltage of 100 volts peak is applied to a resistor, it will have the same heating effect as 70.7 volts DC. By formula

$$\text{effective value} = 0.707 \text{ peak value}$$

The effective value of voltage is sometimes called the rms value. The abbreviation rms stands for *root mean square*. In later units, the relation between the heat developed in a resistor and the square of the current value will be developed.

Voltmeters used to measure electric voltages are designed to indicate the effective, or rms, value of voltage. The air-conditioning technician will have little cause for concern with peak values. This presentation on peak and effective (rms) values was given for information purposes only.

SUMMARY

- A voltage is induced in a wire whenever the wire is passed through a magnetic field.
- The motion of an electron in a magnetic field is covered by the left-hand rule.
- In a generator, voltage is induced in a rotating armature as the armature rotates in a magnetic field.

- In an alternator, voltage is induced in a fixed armature through the use of a rotating magnetic field.
- In both a generator and an alternator, the induced voltage is in the form of a sine wave.
- In many generators and alternators, three sets of coils are used in the armature. The induced voltage is called three-phase voltage.
- Three-phase alternator connections are either wye or delta.

PRACTICAL EXPERIENCE

Required equipment   Voltmeter (AC) and access to a three-phase voltage panel

Procedure

1. Using the voltmeter, measure the voltage between the three phases of voltage.

   Phase 1 _____ volts

   Phase 2 _____ volts

   Phase 3 _____ volts

2. Measure the voltage from each phase to ground.

   Phase 1 _____ volts

   Phase 2 _____ volts

   Phase 3 _____ volts

3. Is the transformer supplying the system connected in wye or delta?

4. If this is a delta system, identify the high phase.

Conclusions

1. Whenever electrons are in motion in a magnetic field, the magnetic field will distort the direction of the original motion of the electrons.

2. Because of this relationship of magnetic fields and electrons, the generation of voltage and current is possible.

3. Practical forms of electromechanical generators of electricity provide alternating current.

4. Three-phase generators may be connected in either the delta or wye configuration.

REVIEW QUESTIONS

1. What is the most common form of electrical energy in the United States?

2. What is the frequency of the AC produced in the United States?

3. What two factors determine the voltage induced in a wire?

4. What is the left-hand rule for generator action?

5. The abbreviation DC stands for _____.

6. The abbreviation AC stands for _____.

7. In the two-phase system, the voltages are out of phase by _____ degrees.

8. In the three-phase system, the voltages are out of phase by _____ degrees.

9. In the wye system with a winding voltage of 100 volts, the line voltage is _____.

10. In the delta system with a line current of 34.4 amperes, the winding current is _____ amperes.

UNIT 9

ELECTRICAL SAFETY

OBJECTIVES

After completion and review of this unit, you should be able to understand:

- Effects of current flow through the body.
- Factors affecting the amount of current flow.
- Dangerous conditions and equipment.
- First aid for shock victims.

EFFECTS OF CURRENT FLOW THROUGH THE BODY

Many tests have been made attempting to determine the effects of current levels on the human body. The general effects are shown in Table 9–1. Keep in mind that individuals will have different responses to current flow through their bodies. The chart shows the approximate amount of current causing a given body response.

Besides the responses given in Table 9–1, the secondary reactions to electrical shock are often the really serious part of the encounter. Any electrical shock that causes an involuntary response can cause a serious accident.

For example, a technician might be on a ladder or scaffolding while using a drill. If she or he receives a shock, the technician might involuntarily drop the drill. This could cause a serious accident to anyone who happened to be standing below.

TABLE 9–1 Effects of current on the body

Amount of Current (mA)*	Response
0–0.5	No response
0.5–2	Slight tingling to mild shock; quick withdrawal from body contact
2–10	Mild to heavy shock; muscular tightening
10–50	Painful shock; cannot let go
50–100	Severe shock; breathing difficulties
100–200	Heart convulsion; death
Over 200	Severe burns; breathing stops

1 mA = 0.001 ampere

Another example is a technician working on a roof of a building. If she or he receives a strong electrical shock, a combination of voluntary and/or involuntary muscle contractions could cause the technician to jump off (or be knocked off) the roof. The technician might live through the shock, but the effects of the fall from the roof might be another matter. It is important that the technician always be aware of the electrical dangers faced in the service area.

FACTORS AFFECTING BODY CURRENT FLOW

It was stated earlier in this textbook that Ohm's law is true for the whole circuit or any part of the circuit. When the human body becomes the circuit (the path for current flow), Ohm's law still holds true. The factors needed in determining the amount of current are the applied voltage and the body resistance. The applied voltage is the voltage with which the body comes in contact. The resistance of the body is the variable factor.

If the resistance of the human body were high enough, for example, 1 million ohms, electrical shock would not be a problem.

EXAMPLE 1

Air-conditioning and refrigeration service technicians normally work with AC voltages under 600 volts. If the body resistance were 1 million ohms, the current flow by Ohm's law would be

$$I = \frac{E}{R}$$

$$I = \frac{600}{1,000,000} = 0.6 \text{ mA or } 600 \text{ microampere}$$

The 0.6 milliampere of current is barely enough to cause a human to respond or even recognize that a shock was received. ■

The problem is that the resistance of the human body is not a fixed value. In addition, the resistance of the human body is mainly the skin resistance. Internally the body is mostly water, and salty at that. Saltwater is a good conductor of electricity. It has low resistance.

Since it is the skin resistance that controls the overall body resistance, let's take a closer look at skin resistance. Dry skin has a resistance of about 200,000 to 700,000 ohms. A resistance at this level would be sufficiently high, keeping current flow below a value that would cause a problem.

Wet skin results in a different situation. The resistance of skin wet with perspiration or other moisture can go as low as 300 ohms. Any contact with line voltage when the skin is wet can cause serious injury or death. Remember, Ohm's law is always true.

$$I = \frac{E}{R}$$

$$I = \frac{120 \text{ volts}}{300 \text{ ohms}}$$

$$I = 400 \text{ milliamperes}$$

According to Table 9–1, a current flow of 400 milliamperes through the body can cause death.

Consider the conditions in which a service technician might be working. One area in which a technician is often required to work is a kitchen. Here the high temperature can easily cause a body to perspire. In a short time, the body and clothing—shirt, pants, socks, and shoes—become damp with perspiration. Body resistance, as well as contact resistance through clothes, rapidly reduces. Any contact with normal working voltages could result in a dangerous shock and burns.

Keep in mind that dry clothes provide good insulation. Clothes wet with perspiration or other moisture are usually low-resistance paths for electricity. When you are wet, it is dangerous to come in contact with line voltages.

DANGEROUS WORKING CONDITIONS AND EQUIPMENT

A dry body in dry clothes and dry shoes has sufficient resistance so that contact with a live wire should not cause a dangerous shock. The major difficulty is that it is nearly impossible to maintain these conditions when working. Dangerous working conditions exist whenever a technician works around live line voltages and his or her body is damp with perspiration or other moisture. It is also dangerous to work around wet equipment because wet metal surfaces provide a better contact to the body. Whenever water is present, whether it is a pool of water on a roof or floor or dampness due to perspiration or moisture on metal cabinets, consider the situation dangerous from an electrical standpoint.

HAND TOOLS

The service technician often encounters problems when power hand tools must be used. Whenever a power hand tool is picked up, line voltage (120 volts AC) is close to the gripped hand tool. This should always be considered a potentially dangerous situation.

The first rule in using power hand tools is a matter of common sense but is often ignored until someone receives a dangerous shock:

1. If anyone receives a slight tingling shock while using a power hand tool, the tool

should be discarded immediately. Do not suggest that the job be finished and then the tool replaced. Stop using the tool at the first sign of a shock.

The second rule regarding power hand tools is an extension of the first rule:

2. When discarding electrically dangerous power hand tools, destroy the hand tool completely, thus ensuring that an unsuspecting individual does not find the tool and receive a dangerous or fatal shock while attempting to use it.

The electrical safety problem with power hand tools is especially significant when older model, two-wire, metal-case power tools are used. The same problem exists when the three-wire power hand tools are used if a good ground does not exist. When a shock occurs, the electrical circuit causing the possibly dangerous shock must first be investigated. Then check for a lack of good grounding with the three-wire, grounded case tool.

In Figure 9–1, a schematic for a hand-drill motor with a motor winding short to the metal case is shown. A stick drawing of the individual holding the tool is also shown. This is a diagram of a dangerous situation. A number of serious results could develop from this situation. Let's consider the two extremes.

1. The work location is dry; the technician is dry; this is the first task of the day. The resistance of the technician's hand to the drill case is high, and grounding through the technician's left hand and feet is through high resistance.

In this situation, the technician might not feel any shock at all, or perhaps just a slight tingling shock. The technician actually would be lucky to feel the slight tingling shock and therefore know that the drill was dangerous and should be discarded. Otherwise, the drill might continue to be used, resulting in situation 2 at some later time.

2. The work location is damp; the technician is wet with perspiration. The work site is cramped between two damp metal cabinets, one of which needs a hole drilled through its case. Since the metal case of the cabinet to be drilled is thick, pressure will be needed on the drill. If the technician braces his or her back against the case of the second cabinet and presses the hand trigger switch on the shorted drill, the chances are the technician will die right there from severe electrical shock.

It is obvious that if the drill were repaired or replaced immediately after the first indication of a tingling shock, the technician would not have been killed in the second situation.

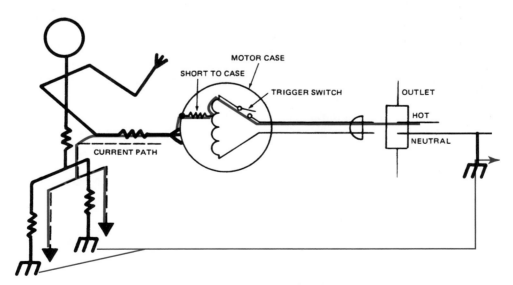

FIGURE 9–1 Two-wire drill with a short to case.

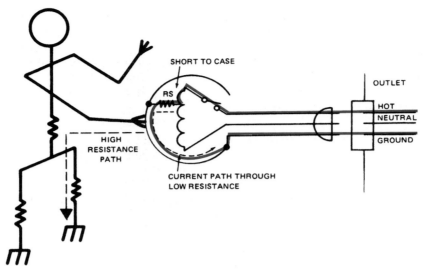

FIGURE 9–2 Three-wire drill with a short to case.

Grounding of Tool Housing

If the metal case of a three-wire hand tool is properly grounded, a short within the tool will most likely result in a blown fuse. Even if the fuse does not blow, the possibility of a dangerous electrical shock is lessened to a great extent since the case of the tool will be at close-to-ground potential, as shown in Figure 9–2.

One of the main points for electrical safety when using hand tools with three-wire power cords is to ensure that the ground connection is complete to ground. One problem encountered with three-wire power cords is to find the grounding prong broken off. This usually occurs when someone has used the hand tool previously in a situation where only the old type of two-wire receptacle was available. The hand tool should not be used until a new three-wire plug or complete cord has been installed.

Cheater plugs adapting the old two-hole receptacles to the three-wire plug are available. The ground pigtail must be connected to a good ground to create a safe situation. Do not be satisfied with a connection to a screw holding the outlet cover plate. Make sure the outlet box is grounded.

The use of double-insulated tools is making inroads in the air-conditioning field. Since the tools use the two-prong plug, the need for a three-prong grounding receptacle is eliminated. The problems caused by ungrounded three-prong receptacles are also eliminated.

Since safety engineers have not yet come to a definite conclusion about the safety of double-insulated

tools, technicians should keep abreast of this changing situation.

Ground Fault Current Interrupter

A standard appliance plug has three prongs. Two are flat, being the hot and the neutral legs of the power supply. The U-shaped prong is the ground. The only appliances that are manufactured today without ground wires are ones that are **double insulated**. They are designed with plastic casings, so that even if the wiring inside the device breaks and touches the casing, the casing itself is an insulator and will not become electrically pressurized.

In some locations (especially wet locations such as bathrooms and kitchens), conventional grounded outlets are no longer permitted by code. A **ground fault current interrupter (GFCI)** shown in Figure 9–3

FIGURE 9–3 Ground fault current interrupter.

provides an extra safety feature. It senses if the current being supplied in the hot leg equals the current being returned in the neutral leg. If they are at all different, it means that there is some current leakage, and the GFCI will open, stopping the flow of current. The GFCI provides ground fault protection only. It does not provide any protection against a short circuit.

During normal operation, no current flows through the ground wire. It is only provided as a personnel safety precaution in the event of a circuit malfunction. Although a device does not require that the ground wire be connected in order to run, it would be a serious mistake to not connect the casing ground wire.

Some people use the term *ground* when they are referring to a *neutral* wire. Although in some circumstances (such as automotive wiring), ground and neutral are the same, they are not the same in building wiring. The neutral wire normally carries current, whereas the ground wire carries current only in the event of a malfunction.

FIRST AID FOR ELECTRIC SHOCK

In cases of electric shock, the first and most important step is to disconnect the victim from the electric power source. If possible, turn the power off. If this is not possible, use a dry board or other nonconducting material to separate the victim from the electric connection.

Start artificial respiration as soon as it is safe to touch the victim. Speed is essential when dealing with victims of electric shock. Do not stop the administration of the artificial respiration, even if the victim appears dead. Long periods of time—up to eight hours—have elapsed before some victims have responded. Only a physician should determine if the victim has died.

Most importantly, technicians should take the official first-aid course offered in their areas. It is extremely important to know the best methods of administrating first aid to shock victims.

Keep yourself aware of accidents as they happen by reading trade periodicals and newspapers. A recent accident reported in a trade paper described a situation in which a technician was working in the attic of a manufacturing plant. Being safety conscious, he had disconnected power from the air-conditioning system he was working on. The attic was not well lighted, it was hot, he was perspiring, and his clothes were damp. Near the electrical conduit feeding the air-conditioning systems were commercial gas lines providing service to the building.

The technician had just finished placing a blower motor in the air distribution system. He raised himself up and possibly stretched to relieve the tension in his back. At head level behind him was an open junction box. The cover had been left off by a previous worker. A wire nut had loosened and fallen off a pigtail connection, or perhaps it had never been installed.

As the air-conditioning technician stretched back, his head came in contact with the hot pigtail. The pigtail was the junction in a 278-volt lighting system. The air-conditioning technician had his hand and back against the gas line, which is an excellent ground. The attempts to revive him were not successful.

Be aware of what you are doing, where you are, and what is around you. Be safety conscious at all times. Do not be satisfied with a halfhearted approach to an understanding of safe working conditions.

SUMMARY

- With electrical shock, it is the amount of current flow that determines the effect of the shock on the body.
- Skin resistance is the main control of body resistance. Wet skin means low resistance. Low resistance means high current. High body current is very dangerous.
- If you feel a shock when using a hand tool, do not use the tool again or allow any other person to use that tool.
- Always be aware of safety considerations. Your life depends on it!

REVIEW QUESTIONS

1. The only electrical shocks that are dangerous are the ones that cause harsh burns.

   T _____   F _____

2. A victim of electrical shock is always knocked away from the electric connection by involuntary muscle contractions.

   T _____   F _____

3. Most shocks are the result of carelessness.

   T _____   F _____

4. No voltage should be considered safe, since it is the current that causes damage, and body resistance is changeable.

   T _____   F _____

5. Hand power tools are not dangerous if grasped with only one hand.

   T _____   F _____

6. A slight tingling shock from a hand tool is not important.

   T _____   F _____

7. A technician can feel safe as long as a three-wire plug is on his or her hand drill.

   T _____   F _____

8. Two wire connections are needed on a panel tap.

   T _____   F _____

9. The voltage between the black wire in a panel box and ground is 120 volts.

   T _____   F _____

10. In case of electric shock, the first thing to do is to remove the victim from the electrical connection.

    T _____   F _____

UNIT 10

CAPACITANCE AND INDUCTANCE

OBJECTIVES

After completion and review of this unit, you should be able to understand that:

- Capacitance relates to the storage of electric energy in electrostatic fields.
- Inductance relates to the storage of electricity in magnetic fields.
- Capacitance opposes a change in voltage.
- Inductance opposes a change in current.
- Current and voltage in an AC circuit are not always in phase with each other.

The first hint of capacitive action was given in Unit 1. A generator was shown pulling electrons from a wire connected to the positive terminal of the generator and pushing them out on the wire connected to the negative terminal. A similar situation is shown in Figure 10–1; the generator is running and wires are connected to the positive and negative terminals. For a voltage to appear between the positive and negative wires, electrons have to be pulled from the positive side and forced out on the negative side.

If a circuit constructed with a sensitive meter is connected as shown in Figure 10–2, the meter will indicate electron movement or current flow during the charging of the wires. The meter needle will move up as the switch is closed and immediately drop back to zero as the wires became charged. The number of free electrons that move would control the meter needle movement.

The number of electrons that have to move to "charge" the wires could be increased by replacing the external wires with large, flat plates. Increasing the surface area enlarges the charge capability.

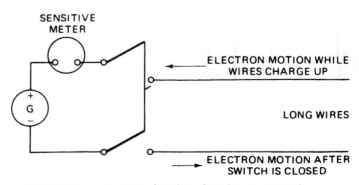

FIGURE 10–2   A meter showing charging current when switch is closed.

FIGURE 10–1   A DC generator charging the wires.

73

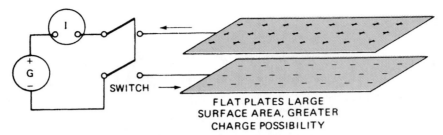

FIGURE 10–3 Increased charging current when large plates are used.

Metal plates with a large surface area separated by a high-resistance insulating material make up a **capacitor**. In Figure 10–3, air is the insulating material.

DC CURRENT FLOW IN CAPACITIVE CIRCUITS

The explanatory text in Figure 10–2 indicates that after the switch is closed there is a momentary movement of electrons. With the DC generator connected to the wires, some of the free electrons move from the top wire through the generator and out on the bottom wire. After the wires become charged, voltage appears between all points on the top and bottom wires. Electron movement then stops.

A similar situation exists if a capacitor is connected across a DC generator or any other DC power source. As the capacitor is connected across the generator, the current jumps up. As the capacitor charges, the current flow drops to zero, and the source voltage appears across the capacitor terminals. A graph of voltage and current as related to time is shown in Figure 10–4. The following explanations describe what happens to the current and the voltage as shown in the graph.

Current

1. As the switch is closed at time zero, the current rises immediately to maximum.
2. The current decreases as the capacitor charges.
3. The capacitor is fully charged; the current drops off to 0 amperes.

Voltage

1. As the switch is closed, the voltage across the capacitor rises rapidly from 0 volts.
2. The voltage rise starts to taper off as the voltage across the capacitor approaches the source (generator) voltage.
3. The voltage across the capacitor reaches the source voltage; the capacitor is fully charged. The voltage across the capacitor remains constant at the source voltage level.

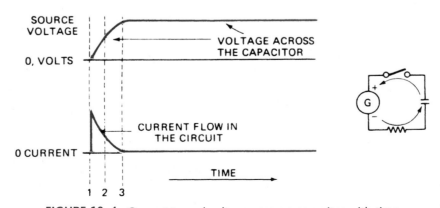

FIGURE 10–4 Current to and voltage across a capacitor with time.

AC CURRENT FLOW IN CAPACITIVE CIRCUITS

When an AC generator is connected across a capacitor, the relation of current and voltage is similar to what was described for a DC source. The current and voltage relationship with an AC source is shown in Figure 10–5.

1. At time position 1, the voltage is zero and current flow is maximum (Figure 10–6a).

2. At time position 2, the voltage across the capacitor is rising and the current flow is decreasing (Figure 10–6a).

3. At time position 3, the voltage is at maximum while the current is at zero.

In steps 1 through 3, current was flowing into the capacitor while the capacitor was charging up. This is shown in Figure 10–6a. Electrons move into the bottom plate of the capacitor and out of the top plate.

The voltage across the capacitor is increasing. As step 3 is passed, the current flow changes direction. Electrons start to move back out of the bottom plate and into the top plate of the capacitor. This is shown in Figure 10–6b. The voltage across the capacitor is decreasing.

4. At time position 4, the voltage across the capacitor decreases as the current flow increases.

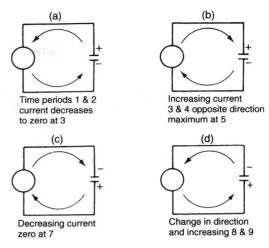

(a) Time periods 1 & 2 current decreases to zero at 3

(b) Increasing current 3 & 4 opposite direction maximum at 5

(c) Decreasing current zero at 7

(d) Change in direction and increasing 8 & 9

ARROWS INDICATE DIRECTION OF ELECTRON MOVEMENT

FIGURE 10–6   Current flow, as related to Figure 10–5.

5. At time position 5, the current flow is maximum. The voltage across the capacitor falls to zero, and the capacitor starts to charge in the opposite direction (Figure 10–6c). The current flow is in the same direction as in Figure 10–6b, but the voltage polarity across the capacitor has changed.

6. At time position 6, current flow is dropping off. The voltage across the capacitor approaches maximum negative.

7. At time position 7, current flow is back to zero. The voltage is at maximum negative.

8. At time position 8, current flow has again changed direction. The voltage across the capacitor starts to decrease (Figure 10–6d).

9. At time position 9, current flow is maximum. The voltage across the capacitor rapidly passes through zero, and the capacitor starts to charge. This completes one cycle of AC in a span of a second. This means that there are 60 repeats each second in the graph in Figure 10–5.

CAPACITOR CONSTRUCTION

For practical reasons, capacitors use thin metal sheets, usually aluminum, for the plates. Wax paper, oil-saturated paper, or a chemical electrolyte is used as the high-resistance material between the plates. For air-conditioning and refrigeration service, the two common types of capacitors

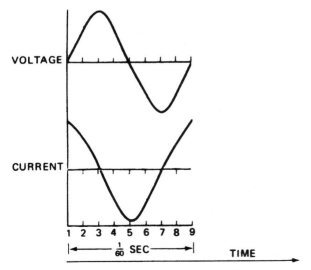

FIGURE 10–5   Current and voltage are 90° out of phase with AC voltage; current leads voltage.

FIGURE 10–7(a)   Oil-filled capacitors (run capacitors). (Courtesy of BET Inc.)

FIGURE 10–7(b)   Electrolytic capacitors (start capacitors). (Courtesy of BET Inc.)

are oil-filled and electrolytic capacitors (Figures 10–7a and b).

Oil-filled capacitors are used in series with the start winding of AC motors in order to increase the torque of the motor. Oil-filled capacitors are also used when the application requires the capacitor to continuously be in the circuit.

Electrolytic capacitors are used in the start circuits of high-horsepower AC motors. An electrolytic capacitor provides a high capacitance in a small space. Electrolytic capacitors are usually connected for intermittent use.

CAPACITOR RATINGS

Capacitors are usually rated by capacity and a maximum voltage rating. This capacity unit is the farad, which is a very large unit. Capacitors are usually rated in microfarads (one-millionth of a farad),

which is a practical unit for most applications. Some capacitors are marked 10 MFD, 10 mfd, or 10 MF, each of which means 10 microfarads. The common abbreviation for microfarad is μF. Whenever a capacitor is replaced in an electrical device, the capacitance values should be matched as closely as possible. The voltage rating of a capacitor is also an important consideration. The voltage rating of the replacement capacitor should equal or exceed that of the original capacitor.

Standard oil-filled capacitors for continuous use are usually available in capacitance values up to 50 μF, with voltage ratings up to 500 V. Electrolytic capacitors for intermittent use are usually available in capacitance values of 70 to 800 μF at 125 V and up to 200 μF at 440 V.

The most common purpose for capacitors in air-conditioning and refrigeration circuits is to provide starting torque and to maintain running torque in AC motors. Capacitive circuits will be covered more completely in Unit 13.

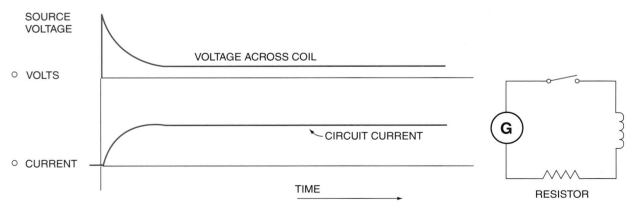

FIGURE 10–8   Current and voltage, inductive circuit.

FIGURE 10–9 Alternating current.

INDUCTANCE

Some of the factors governing inductance in electrical circuits were covered in Unit 2, Magnetism. One important factor not yet considered is the relationship between current and voltage in an inductive circuit when a DC voltage is applied. A graph of this relationship is shown in Figure 10–8.

Current

1. As the switch is closed at time zero, the current starts to rise from zero.
2. The current rises to maximum.
3. The current stops rising when the total current is equal to the voltage divided by the resistance of the circuit (Ohm's law).

Voltage

1. As the switch is closed, the voltage across the inductance coil is maximum (supply voltage).
2. As the current rises, the voltage across the coil decreases.
3. The voltage across the coil decreases to a fixed level ($I \times R$ of coil) as the current increases to its maximum value.

Compare the relationship of current and voltage in an inductive circuit with that of a capacitive circuit shown in Figure 10–8. Make note of, and remember, that in the capacitive circuit, the current was maximum before the voltage, and in the inductive circuit, the voltage was maximum before the current.

Laws governing inductance in electrical circuits have been presented in earlier units. One of the most important of them is that any time a conductor (wire) is cut by a magnetic field, a voltage is induced in the wire.

Consider a coil of wire with an AC voltage applied to it. The current flow through the coil will be continuously variable: starting at zero, rising to a maximum, returning to zero, then rising to a maximum in the opposite direction before returning to zero again (Figure 10–9).

When a continuously varying current (AC) flows through a coil of wire, it produces a continuously varying magnetic field around the coil. This varying magnetic field cuts through the coil and produces a voltage in the coil. The voltage that is induced is in direct opposition to the supplied voltage. The induced voltage is called back electromotive force, abreviated as back EMF or BEMF.

In Figure 10–10, a coil of wire is shown with a voltage of 100 volts at 60 hertz AC supplied to it. The ammeter indicates 1 ampere of current flow. The DC resistance of the coil is given as 1 ohm. According to Ohm's law, only 1 volt is needed to force 1 ampere through the 1-ohm resistance of the coil. The remaining 99 volts overcome the 99 volts of back EMF developed in the coil. The back EMF of the coil has a greater effect in the control of current than in the resistance of the coil.

Inductive Effects

Coils of wire are used in many applications in air-conditioning and refrigeration electrical systems. The most common uses are in motors, transformers, relays, and solenoids. In each type of device, back EMF is a major factor in the control of current.

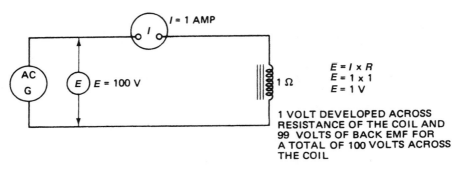

FIGURE 10–10 Back EMF generated in a coil.

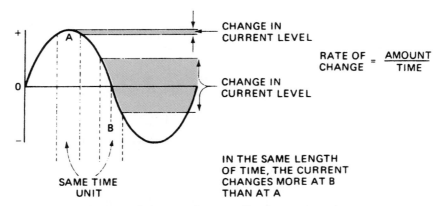

FIGURE 10–11 Rate of change of current flow is greatest when current passes through zero.

Current Flow in Inductive Circuits

The AC current flow through a circuit containing a coil varies (Figure 10–10). The magnetic field that develops around a coil carrying current is proportional to the current. As the current changes, so does the magnetic field. Around point A in Figure 10–11, the current is changing very little. The current rises slowly to a maximum and then decreases to a low level. At the maximum value of current, point A, the current is not increasing or decreasing; the rate of change is zero. At this same point, the change in the magnetic field is zero. Without a changing magnetic field, the induced voltage (back EMF) is zero.

The maximum voltage is induced in a coil when the rate of change of the magnetic field is greatest. In Figure 10–11, the maximum rate of change of current is around point B when the current passes through zero. The maximum rate of change in the magnetic field occurs at the same time. The maximum voltage (back EMF) also occurs at this time. A plot of this relationship is shown in Figure 10–12.

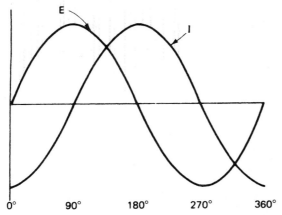

FIGURE 10–12 Current lags voltage by 90° in a pure inductance.

This is a graph of the current through and voltage across a pure inductance coil.

1. The voltage is maximum as the current passes through zero.
2. The voltage is zero as the current reaches maximum.

It is not necessary for current and voltage to rise and fall together in AC circuits. Actually, it is more common for them not to do so, since so many electrical circuits contain coils. Further treatment of coils and the angular relationship between current and voltage will be covered in Unit 13.

SUMMARY

- In a capacitor, energy is stored in the form of an electrostatic field.
- In an inductive circuit, energy is stored in the magnetic field.
- Capacitance in a circuit opposes a change in voltage.
- Inductance in a circuit opposes a change in current.
- If a circuit contains inductance and/or capacitance, the current and voltage in the circuit will likely be out of phase with each other.

PRACTICAL EXPERIENCE

Required equipment Analog ohmmeter, various capacitors, special bleed resistor (see Figure 10–13)

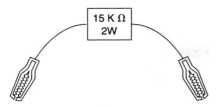

FIGURE 10–13   15 K Ω 2-watt bleed resistor with wire and insulated clip leads attached.

Procedure

1. Using the bleed resistor with clip leads, carefully connect the resistor across each capacitor. Make connection to one terminal of the capacitor at a time.

2. List the sizes in MF of each capacitor to be tested.
   - a. _____ MF
   - b. _____ MF
   - c. _____ MF
   - d. _____ MF
   - e. _____ MF

(**NOTE:** Use a standard speed count sequence to estimate the time it takes for each capacitor to charge.)

3. Place the ohmmeter selector switch on the highest R position (probably $R \times 10,000$).

4. In turn, connect the ohmmeter across each capacitor. Start counting when the connection is made.

5. Record the count (time) it takes for each capacitor to charge to the meter's internal battery voltage.

	Size		Count	
1.	_____	MF	_____	Time
2.	_____	MF	_____	Time
3.	_____	MF	_____	Time
4.	_____	MF	_____	Time

Question Does it take a longer time to charge large MF values or small MF values?

Conclusions

1. Capacitance and inductance play many important roles in air-conditioning systems.
2. In air-conditioning circuits, the current and voltage are nearly in phase with each other.
3. Back EMF (BEMF) is generated in coils.

REVIEW QUESTIONS

1. Name the two common types of capacitors used in air-conditioning and refrigeration systems.

2. A capacitor is made up of two metal plates separated by some _____ material.

3. Which type of capacitor is generally used intermittently in compressor motor-starting circuits?

4. Two important capacitor ratings are _____ and _____.

5. What is the relationship between the voltage induced in a coil-carrying current and the voltage that produced the current?

6. With AC current, the rate of change is greatest when the current passes through _____.

7. When the AC voltage applied to a capacitor is at maximum, the current flow into the capacitor is _____.

8. The (electrolytic, oil-filled) _____ capacitor usually has the higher microfarad value in a motor circuit.

9. Refer to Figure 10–11, which illustrates current flow in an inductive circuit. The greater induced voltage across the coil takes place at point (A or B) _____.

10. Refer to Figure 10–12. In this voltage and current relationship, the current is shown (leading, lagging) _____ the voltage by _____ degrees.

UNIT 11
ELECTRICAL POWER AND ENERGY

OBJECTIVES

After completion and review of this unit, you should be able to:

- Calculate electrical power using voltage current and resistance.
- Calculate electrical energy using electrical power and time.
- Control electrical power using relays and contactors.

Power is a measure of the rate of doing work. Mechanical power is measured in horsepower (hp), and electrical power is measured in watts (w). The unit of electrical power is named for James Watt (1736–1819), a Scottish engineer who developed the modern condensing steam engine.

ELECTRIC POWER

In an electrical system, power is equal to voltage times current.

$$W = I \times E$$

The electric heat strip unit of an air-conditioning system may have a hot resistance of about 6 ohms (Figure 11–1). (The cold resistance of this unit would be only slightly lower.) When in use, 9600 watts of power is produced in the unit. The unit is designed for 240 volts at 40 amperes.

$$I = \frac{E}{R}$$

$$I = \frac{240}{6}$$

$$I = 40 \text{ amps}$$

$$W = I \times E$$

$$W = 40 \times 240$$

$$W = 9600 \text{ watts}$$

This may be seen written on a nameplate as 9.6 kW, which is 9.6 thousand watts. The k stands for one thousand.

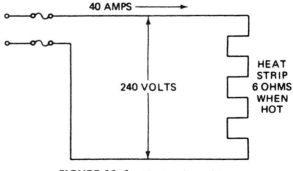

FIGURE 11–1   Heat strip resistor.

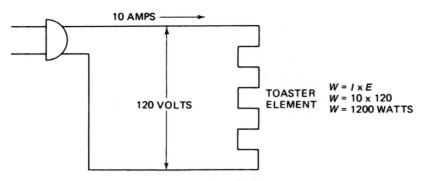

FIGURE 11–2   Standard toaster.

A good example of a power-consuming device is an electric bread toaster. In Figure 11–2, an electric toaster is shown plugged into a 120-volt electrical source. Ten amperes of current is being drawn from the power source. The power developed in the toaster may be determined by using the following formula: watts = amperes × volts.

$$W = I \times E$$

$$W = 10 \times 120$$

$$W = 1200 \text{ watts or } 1.2 \text{ kW}$$

Another formula is sometimes used to determine power in electrical circuits.

$$W = I^2 \times R$$

As a statement, the formula is that power (W) is equal to amperes (I) squared times resistance (R). Another way of stating the formula is that power (W) is equal to amperes (I) times amperes (I) times resistance (R).

Since in Ohm's law $E = I \times R$, the factors $I \times R$ can be substituted for E in the original power formula. Substituting ($I \times R$) for E

$$W = I \times (I \times R)$$

$$W = I \times I \times R$$

$$W = I^2 \times R$$

In the toaster example, the resistance of the toaster heating element, according to Ohm's law, is

$$R = \frac{E}{I}$$

$$R = \frac{120}{10}$$

$$R = 12 \text{ ohms}$$

When this factor is used in the formula $W = I^2 \times R$,

$$W = 10 \times 10 \times 12$$

$$W = 100 \times 12$$

$$W = 1200 \text{ watts}$$

This same answer can be obtained using the formula $W = I \times E$.

ELECTRIC MOTORS

Electric **power** is required to operate electric motors. In a motor, the desired mechanical output is a rotation of the motor shaft. Some electrical power is continually lost in a motor because of the heat produced. A motor is, therefore, not 100% efficient.

Consider an electric motor used to drive a refrigeration compressor (Figure 11–3). The compressor

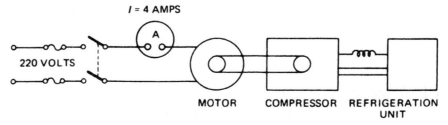

FIGURE 11–3   Motor rated in output horsepower.

is rated as 1 **horsepower (hp)**. The motor, also rated as 1 hp, is delivering 1-hp mechanical power to the compressor. With 4 amperes of current being drawn from the power line (at 220 volts), 880 **watts** of electrical power is being delivered to the motor.

$$W = I \times E$$

$$W = 4 \times 220$$

$$W = 880 \text{ watts}$$

If the motor is operating at about 85% efficiency, about 134 watts are converted into heat in the motor. The remaining 746 watts is converted into mechanical rotating power at the motor shaft (output). (The conversion between horsepower and watts is 746 watts equals 1 horsepower.)

Therefore, with 746 watts converted to rotating power, 1 hp is delivered to the motor shaft. Motors are rated in shaft horsepower, that is, output, not power input. As an aid in remembering the hp/watt conversion factor, remember that Columbus discovered America in 1492; 1492 ÷ 2 = 746; 746 watts = 1 horsepower.

ELECTRIC ENERGY

Electric power was defined earlier as a measure of the rate of doing work. Electric **energy** relates to the total amount of work done. Stated a different way, energy equals the rate of doing work (power) multiplied by time.

$$\text{energy} = \text{rate} \times \text{time}$$

The power company charges its customers for the total amount of electrical energy it supplies.

Since 1 **watt-hour** (Wh) is a very small unit of energy, the power company charges per kilowatt-hour. A kilowatt-hour (kWh) is equal to 1000 watts for 1 hour, 500 watts for 2 hours, 2000 watts for ½ hour, and so on. For example:

$$1000 \text{ w} \times 1 \text{ h} = 1000 \text{ Wh} \,(1 \text{ kWh})$$

$$500 \text{ w} \times 2 \text{ h} = 1000 \text{ Wh} \,(1 \text{ kWh})$$

$$2000 \text{ w} \times 0.5 \text{ h} = 1000 \text{ Wh} \,(1 \text{ kWh})$$

where kilo (k) stands for one thousand. The power company bases its charge for providing electrical energy on its cost of production plus a small profit. The cost varies by geographic location and can be obtained from your local company. The following example, based on $0.06 per kWh, illustrates how the cost of operation may be determined when power consumption is known.

EXAMPLE 1

An electric power company charges its customers 6 cents for 1 kilowatt hour (kWh) of energy. How much does it cost the customer to heat for eight hours if the heat is obtained from a heat strip of 22 ohms connected to a 220-volt source (Figure 11–4)?

Solution

Solve for current.

$$I = \frac{E}{R}$$

$$I = \frac{220}{22}$$

$$I = 10 \text{ amperes}$$

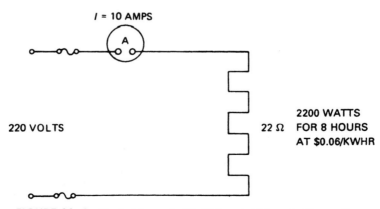

FIGURE 11–4   Heat strip costs $1.06 for eight hours of operation.

FIGURE 11–5   Watt-hour meter. (Courtesy of BET Inc.)

Solve for power.

$$W = I \times E$$
$$W = 10 \times 220$$
$$W = 2200 \, watts$$

Solve for energy.

$$Wh = watts \times time$$
$$Wh = 2200 \, watts \times 8 \, hours$$
$$Wh = 17,600 \, watt\text{-}hours \text{ or } 17.6 \, kWh$$

Solve for cost at 0.06 cents per kWh.

$$Cost = kWh \times 0.06$$
$$Cost = 17.6 \times 0.06$$
$$Cost = \$1.06$$

For the power company to keep track accurately of the total amount of electrical energy being supplied to a customer, a watt-hour meter is usually installed where the electric service is delivered to a building. A watt-hour meter is an electromechanical device that rotates dials, recording the energy delivered to the customer (Figure 11–5). This meter is read monthly by the power company, and the customer is billed for the energy used.　■

CONTROL OF ELECTRIC POWER

The control of electric power is a problem that must be dealt with every day. Electric power must be turned off and on either manually or automatically as needed for lights, motors, heaters, and other power-consuming devices.

In low-power units, such as lamps, electric drills, and electric toasters, simple switches may be used to provide the required control. In larger power-consuming devices, such as air-conditioning compressor units or electric-heat strip units, high power (usually at higher voltages) must be controlled. In these types of units, power must be controlled at a point remote from where it is being consumed. For example, an air-conditioning compressor (often outside the house) is controlled by a thermostat in the living area inside the house. A low-voltage signal (usually 24 volts) is fed to a relay (an electrically operated switch) in the air-conditioning unit. The relay, when energized, supplies high voltage (usually 220 volts) at high current levels to the compressor motor.

Relays are electromechanical devices used as controls for electric power. A relay has a movable metal part called the **armature**. An electrical contact or contacts may be connected to the armature; when the armature moves, the contacts also move. Depending on the application, this action may open some electric circuits while closing others.

In Figure 11–6, the relay is in the de-energized (no coil power applied) position. When no power is supplied to the relay coil, the spring pulls the armature down at the right end. The armature pivots at the point, moving the left end of the armature up. An electric circuit is complete from the normally closed contact through the movable contact to the connecting wire to the terminal output.

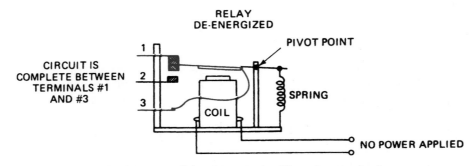

FIGURE 11–6   Relay armature position, de-energized.

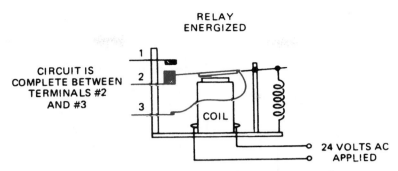

FIGURE 11–7   Relay armature position, energized.

In Figure 11–7, the same relay is shown, but electric power has been applied to the relay coil. The magnetic field developed around the coil attracts the iron in the armature. The armature moves down, making electrical contact between the normally open contact and the movable contact. As long as power is supplied to the coil, the relay remains energized. When power is removed from the coil, the spring pulls the armature to its original position, returning the connection to the normally closed contacts.

Relays come in various shapes and sizes and are designed for special purposes. Sometimes the name of the device is changed, because of its application, so it may be called a *contactor*. The mechanical construction of a contactor (Figure 11–8) is somewhat different from that of a relay, but it operates on the same principle. Low power supplied to the coil of the contactor produces a magnetic field that attracts an armature. When the armature moves, it closes electric contacts that provide for the control of high-power circuits. More will be said about relays and contactors in Unit 16.

MEASURING ELECTRIC POWER

One method of determining electric power in a circuit is to measure the voltage across the circuit and the current through the circuit. The power developed is found by the formula

$$W = I \times E$$

In Figure 11–9, the circuit of an electric heat strip is shown. The voltage across the heat strip is 220 volts, as indicated by the voltmeter. The ammeter indicates 43.6 amperes.

$$W = I \times E$$
$$W = 43.6 \times 220$$
$$W = 9600 \text{ watts}$$

Another method of measuring power is to use a wattmeter. A **wattmeter** is a device that uses both the current in the circuit and the voltage across the circuit to provide an indication of the power developed

FIGURE 11–8   Contactor. (Courtesy of BET Inc.)

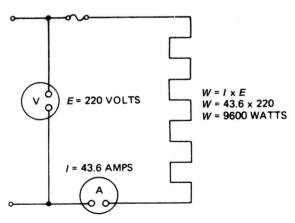

FIGURE 11–9   Power measurement using a voltmeter and an ammeter.

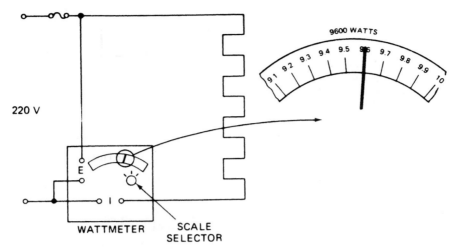

FIGURE 11–10   Power measurement using a wattmeter.

in the circuit. The heat strip of Figure 11–9 could have a wattmeter connected rather than a voltmeter and an ammeter. The power developed would be read directly from the wattmeter, as shown in Figure 11–10.

It is always better to use a wattmeter for determining power if one is available. Since only one meter reading is needed, the chance of error is lessened. In AC circuits, it is always better to use a wattmeter for power measurement because of other factors that can effect AC. Further information on power in AC circuits will be covered in Unit 13.

SUMMARY

- Electric power may be calculated using the following formulas:

$$W = I \times E$$
$$W = I^2 \times R$$
$$W = \frac{E}{R}$$

- Wattmeters indicate electrical energy through the use of current measurement and voltage measurement in the same instrument.
- Electrical energy is electrical power times time. It indicates the total work done.

PRACTICAL EXPERIENCE

Required equipment   Heat strip, AC voltmeter, ohmmeter, and clamp-on ammeter

Procedure

1. Observe the nameplate specifications of the heat strip.
2. Using the ohmmeter, measure the cold resistance of the heat strip.
3. Connect the heat strip to the proper voltage as indicated on the nameplate.
4. Measure the voltage at the heat strip terminals.
5. Use the clamp-on ammeter to measure the current flow through the strip.
6. Calculate the power developed in the heat strip.

$$W = I \times E = \underline{\hspace{2cm}} \text{watts}$$

7. Calculate the hot resistance of the heat strip.

$$R = \frac{E}{I} = \underline{\hspace{2cm}} \text{ohms}$$

8. Is the hot resistance higher or lower than the cold resistance of the strip?

REVIEW QUESTIONS

1. The formula for power using voltage and current is $W = \underline{\hspace{3cm}}$.
2. The formula for power using current and resistance is $W = \underline{\hspace{3cm}}$.
3. A strip heater draws 30 amperes of current at 220 volts. How much power is developed in the unit?
4. If a strip heater of 8600 watts is operated for 100 hours, how much electrical energy will be converted to heat?

5. A power company is selling electric energy at $0.04 per kWh. How much will it cost to operate a 10,000-watt heater for 60 hours?

6. The movable part of a relay is called an

   _____.

7. A contactor is a special-purpose _____.

8. In Figure 11–11, if the power developed in the heater is 12,000 watts and the supply voltage is 220 volts, what is the current flow through the heater?

9. In Figure 11–11, the power developed in the heater is 8000 watts. The power source is 220 volts. What is the resistance of the heater?

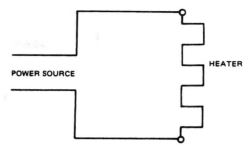

FIGURE 11–11   Strip heater.

10. The circuit of Figure 11–11 is to supply 18,000 watts of heat. The power source is 220 volts. What is the resistance value of the heater?

UNIT 12
TRANSFORMERS

OBJECTIVES

After completion and review of this unit, you should be able to understand:

- Transformer theory; the relation between transformer turns; and ratio, voltage, and current.
- Power losses in transformers.
- Delta and wye connections of transformers.
- How transformers are rated.

TRANSFORMERS

Transformers are nonmechanical electric devices used in many applications for the control of electrical power. High voltages used on transmission lines for electrical power transfers are dropped (reduced) with the use of transformers. Low-level voltages used in air-conditioning and refrigeration control systems are obtained through the use of transformers.

For example, the transmission cable may be 2400 volts, whereas utility service requirements are 240 volts. A transformer is used to reduce the 2400-volt transmission to a usable 240 volts. This 240 volts supplies power to the air-conditioning system's high-voltage electrical components. Another transformer within the electrical system steps this 240 volts down to 24 volts for the low-voltage control system.

Volts per Turn

In Unit 10, the back EMF developed in a coil of wire carrying an AC current was discussed. It was shown that the inductance of the coil is the major effect in the control of current. The varying magnetic field around a coil carrying an AC current cuts through the coil and produces a back EMF that limits the total current. The small current that flows through the coil is called the *magnetizing current*.

Figure 12–1 shows a coil of wire wound on an iron core. The coil of wire has 600 turns in it, connected in series. The back EMF developed in the coil is very close to 100 volts. Another way of stating the situation is that for every six turns of wire on the coil, approximately 1 volt is developed.

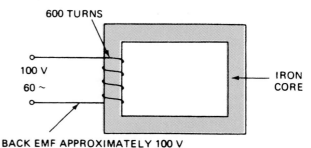

FIGURE 12–1 Back EMF of 0.166 volts per turn in the 600-turn coil.

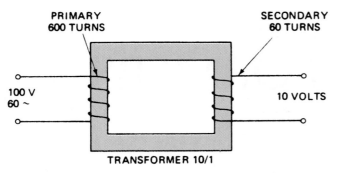

FIGURE 12–2 Voltage is developed in the secondary at 0.166 volts per turn.

A second coil of wire is wound on the same iron core (Figure 12–2). The second coil is called the *secondary winding;* the first coil is called the **primary winding**. The same magnetic field that cut through the primary winding, developing back EMF, cuts through the secondary winding. A voltage is developed in the secondary winding at 1 volt per six turns. Since there are 60 turns in the secondary, a total of 10 volts is developed across the secondary winding (60 ÷ 6 = 10).

There is a voltage relationship between the primary and secondary windings of a transformer. Simply stated, the turns ratio is equal to the voltage ratio. The transformer shown in Figure 12–2 has a turns ratio of 600 to 60 or 10 to 1; the voltage ratio is also 10 to 1.

Figure 12–3 shows the same transformer with a load resistance of 1 ohm connected across the secondary. The 10-volt output across the secondary winding will be impressed across the 1-ohm resistor after the switch is closed. According to Ohm's law

$$I = \frac{E}{R}$$

$$I = \frac{10}{1}$$

$$I = 10 \text{ amperes}$$

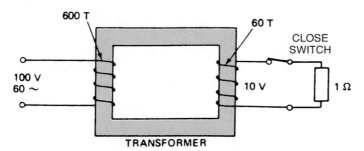

FIGURE 12–3 Current flow of 10 amperes developed 100 watts in the load resistor.

The 10 amperes of current flow through the I-ohm resistor.

Electrical power is supplied by the secondary of the transformer. This power must be coming through the primary of the transformer from the 100-volt, 60-hertz power source. The power developed in the **secondary circuit** is

$$W = I \times E$$

$$W = 10 \times 10$$

$$W = 100 \text{ watts}$$

When the primary of a transformer is connected to a source voltage, a small current called the magnetizing current flows through the primary, as mentioned previously. The magnetic field set up by this magnetizing current is indicated in Figure 12–4 as 1. When a load is connected across the secondary, current flows in the secondary. This current produces a magnetic field in the core of the transformer in direct opposition to the field set up by the magnetizing current. In Figure 12–4, this magnetic field is indicated as 2. The field strength is based on 600 ampere-turns and tends to reduce the back EMF of the primary. The decreased back EMF allows for an increase in current in the primary.

A current flow producing a magnetic field based on 600 ampere-turns is required in the primary to overcome the 600 ampere-turns in the secondary. With 600 turns in the primary, 1 ampere of current is needed to produce this field. The field is shown as 3 in Figure 12–4.

The relationship between the currents in the primary and the secondary can now be determined. The currents are inversely proportional to the turns ratio. For example, in Figure 12–3

1. The turns ratio is 600/60 or 10/1.
2. The voltage ratio is 600/60 or 10/1.
3. The current ratio is 60/600 or 1/10.

The power developed in the secondary has been shown to be 100 watts. The power developed in the primary is

$$W = I \times E$$

$$W = 1 \times 100$$

$$W = 100 \text{ watts}$$

Actually, the current flow in the primary is slightly more than 1 ampere since the magnetizing current also flows in the primary. The transformer is not 100% efficient; there is some power loss.

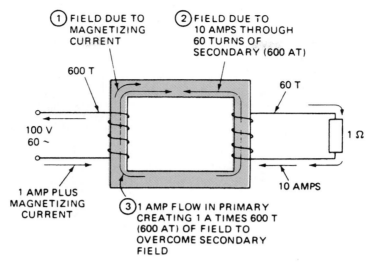

FIGURE 12–4   The power developed in the secondary is equal to the power supplied to the primary.

Power Loss in Transformers

There are three main reasons for power loss in a transformer:

1. Resistance of the windings
2. Hysteresis
3. Eddy currents

Resistance of the Windings   The power loss in the resistance of the windings needs no further explanation. Whenever current flows through resistance, heat is developed and power is lost. This power loss is known as the I^2R loss.

Hysteresis   When alternating current flows in a coil, the magnetic field around the coil changes direction with the current. If the coil has an iron core, as in a transformer, the direction of the magnetism of the iron core must also change with the current. Each magnetic molecule of the iron core must change direction with each alternation of the current. This continuous motion of the magnetic molecules produces heat due to friction, which is a power loss in the transformer.

Eddy Currents   Eddy currents are set up in the iron core of a transformer. The core of the transformer is a metal, and it acts the same as a secondary winding. A voltage is induced in the core and current flows in the core. Eddy currents flow throughout the core. If solid iron cores were used in transformers, the losses due to eddy currents would be quite high.

To overcome the high-eddy current losses, the cores of transformers and most other magnetic devices using alternating current are laminated. Laminations are thin, sheet-metal parts that may be mechanically fastened together to produce the required core. Through the use of thin sheets of metal, the eddy current is confined to the individual lamination. Because the resistance path for eddy currents is increased, the total eddy current draw is decreased. Laminated iron cores are found in most AC devices, such as transformers, motors, solenoids, relays, and contactors.

POWER TRANSFORMERS

Power transformers are used throughout the world to step up voltages for transmission and distribution. They are also used to step down voltages for use at the destination point.

For example, a power company may generate electric power at a voltage level of 13,800 volts at the generating station. The main factor in determining the voltage level at the generator is the insulation problem related to operating generators at higher voltages. Before the electrical power is distributed on transmission lines across the country to individual localities, the voltage level is stepped up, using transformers. A standard transmission line voltage level is 69,000 volts. Substations (transformers) are used to reduce the voltage to safer levels (about 2400 volts) for distribution around populated areas. Finally, at the end point of use, the voltage is further reduced by a transformer to the

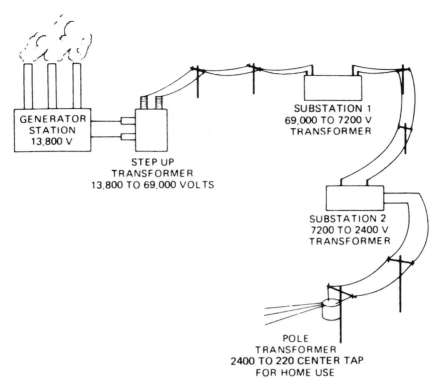

FIGURE 12–5 Power distribution using transformers to step voltage up and down.

level required by the consumer, usually 220/110 volts. Figure 12–5 shows an example of a full system from generation to final use of electrical power by the residential customer. The voltage levels shown are representative of industry standards.

Power transformers at substations are usually large, bulky items (Figure 12–6). High-voltage insu-

lators of ceramic material are usually visible at the terminals of the transformers. Power transformers used to step voltage down from a final substation (2400 volts) to the voltage needed in residential use (220/110 volts) are located on power-line poles (Figure 12–7). A single transformer will usually supply several homes. The final step-down transformer

FIGURE 12–6 Transformer substation. (Courtesy of Florida Power & Light Company)

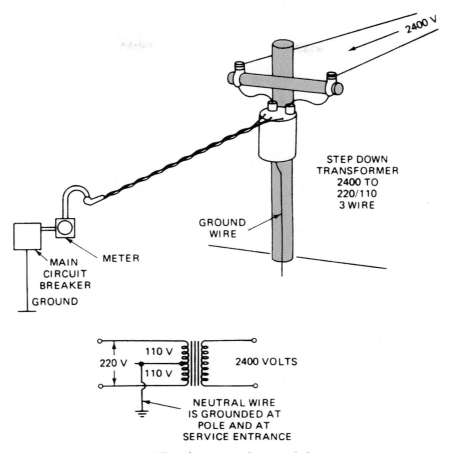

FIGURE 12–7 Transformer used to supply home power.

is center tapped and therefore can provide the required 220 volts for power circuits and also the 110 volts required for lighting and small appliances.

Transmission Losses

Voltages are stepped up whenever power is distributed over long distances in order to decrease power losses. For example, consider a manufacturing plant located approximately 10 miles from a power-

generating station (Figure 12–8). The manufacturing plant requires 100,000 watts of electrical power. The generating station is producing power at a 10,000-volt level. A current of 10 amperes is needed at the 10,000-volt level to provide the 100,000 watts.

Two wires are required to get the electrical power to the manufacturing plant; each wire is 50,000 feet long. If size 0 wire is used, the total resistance of the two wires will be about 10 ohms, because size 0 wire

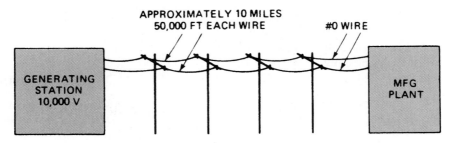

FIGURE 12–8 Power distribution at 10,000 volts (high voltage and power loss in lines).

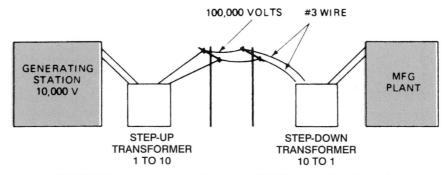

FIGURE 12–9 Power distribution at 100,000 volts (low losses).

has 0.1 ohm resistance per 1000 feet. According to Ohm's law, the voltage developed on the resistance of the transmission line would be 100 volts.

$$E = I \times R$$
$$E = 10 \times 10$$
$$E = 100 \, \text{volts}$$

This would leave only 9900 volts to be delivered to the manufacturing plant. The power company would continually waste the power developed in the resistance of the transmission lines. The power loss is

$$W = I \times E$$
$$W = 10 \times 100$$
$$W = 1000 \, \text{watts lost}$$

The losses can be reduced and additional savings can be realized by employing transformers, thus making feasible the use of smaller wire in the transmission lines. In Figure 12–9, the power station and the manufacturing plant are located at the same distance as in Figure 12–8. A step-up transformer has been added at the generator station and a step-down transformer has been added at the manufacturing plant. The transmission lines have been changed to size 3 wire, which has about 0.2 ohm resistance per 1000 feet. The resistance of the total transmission line, two lengths of 50,000 feet each, is about 20 ohms. With the step-up of the voltage at the power plant from 10,000 to 100,000 volts, the current is stepped down from 10 to 1 ampere. The power loss in the lines has been reduced considerably.

The voltage developed in the resistance of the transmission lines is

$$E = I \times R$$
$$E = 1 \times 20$$
$$E = 20 \, \text{volts}$$

The power loss is

$$W = I \times E$$
$$W = 1 \times 20$$
$$W = 20 \, \text{watts lost}$$

The voltage at the primary of the step-down transformer is

$$100{,}000 - 20 = 99{,}980$$

At the secondary, the voltage is 9998 volts, which is much closer to the 10,000 volts being delivered by the generating station.

In actual practice, the voltage required at a manufacturing plant is seldom as high as that generated at a power station. The examples given of using transformers to step up and step down voltages to save power are similar to the methods used in actual situations to adjust voltage at the receiving user.

POWER DISTRIBUTION: DELTA AND WYE

Almost all electrical power is generated using three-phase alternators. Power transformers are used to step the voltages up or down as required by the distribution system. The turns ratio of the transformers is not the only factor to consider when three-phase power is transformed. The transformer connections, delta or wye, must be taken into account.

Three-Phase Delta-Wye Combinations

The primaries and secondaries of transformers used in three-phase systems may be connected in delta or wye configuration. The output voltage is determined not only by the input voltage and the turns ratio; the delta or wye connection of the primaries

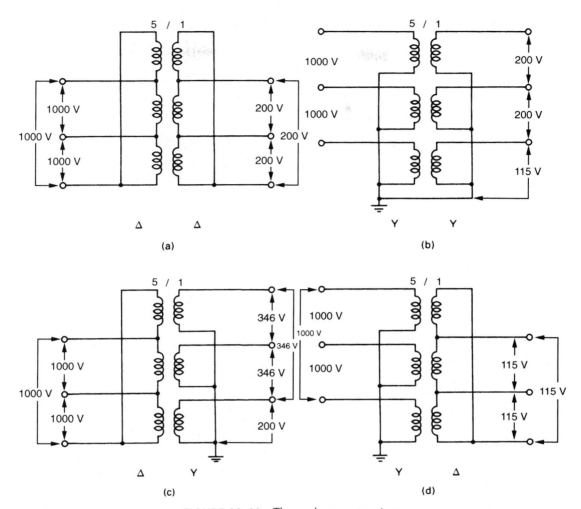

FIGURE 12-10 Three-phase connectors.

and secondaries also affects the output voltage. Consider the four possible connections of the three transformers in Figure 12–10. The transformers are stepped down, with a 5/1 turns ratio. The input voltage is 1000 volts.

In Figure 12–10a, the transformers are connected delta to delta. The primary coil voltage is equal to the line voltage. The secondary line voltage is equal to the secondary coil voltage.

In Figure 12–10b, the transformers are connected wye to wye. The primary coil voltage is equal to the line voltage divided by the square root of 3 (1.732). There are 577 volts across each primary coil. The secondary coil voltage is reduced by a factor of 5 due to the turns ratio. There are 115 volts across each secondary coil. The voltage from any line to ground is 115 volts. The voltage between any two lines is the coil voltage times the square root of 3, or 200 volts (115 V × 1.732 = 200 V).

In Figure 12–10c, the transformer primaries are delta connected. The secondaries are wye connected. There are 1000 volts across each primary coil. The step-down turns ratio provides 200 volts across each secondary coil. Since the secondaries are wye connected, the voltage line to ground is the voltage of the coil, or 200 volts. The voltage between any two lines is the coil voltage times the square root of 3, or 346 volts (200 V × 1.732 = 346 V).

In Figure 12–10d, the primaries are connected in wye. The transformer primary coil voltage is equal to the line voltage divided by the square root of 3, or 577 volts. The 5/1 turns ratio provides a secondary coil voltage of 115 volts. The secondaries are delta connected. The line voltage is equal to the coil voltage, 115 volts.

With the delta connection, the line voltage equals the coil voltage and the coil voltage equals the line voltage. With the wye connection, the line voltage equals the coil voltage times 1.732 and the coil voltage equals the line voltage divided by 1.732.

With the delta connection, the line current equals the coil current times 1.732, and the coil current equals the line current divided by 1.732. With the wye connection, the line current equals the coil current, and the coil current equals the line current.

Standard Delta-Connected Secondaries

Figure 12–11 shows the secondary connection in a standard delta system. The voltage between any two lines is 240 volts. The voltage between line A and neutral or line C and neutral is 120 volts. The voltage between line B and neutral is approximately 208 volts.

In large commercial and office installations, the 240 volts could be used to operate power equipment such as air conditioners. The 120 volts could be used to supply the outlets. The phase B to ground voltage is 208 volts. This phase is often called the *high leg*

or *wild leg*. Normally, this phase is identified in panel boxes by orange tape or paint. Phase B is normally placed in the center lug of a panel box. Special care must be taken not to use this phase to ground for 120 volts. The higher voltage (208 V) could damage 120-volt equipment.

Open-Delta System

Three-phase power is sometimes provided using the open-delta system. An advantage of the open-delta system is that only two transformers are required for the three-phase system. The open-delta connection is shown in Figure 12–12. Note that the high leg or wild leg, phase B, is also present in the open-delta system.

SIGNAL TRANSFORMERS

Transformers are used in almost all air-conditioning units to provide safe, low-level AC voltage for control purposes. The input voltages for air-conditioning systems may be 110, 220, or 440 volts; 220 volts is the most common input for residential control systems. A step-down transformer within the air-conditioning unit provides 24 volts AC for use in the control circuit. The control circuit is used to energize high-voltage circuits for the control of motors and other devices.

An example of a low-voltage control circuit in an air-conditioning system using a 24-volt step-down transformer is shown in Figure 12–13. A voltage of 220 volts is fed through the disconnect switch to the primary of the low-voltage transformer and to the top of the two power contactors. With the thermostat switches (covered in Unit 16) as shown, the 24 volts from the secondary of the low-voltage transformer is fed to the coil of the compressor contactor and to the coil of the inside fan motor; both contactors energize. The 220 volts is then fed through both contactors to the motors.

After the room cools, the thermostat contacts open at a preselected temperature. This removes the 24 volts from the contactor coils; the contactors de-energize. The contacts then open and 220-volt power is removed from the motors. The motors will not run until the room temperature rises and the switches (contacts) in the thermostat again close.

The use of a low-voltage transformer creates a safe system of signal transfer. The individual making adjustments at the thermostat is not in close proximity to dangerous high-voltage power circuits.

Another important factor in the use of low voltage in air-conditioning control circuits is the cost of

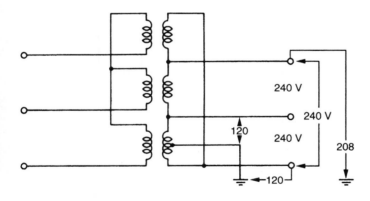

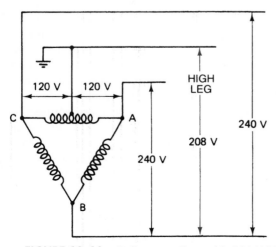

FIGURE 12–11   Delta secondary with 208 V high leg.

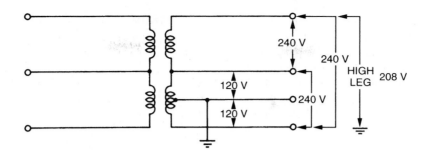

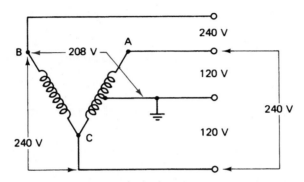

FIGURE 12–12 Open delta. (Note high leg.)

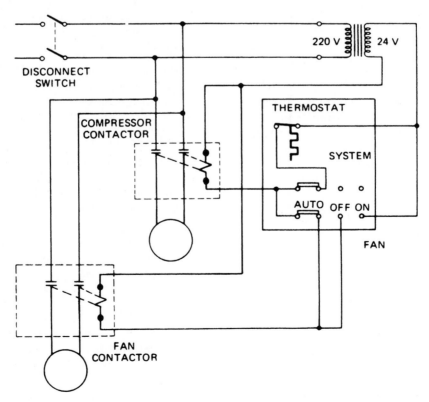

FIGURE 12–13 Low-voltage control of high voltage.

installation of control-circuit wiring. If 110 volts were used in the control circuit, wiring between the thermostat and the evaporator/condenser would have to be treated the same as the power circuits for lighting or outlets. If the local building code called for the use of conduits in all line–voltage circuits, then a conduit would have to be installed between the evaporator/condenser unit and the thermostat. Installation costs could vary greatly, depending on the distance between the thermostat and the condenser.

The same installation using a low-voltage control system is less expensive. Control circuit wiring between the evaporator/condenser and thermostat consists of a standard signal cable that meets building code requirements. If the conduit is not needed, the cost decreases dramatically.

TRANSFORMER RATINGS

Transformers are usually rated with the primary and secondary voltage and the volt–ampere (VA) design of the unit. An example of this is a control transformer. This transformer is designed to operate with 220 volts on the primary, and it will produce 24 volts at the secondary terminals. The volt–ampere rating of the transformer is 48 VA. This means that a maximum of 2 amperes can be drawn from the transformer's secondary (2 amperes $\times$ 24 volts = 48 VA).

SUMMARY

- The turns ratio of a transformer determines the relationship between input and output voltage. The number of turns in the primary divided by the number of turns in the secondary is equal to the voltage of the primary divided by the voltage of the secondary.

$$\frac{N_p}{N_s} = \frac{E_p}{E_s}$$

- Power losses in a transformer are due to three main causes: hysteresis, eddy current, and *IR* losses.
- Transformers are rated in VA (volt–amperes) rather than power.

PRACTICAL EXPERIENCE

Required equipment Three-phase distribution panel (208 V–240 V), voltmeter

Procedure

1. Measure and record the individual voltages phase A to B, phase B to C, and phase C to A.

   A to B _____ volts

   B to C _____ volts

   C to A _____ volts

2. Measure the voltage phase A to neutral. _____ volts

3. Measure the voltage phase B to neutral. _____ volts

4. Measure the voltage phase C to neutral. _____ volts

5. Is the power panel being fed from a delta- or wye-connected secondary?

REVIEW QUESTIONS

1. The voltage ratio of a transformer is proportional to the turns ratio.

   T _____   F _____

2. The current ratio of a transformer is proportional to the turns ratio.

   T _____   F _____

3. The power supplied to the primary is approximately equal to the power taken from the secondary.

   T _____   F _____

4. The amount of power loss in transmission lines will be (higher, lower) if the power is transmitted at higher voltages.

   T _____   F _____

5. Why are low-voltage transformers used in air-conditioning systems?

6. A transformer has 800 turns on its primary and 400 turns on its secondary. The primary voltage is 200 volts. What is the secondary voltage?

7. In the transformer in question 6, if 4 amperes are drawn in the secondary circuit, how much current flows in the primary?

8. What are the three main losses in a transformer?

9. Which transformer loss is related to the motion of molecules?

10. What ratings are usually provided on control transformers?

UNIT 13

PHASE SHIFT AND POWER FACTOR

OBJECTIVES

After completion and study of this unit, you should be able to understand that:

- ■■■ In AC circuits, current and voltage do not necessarily rise and fall together.
- ■■■ Whenever a circuit is acting inductively or capacitively, there will be a phase shift between the current and voltage.
- ■■■ Whenever there is a phase shift between current and voltage, power is no longer equal to current times voltage. A third factor—the power factor (*pf*)—must be considered.

CURRENT AND VOLTAGE PHASE RELATION

In Unit 10 it was shown that current and voltage in certain electric circuits do not rise and fall together. In an inductive circuit, such as a coil of wire, current lags behind (follows) voltage (Figure 13–1). Voltage passes through zero and rises in a positive direction, whereas current is still negative and follows behind the voltage. The current in such a circuit is called a *lagging current*.

To better understand power in inductive circuits, take another look at power in resistive circuits. The current and voltage are in phase with each other. The power developed is equal to $I \times E$. In the graph in Figure 13–2, during the first time period, both the voltage and the current are positive. During the second time period, the voltage and the current are both negative. In algebra, a positive times a positive equals a positive ($+ \times + = +$) and a negative times a negative equals a positive ($- \times - = +$). During both time periods, therefore, the power is positive. This power is developed in the resistor as heat.

The power developed in a resistive circuit is shown in Figure 13–3. In a pure inductance circuit (one that does not contain resistance), current follows the voltage by 90°. Such a circuit is not possible. If it were, no power would be dissipated, and no heat would be developed. To help explain phase shift, a perfect inductor is considered. It must be noted, however, that this is for explanation only, since perfect inductors do not exist. All wires have resistance; all core materials have hystereses and eddy current losses.

In a perfect inductor, the current lags the voltage by exactly 90°. Figure 13–4 shows an iron-core coil and provides a graph indicating that the current lags the voltage by 90°. The electrical energy provided by the 120-volt, 60-hertz source is converted into magnetic energy in the coil. The magnetic energy is converted back into electrical energy and is fed back to the incoming line. No power or energy is used in the coil; no heat is developed.

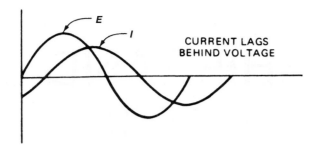

FIGURE 13–1 Voltage and current in an inductive circuit.

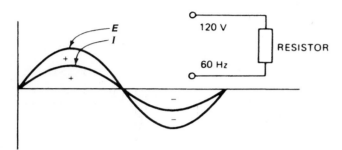

FIGURE 13–2 Voltage and current in a resistive circuit.

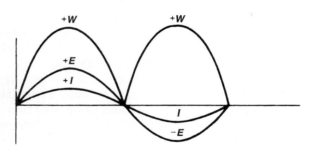

FIGURE 13–3 Power developed in a resistive circuit.

The power that is fed to the coil from the source is positive power. The power that is fed from the coil back to the source is negative power. In a pure inductance coil, they would be equal. An example of this is given in Figure 13–5, where current, voltage, and power are shown on the same graph.

1. In time period 1, a positive voltage times a negative current provides negative power $(+E \times -I = -W)$.

2. In time period 2, a positive voltage times a positive current provides positive power $(+E \times +I = +W)$.

3. In time period 3, a negative voltage times a positive current provides negative power $(-E \times +I = -W)$.

4. In time period 4, a negative voltage times a negative current provides positive power $(-E \times -I = +W)$.

The power fed to the circuit in time periods 2 and 4 is equal to the power fed from the circuit back to the line in time periods 1 and 3. No power is consumed in the inductance coil. This situation would only appear in a pure inductive circuit.

In actual practice, inductive circuits in air-conditioning and refrigeration applications are not pure inductive circuits. Some of the electric power that is fed to the circuit remains either as heat or is converted into mechanical power. A good example of a device converting electric power to mechanical

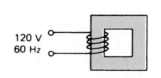

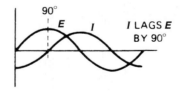

FIGURE 13–4 Current and voltage in a perfect inductor.

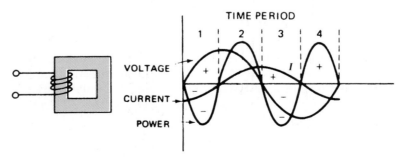

FIGURE 13–5 Positive and negative power in a pure inductance.

TABLE 13–1 Cosine Values

Angle	Cosine	Power factor (%)
0	1.000	100.0
5	0.996	99.6
10	0.985	98.5
15	0.966	96.6
20	0.940	94.0
25	0.906	90.6
30	0.866	86.6
35	0.819	81.9
40	0.766	76.6
45	0.707	70.7
50	0.643	64.3
55	0.574	57.4
60	0.500	50.0
65	0.423	42.3
70	0.342	34.2
75	0.259	25.9
80	0.174	17.4
85	0.087	8.7
90	0.000	0.0

power is a **motor**. In a circuit containing an electric motor, the current could lag the voltage by about 30°. This means that some of the power that is fed to the motor is not used but is fed back to the line. Actually, only 86.6% of the power fed to the circuit is used. This figure, 86.6%, is the power factor (*pf*) of the circuit. It is the cosine of the angle between the current and the voltage. A list of cosine values is presented in Table 13–1.

Graphically, this relationship is shown in Figure 13–6. Since current and voltage are not necessarily in phase with each other in circuits, this must be considered in the calculations of power. Power factor must also be considered in the calculations of power. The formula is

$$W = I \times E \times pf$$

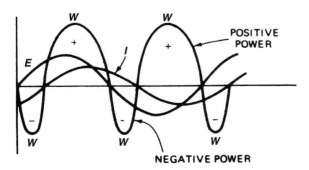

FIGURE 13–6 Power in a circuit containing resistance and inductance.

Since power factor is the cosine of the angle between the current and the voltage, the greater the angle is, the smaller the power factor; the smaller the angle is, the larger the power factor. When current and voltage are in phase, the angle is 0°. Since the cosine of 0° is 1.0, the power formula when the current and voltage are in phase is

$$W = I \times E \times pf$$

$$W = I \times E \times 1$$

$$W = I \times E$$

If all the power fed to the circuit were to be used in the circuit, no power would be returned to the source. This situation was shown in Figure 13–3.

When current and voltage are 90° out of phase with each other, as in a pure inductive circuit, the power factor (*pf*) is 0 (the cosine of 90° is 0). The power formula becomes

$$W = I \times E \times pf$$

$$W = I \times E \times 0$$

$$W = 0$$

The power used in the circuit would be 0, since any number multiplied by 0 is 0. An example of this was shown in Figure 13–5, in which the same amount of power supplied by the source as positive power is returned to the source as negative power.

In actual practice, of course, the power factor is seldom, if ever, as low as 0 or as high as 1.0. It is a percentage, such as those given in Table 13–1. If the current lags the voltage by 30°, the power factor for 30° is 86.6%, or 0.866.

The air-conditioning and refrigeration technician will seldom have to calculate or work with the power factor. It is necessary that the technician understand that the power formula does contain the *pf* component, in order that he or she not be confused when measuring voltage, current, and power.

CAPACITANCE CIRCUITS

The current in a capacitive circuit has characteristics exactly opposite the current characteristics of inductive circuits. Current leads the voltage. In a perfect capacitor, current would lead the voltage (across the capacitor) by 90°. There is no such thing as a perfect capacitor, but nearly perfect capacitors in which current and voltage relations approach

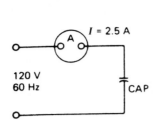
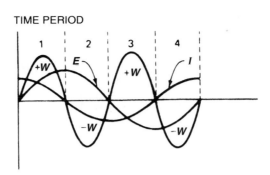

FIGURE 13–7 Current, voltage, and power relationship in a pure capacitive circuit.

90° are common. Figure 13–7 shows the current, voltage, and power relationships in a capacitive circuit.

1. In time period 1, current and voltage are positive; power is positive ($+I \times +E = +W$).
2. In time period 2, current is negative but voltage is positive; power is negative ($-I \times +E = -W$).
3. In time period 3, current and voltage are negative; power is positive ($-I \times -E = +W$).
4. In time period 4, current is positive but voltage is negative; power is negative ($+I \times -E = -W$).

The power fed to the capacitor in time periods 1 and 3 is equal to the power fed back to the line in time periods 2 and 4. No power is consumed in the capacitor.

Observe the relationship between Figures 13–5 and 13–7. In time period 1, the coil (Figure 13–5) is sending power back to the line (negative power). During this time period, the capacitor (Figure 13–7) is taking power from the line. During time period 2, the capacitor is sending power back to the line (negative power), whereas the coil is receiving power.

An interesting combination may be made with coils and capacitors. Assume that in their individual circuits, both the coil and the capacitor draw 2 amperes. If the two circuits are connected together in parallel, the line (source) has to supply very little power (Figure 13–8). Current circulates back and forth between the capacitor and coil, as indicated on ammeter 2; little current is flowing in the line, as indicated on ammeter 1.

The power necessary to overcome the losses in the circuit is the only power supplied by the line.

If perfect inductance coils and perfect capacitors were possible, no power at all would have to be supplied by the line. Current would be indicated in ammeter 2, whereas no current would be indicated in ammeter 1.

Capacitors are used by large manufacturing plants to reduce the incoming line current. Most manufacturing plants use a large number of motors in their operations. The motors are inductive and draw a lagging current. With a lagging current, some of the power that the power company is supplying is fed back to them. Because of losses in the transmission lines, the power company is losing power, and therefore money, because of the lagging current drawn by the manufacturing plant. To overcome this loss, the power company charges more for power supplied at lower power factors. This is not true for electrical power supplied to homes; the power company usually corrects for power factor at distribution points.

Because of the higher charges presented to large manufacturing plants, it is to the plant owner's advantage to correct the power factor and thus to reduce the cost of operation. As an example, consider an electric motor drawing 900 volt–amperes from a 208-volt power source (Figure 13–9).

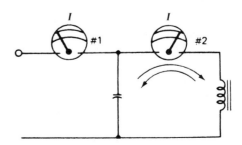

FIGURE 13–8 Low-line current with high-circulating current between capacitor and coil.

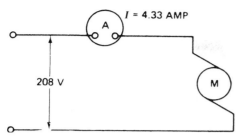

FIGURE 13–9 Motor drawing current with a lagging power factor.

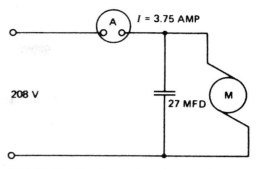

FIGURE 13–11 Correction of power factor by the addition of a capacitor.

The uncorrected line current would be 4.33 amperes.

$$I = \frac{VA}{E}$$

$$I = \frac{900}{208}$$

$$I = 4.33 \text{ amperes}$$

By measurement, it is found that the power factor is equal to 0.866, indicating that the current is lagging the voltage by 30°. The power factor can be determined by measuring the current with an ammeter, the voltage with a voltmeter, and the power with a wattmeter; the wattmeter reading is then divided by $I \times E$ or VA.

$$pf = \frac{W}{VA}$$

The circuit connections for determining the power factor are shown in Figure 13–10.

$$pf = \frac{W}{VA}$$

$$pf = \frac{780}{208 \times 4.33}$$

$$pf = 0.866$$

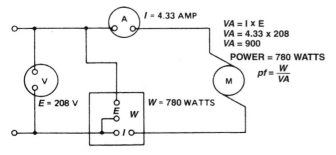

FIGURE 13–10 Measuring true power and volt–amperes.

The power factor of this circuit could be improved and the line current reduced by adding a capacitor in parallel with the motor (Figure 13–11). With only the addition of the capacitor, the line current could be reduced from 4.33 to 3.75 amperes; there would be no difference in the operation of the motor. The power company would be supplying 780 volt–amperes, and the motor would be using 780 watts. This example covers the calculations of a single, low-horsepower motor. The example is typical for all the motors in a complete manufacturing plant. The explanation is given in order that the air-conditioning technician might have a better understanding of power factor and power-factor correction. It is not the job of an air-conditioning technician to correct power factor.

The capacitors that are used with compressor motors contribute to a better power factor although they are not included for this purpose. Capacitors are used in the start-winding circuits to increase starting torque; the power-factor improvement that is obtained is of secondary importance.

VOLTAGE AND CURRENT MEASUREMENTS

Whenever the voltage and current are out of phase with each other in an AC circuit, their indications do not seem to follow the laws of series or parallel circuits. The laws of series and parallel circuits are true, but electrical measuring instruments provide a reading of effective values, rather than the instantaneous values on which the laws are based.

Consider the circuit in Figure 13–12: A resistor of 43 ohms and a capacitor of 100 microfarads are connected in series. The voltmeter across the capacitor indicates 53.2 volts, and the voltmeter across the resistor indicates 86 volts. It seems that the sum of these two voltages is not the same as the supply

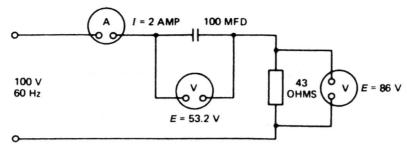

FIGURE 13–12   Resistor capacitor circuit $E_C + E_R$ is greater than the supply voltage.

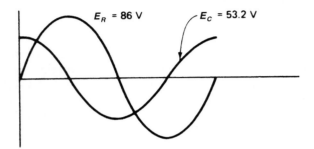

FIGURE 13–13   Voltage relationship of E_R and E_C.

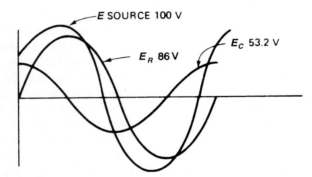

FIGURE 13–14   Voltage relationship of E_R, E_C, and E source.

voltage of 100 volts (53.2 + 86 = 139.2). The waveforms of the two voltages are shown in Figure 13–13. Note that when the voltage across the resistor E_R is positive, the voltage across the capacitor E_C is first positive and then negative (one alternation of E_R). If these two waveforms are added together, the result is the supply voltage of 100 volts (Figure 13–14).

In a series circuit, the same current flows through each component. This law is true even if the current and voltage are not in phase with each other. In parallel circuits, the same voltage appears across each component. This law is true for circuits where the current and voltage are not in phase with each other.

In parallel circuits, the line current equals the sum of the currents in the branches. This law is also true in AC circuits, but instantaneous values must be used.

Consider the circuit in Figure 13–15. A resistor, a capacitor, and a coil are connected in parallel. Ammeters connected in series with each component indicate the current flow through the component.

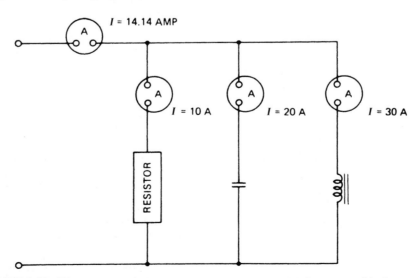

FIGURE 13–15   Parallel circuit containing resistance, capacitance, and inductance.

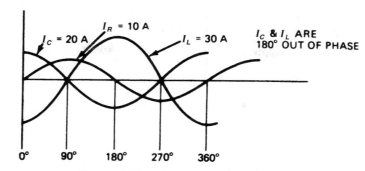

FIGURE 13–16 Phase relation of current in the resistor, capacitor, and coil.

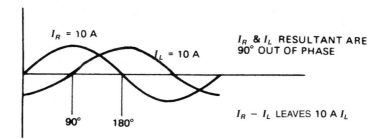

FIGURE 13–17 Resistive current and resultant inductive current.

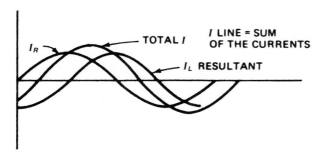

FIGURE 13–18 The line current is the sum of the reisitive and resultant inductive current.

The ammeter in series with the line indicates the total line current. It is not possible simply to add up the currents in the branch circuits to get the line current. In Figure 13–16, the waveforms of the individual currents are shown. The total line current is the sum of the three waveforms of current. Since the current in the capacitor and the current in the coil have polarities exactly opposite each other, the capacitive current can be added to the coil current arithmetically. This leaves a current of 10 amperes acting inductively: $+(20\ I_C) - (30\ I_L) = -10\ I_L$.

The current waveforms are now as shown in Figure 13–17. When the two currents are added together, the final result is as shown in Figure 13–18.

MOTOR CIRCUITS

The compressor motor start-winding circuit of most large air-conditioning systems includes a capacitor in series with the start winding (Figure 13–19). The voltage across the capacitor might be as high as 320 volts. The capacitor voltage and start-winding voltage do not add up to the supply voltage of 220 volts.

An air-conditioning and refrigeration technician will often make voltage and current measurements in circuits where there is a phase difference between the current and the voltage. It is important that the meter indicators do not cause any confusion.

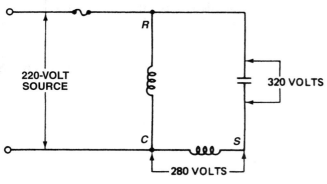

FIGURE 13–19 Voltages across compressor motor components.

The service technician will not have to deal with current or voltage waveforms. This information has been presented so that the technician will have a better insight into circuit operations for trouble-shooting procedures.

SUMMARY

- In AC circuits, current and voltage seldom are in phase with each other.
- Power in AC circuits is equal to $I \times E \times pf$.
- The power factor (*pf*) is equal to the cosine of the angle between the current and voltage.

PRACTICAL EXPERIENCE

Required equipment Air conditioner with a capacitor-run compressor motor, voltmeter. (Access to the compressor motor components is required.)

Procedure

1. Connect the A/C unit to the voltage specified.
2. Measure and record the input voltage. _____ volts
3. Measure and record the voltage across the run winding. _____ volts
4. Measure and record the voltage across the start, or AUX, winding. _____ volts
5. Measure and record the voltage across the run capacitor. _____ volts
6. Do the voltage readings in steps 5 and 6 indicate a phase relation between the two voltages of a series circuit?

REVIEW QUESTIONS

1. In a circuit containing resistance and inductance, the line current will lag behind the voltage.

   T _____   F _____

2. Whenever the current and voltage are out of phase with each other, the total power developed in a circuit is equal to $I_T \times E_T$.

   T _____   F _____

3. The cosine of a 45° angle is 0.866.

   T _____   F _____

4. The current is always in phase with the voltage across a resistance.

   T _____   F _____

5. The current flow in a capacitor leads the voltage across the capacitor by 90°.

   T _____   F _____

6. Pure inductance circuits are commonly found in air-conditioning circuits.

   T _____   F _____

7. The formula for power in an AC circuit is $W = I \times E \times 2$.

   T _____   F _____

8. To find the total applied voltage in an AC series circuit containing inductance and capacitance, the voltages across the components are simply added together.

   T _____   F _____

9. The power factor in an AC circuit is often over 200%.

   T _____   F _____

10. The power factor of a circuit containing only pure capacitance would be 0.50.

    T _____   F _____

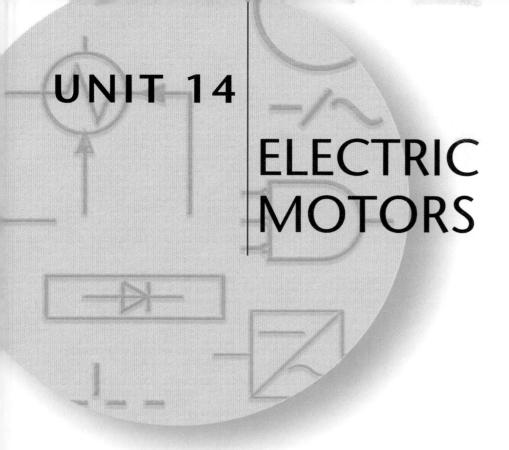

UNIT 14

ELECTRIC MOTORS

OBJECTIVES

After completion and study of this unit, you should be familiar with the principles of and the differences and similarities between:

- Induction motors.
- Shaded-pole motors.
- Split-phase motors.
- Capacitor motors.
- Polyphase motors.
- Two-speed, single-phase systems.
- Two-speed, three-phase systems.
- Variable-speed, three-phase systems.

INDUCTION MOTORS

Most air-conditioning and refrigeration systems operate through the use of a compressor of one type or another, driven by an electric motor. The most common type of motor used to drive these compressors is the **induction** motor.

An induction motor operates on a basic principle of electricity: An induced current opposes the field that produced it. This was covered in Unit 12, in which the secondary current of a transformer opposed the field of the primary. The back EMF of the primary tends to be reduced, allowing for more current in the primary. In this way, power in the secondary is provided for by increased power in the primary. (Refer to Figure 12–4.)

An induction electric motor is actually nothing more than a special type of transformer. The primary of the motor transformer is wound on the stationary parts, called the **stator**. The secondary is made up of copper bars shorted together on the rotating part of the motor, called the **rotor**.

The primary windings on the stator look very much like most transformer windings. The secondary winding on the rotor consists of copper bars on an iron core. The copper bars are all connected together. This essentially provides a combination of short-circuited single-turn windings. Figure 14–1 shows a typical rotor in which the core is made up of die-cut laminations. These laminations reduce eddy currents in the iron core of the rotor.

FIGURE 14–1 Rotor. (Courtesy of BET Inc.)

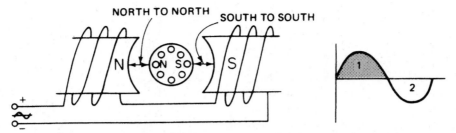

FIGURE 14-2   Magnetic field set up in rotor due to stator magnetic field.

Whenever an AC voltage is connected to the primary of a transformer, a voltage is induced in the secondary. The secondary-induced voltage is in direct opposition to the primary supply voltage. This opposition usually results in a rotation of the motor shaft.

Without some form of electric direction, the rotor may turn in either direction, **clockwise (cw)** or **counterclockwise (ccw)** (Figure 14-2). If the rotor were not started in some direction, the motor would sit like a transformer with a shorted secondary—it would hum but not move. In part of alternation time period 1, current flows through the stator to provide for north and south poles. The induced voltage and the current flow in the rotor produce a magnetic field in the secondary winding in opposition to the magnetic field in the primary winding. The system is in balance; the rotor does not rotate. Some direction action is needed.

In part of the second alternation of the applied voltage time period 2, the current flow and the polarity of the magnetic field of the primary change. The current flow and magnetic polarity in the secondary also change (Figure 14-3). Although the magnetic field of the stator and rotor change, the rotor still does not turn since the opposition of magnetic poles is direct, trying to force the rotor toward the shaft.

When the motor shaft is started in either direction, a time-delay imbalance is created, allowing the motor to continue to turn in the direction in which it was started. The motor continues to turn after it is started because of the repulsion between the stator poles and the poles set up in the rotor (Figure 14-4). As long as electric power is applied to the stator winding of the motor, it continues to rotate in the direction in which it was started. A continuous shift in the magnetic field around the rotor takes place. Current flow in the rotor results from two actions: transformer action of the primary, as previously discussed, and generator action as the copper bars in the rotor cut through the primary magnetic field.

SHADED-POLE MOTORS

One method used to get small motors (¼ horsepower or less) started is to provide shaded poles in the motor (Figure 14-5). There is a slotted area in the two pole pieces with shorted turns of copper wire in the slots. The transformer action involving these shorted turns starts the motor in the desired direction.

The magnetic polarity relationships of the rotor and stator of a shaded-pole motor are shown in Figure 14-6. This relationship exists just after the current flow in the motor coil passes through zero.

The current flow in the shaded pole (shorted turn) sets up a magnetic field in opposition to the magnetic field that produced it. The magnetic field

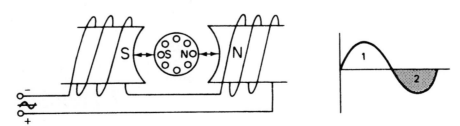

FIGURE 14-3   In negative alternation, the magnetic field of the stator and rotor both change direction.

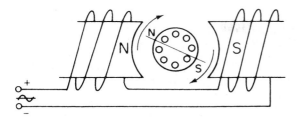

FIGURE 14-4 Turning of rotor causes a shift in the rotor magnetic field.

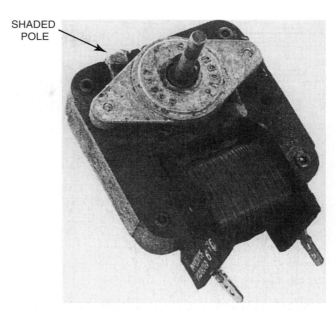

FIGURE 14-5 Shaded-pole motor. (Courtesy of BET Inc.)

set up by the shaded pole is of the same polarity as the magnetic field set up by the rotor currents. The magnetic field of the shorted turn is not as strong as the magnetic field of the main coil, but it is sufficient to cause a delay in the magnetic field set up in the shaded area.

The end result is a rotating magnetic field around the stator of the motor. The relationship is shown in Figure 14-7 and is as follows:

1. The current in the main winding passes through zero and starts to increase. A north pole is started in the main pole area. The opposing effect in the shaded area produces a south pole at the shaded-pole area.

2. As the main field builds up toward a maximum north pole, the shaded area becomes a weak north pole. The effect of the shaded pole is decreasing.

3. The main pole is at maximum. The effect of the shaded pole is zero.

4. The strength of the main pole decreases from a maximum north pole. The shaded area becomes a north pole.

5. The main pole decreases in strength to zero. The shaded area stays a north pole.

6. The main pole changes polarity to a south pole. The shaded area remains a north pole.

As long as the magnetic field is rotating around the stator, the resulting current flow in the rotor causes the rotor to follow. The direction of rotor rotation is always in the direction of the shaded pole (Figure 14-8).

Motor Direction (Shaded-Pole)

Many shaded-pole motors have a shaft extending out of end bells on both sides of the rotor. With a double-shaft motor, either clockwise or counterclockwise rotation is obtained, depending on which shaft is used to drive the load (Figure 14-9). In many cases, it is possible to obtain reversed rotation of a shaded-pole motor with a single shaft (extending from only one end). If the end bells of the motor are interchangeable, the motor may be reversed

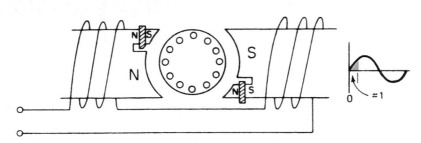

FIGURE 14-6 Magnetic field distribution during time period # 1.

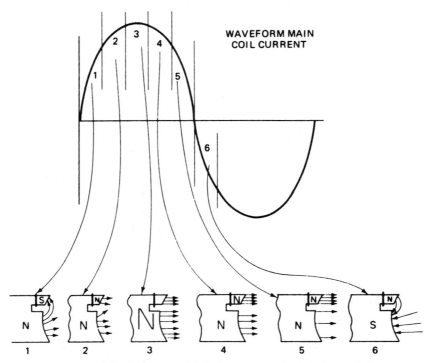

FIGURE 14–7 Stator field during time periods 1 through 6.

simply by reversing the rotor. The end bells are removed, and the motor is reassembled with the shaft
extending from the opposite end of the motor
(Figure 14–10).

SPLIT-PHASE MOTORS

Split-phase motors are usually used for applications
requiring from ⅛ to ¾ horsepower. The starting torque
of split-phase motors is low to moderate compared
with other motor types, except shaded-pole motors.
Split-phase motors are usually found in service for
driving fans, centrifugal pumps, conveyors, and refrigerator compressors. The split-phase motor is the

most common type used in home refrigerator compressor units.

The rotating magnetic field needed to start
split-phase motors is obtained through the use of
two stator windings in the motor, a main winding,
and a start winding. Each of these windings has different inductance and resistance values. Current
flows in the two windings are not in phase with
each other (Figure 14–11). The current in the main
winding (coil A) lags the current in the start winding (coil B). A rotating magnetic field is set up in the
stator and the motor rotates in the direction in
which the field rotates.

After a split-phase motor has started its rotation
and as its speed approaches normal operating speed,
the start winding is no longer needed. A mechanical
centrifugal switch is often used to disconnect electrically the start winding from the power source, usually at about two-thirds of top speed (Figure 14–12).
The motor continues to rotate after the start winding has been electrically disconnected.

Motor Direction (Split-Phase)

The direction of rotation of most split-phase motors
may be reversed by interchanging the connections
of the windings. The motor windings are marked S1

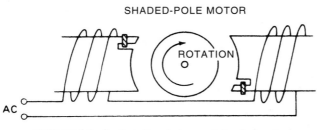

FIGURE 14–8 Rotation is toward shaded pole.

DUAL-SHAFT SHADED POLE MOTOR

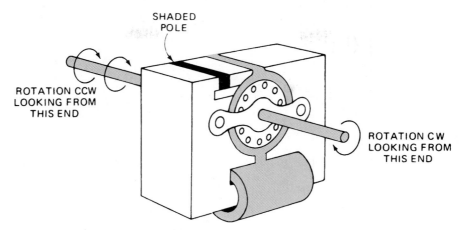

SHADED
POLE

ROTATION CCW
LOOKING FROM
THIS END

ROTATION CW
LOOKING FROM
THIS END

FIGURE 14–9 Dual-shaft motor—shafts rotate in opposite direction when looking at shaft end.

TO REVERSE DIRECTION
DISASSEMBLE AND
REVERSE SHAFT

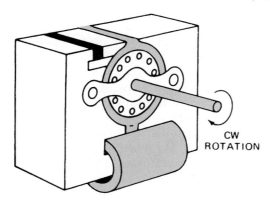

CW
ROTATION

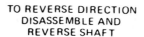

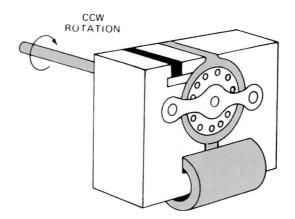

CCW
ROTATION

FIGURE 14–10 With a single-shaft motor, end bells must be interchanged and the rotor installed at opposite side to reverse direction.

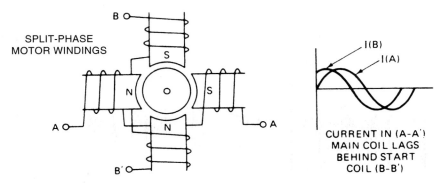

FIGURE 14–11 Split-phase motor winding produces rotating magnetic field.

and S2 for the start winding and R1 and R2 for the run winding. In Figure 14–12a, the winding connection is S1 to R1 and S2 to R2; the motor rotates clockwise. In Figure 14–12b, the connection is S1 to R2 and S2 to R1; the motor rotation is counterclockwise.

The centrifugal switch is in series with the (S) winding of the motor (Figure 14–13). Although bearings often wear out in motors using centrifugal switches, the switch and switch mechanism have proved to be a common source of trouble. The contacts of the switch may become burned and pitted, preventing good electrical contact. Dirt and grit may interfere with the mechanical operation of the switch mechanism, causing the switch to remain open after the motor has been stopped.

If the switch contacts are open and an attempt is made to start the motor, the shaft will not rotate. The motor will hum but will not turn. If this should happen, it is necessary to disassemble the motor and to clean the mechanical portion of the centrifugal switch. The contact surfaces of the electrical switch should be inspected for excessive pitting. Minor pitting may be removed and the contact surface returned to operating condition with fine sandpaper. Care must be taken to remove all sand and sanded contact material from the motor. If in doubt, the centrifugal switch should be replaced as an assembly.

CAPACITOR MOTORS (SPLIT-PHASE)

Capacitor-type, split-phase motors are used in applications in which higher starting torque and/or higher horsepower ratings are required. Common capacitor-type, split-phase motors are available from ½ to 10 horsepower. There are three different types of capacitor split-phase motors common to the air-conditioning and refrigeration industry:

1. Permanent split-capacitor (PSC) motor
2. Capacitor-start motor
3. Capacitor-start, capacitor-run motor

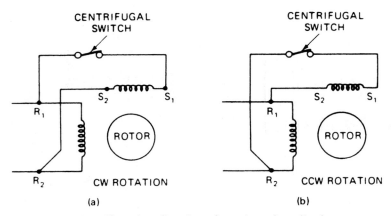

FIGURE 14–12 Changing direction of rotation of a split-phase motor.

FIGURE 14–13 Motor stator-centrifugal switch. (Courtesy of BET Inc.)

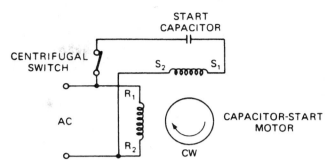

FIGURE 14–15 Centrifugal switch connection in a capacitor-start motor.

Permanent Split-Capacitor Motors

The permanent split-capacitor motor is sized from 1/20 to 10 horsepower. These motors are often used in place of shaded-pole motors because they are somewhat more efficient. The capacitor remains in series with the auxiliary winding for the motor at all times. An electrical circuit for this motor is shown in Figure 14–14.

Capacitor-Start Motors

Capacitor-start motors are readily available from ¼ to 10 horsepower. They are most often used in applications in which high starting torque is required.

Capacitor-start motor operation is very similar to that of the split-phase induction motor with one exception. The capacitor-start motor has an electrolytic capacitor in series with the start winding, sometimes called the *auxiliary winding*. A centrifugal switch or some other control device is used to remove electrically the capacitor and the start

winding from the circuit after the motor reaches approximately two-thirds of operating speed. The diagram of a capacitor-start motor, Figure 14–15, is very similar to that of the split-phase motor, Figure 14–12; the only difference is the addition of the capacitor.

MOTOR DIRECTION (CAPACITOR-START)

The direction of rotation of capacitor-start motors may be changed by reversing the electrical relationship of the two windings. If the motor connections in Figure 14–15 are taken as standard, then the motor direction is reversed by the motor connections shown in Figure 14–16. The method that the technician uses to reverse motor direction depends on what access is available to the winding connections.

A common trouble area in capacitor-start motors is the starting mechanism, which may be a centrifugal switch or some other control device such as a relay. More information on motor starting circuits is given in Units 15 and 16. A second trouble area is the start capacitor. Failure of the start capacitor, usually an electrolytic one, may be due to an open or a short.

Capacitor-Start, Capacitor-Run Motors

The capacitor-start, capacitor-run motor is essentially a combination of the permanent split-capacitor motor and the capacitor-start motor. It is a common motor in air-conditioning systems. The capacitor-start, capacitor-run motor provides both the high starting torque of the capacitor-start motor and the efficiency (as well as good power factor) of the permanent split-capacitor motor.

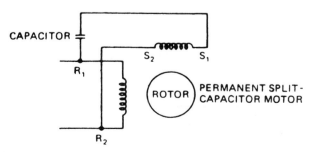

FIGURE 14–14 Connection of components in a permanent split-capacitor motor.

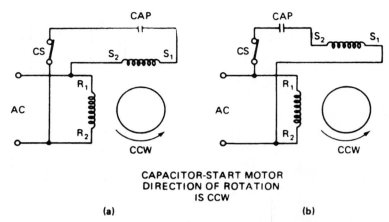

CAPACITOR-START MOTOR
DIRECTION OF ROTATION
IS CCW

(a) (b)

FIGURE 14–16 Changing direction of rotation of a capacitor-start motor.

A diagram of a capacitor-start, capacitor-run motor is shown in Figure 14–17. High starting torque is obtained through the use of an electrolytic (start) capacitor, C2, in parallel with the permanent oil-filled (run) capacitor, C1. When power is applied to the motor and as shaft rotation approaches operating speed, the control switch, S1, opens. The large-value capacitor, C2, is removed from the circuit, and total current draw by the motor is reduced. The permanent capacitor, C1, remains in the circuit, providing for constant torque as well as improving the overall power factor of the motor.

The capacitor-start (capacitor-run) motor is probably the most common type of motor used in the air-conditioning hermetic compressor system. The run and start windings of the motor are connected together internally, and three connections are brought out to motor terminals. These terminals are marked R for run winding, S for start winding, and C for common connection. This arrangement is shown in Figure 14–18.

There are times when a service technician may have to determine the proper terminal marking for a single-phase compressor. An ohmmeter is all that is necessary to make this determination. The following steps should be taken when checking for motor terminal connections:

1. Mark the terminals 1, 2, and 3.
2. Measure the resistance between each set of terminals.
3. The resistance between one set of terminals will be higher than the resistance between the other two sets. The terminal not included in the highest resistance measurement is the common terminal. Mark that terminal C.
4. Measure the resistance between the common terminal and the other terminals.
5. The indication of lower resistance from a terminal to common C determines the run (R) winding.

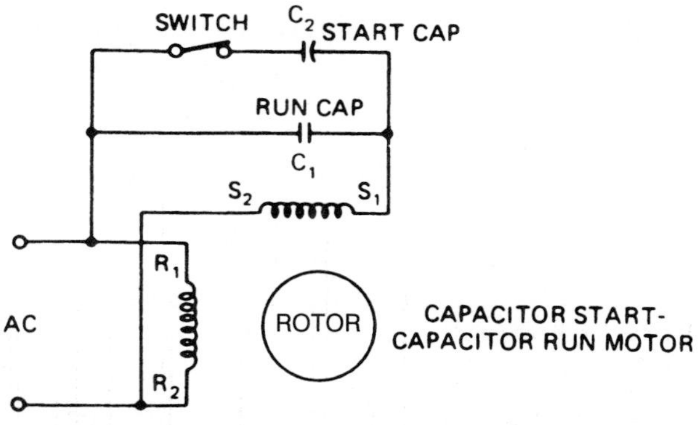

FIGURE 14–17 Connection of components in a capacitor-start, capacitor-run motor.

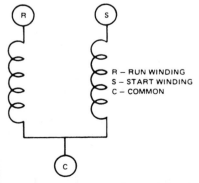

FIGURE 14–18 Run, start, and common marking of a compressor motor.

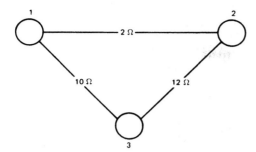

FIGURE 14–19 Measuring resistance in order to determine R, S, and C terminals.

6. The indication of higher resistance from a terminal to common C determines the start (S) winding.

An example of winding resistance is shown in Figure 14–19, which shows the ohmmeter measurements.

1. The highest resistance measured, 12 ohms, is measured between terminals 2 and 3. Terminal 1 must be the common terminal and should be marked C.
2. The resistance between terminals 1 and 2 is 2 ohms, and the resistance between terminals 1 and 3 is 10 ohms.
3. The lowest resistance, 2 ohms, indicates that terminal 2 is the run (R) winding.
4. The next higher resistance, 10 ohms, indicates that terminal 3 is the start (S) winding.

REPULSION-START, INDUCTION-RUN MOTORS

Repulsion-start, induction-run motors are not common in the air-conditioning and refrigeration industry today. Older operating units may be found with this type of motor still in use.

The repulsion-start induction motor includes coil windings on the rotor (Figure 14–20). The individual ends of the coil winding are connected to a **commutator**. A centrifugal mechanism controls the start of the motor, where a set of brushes connects the ends of some of the rotor windings together. As the motor approaches operating speed, the centrifugal device completely shorts the commutator segments together through the use of a shorting ring. The rotor essentially becomes a squirrel-cage rotor. The high manufacturing cost of this motor, compared to other types of motors, ended its popularity.

POLYPHASE MOTORS (THREE-PHASE)

Polyphase motors are simple in construction. They do not require starting windings, centrifugal switches, or start capacitors.

When a three-phase voltage is applied to the stator of a three-phase motor, a rotating magnetic field is automatically present around the stator (Figure 14–21). A set of magnetic pole pairs is indicated by the numbers 1 ①, 2 ②, and 3 ③. With the application of the three phases of voltage to the windings of the poles, a rotating field results around the face of the motor stator.

First pole 1 is a north pole, then pole ③, and then pole ②. Pole ① then becomes a north pole as the current changes direction. Poles ①, 3, and ② are south poles in order as 1, ③, and 2 are north poles. The magnetic field continuously rotates around the stator face. A squirrel-cage rotor rotates in a magnetic field of this type. No other starting system is needed.

Winding Connections

The windings of three-phase motors are connected in either the wye or delta configuration, depending on voltage and current requirements. As a current-limiting scheme, three-phase motors are sometimes started in the wye configuration and are switched into the delta configuration after the motor comes up to speed. In this manner, the starting current is

COMUTATOR SECTION
SHORTED-START

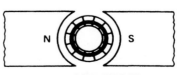

FULL COMUTATOR
SHORTED RUN

FIGURE 14–20 Repulsion-start, induction-run motor.

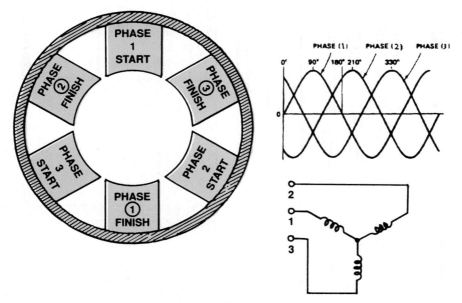

FIGURE 14–21   Three-phase motor.

limited. Figure 14–22 shows examples of wye and delta connections.

Direction Control of Three-Phase Motors

To reverse the direction of a three-phase motor, it is only necessary to interchange any two of the three supply voltage lines. In this manner, the direction of rotation of the magnetic field is reversed and, consequently, so is the direction of rotor rotation.

MOTOR SPEED

The speed of an AC motor is controlled by the number of poles of the motor and the frequency of the applied voltage. By formula, the speed is

$$rpm = \frac{2F \times 60}{N}$$

where

$$rpm = \text{revolutions per minute}$$
$$F = \text{frequency}$$
$$N = \text{number of pairs of poles}$$

The speed of a two-pole motor is

$$rpm = \frac{2F \times 60}{N}$$

$$rpm = \frac{2 \times 60 \times 60}{2} = \frac{7200}{2}$$

$$rpm = 3600$$

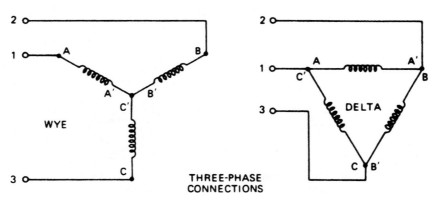

FIGURE 14–22   Wye (Y) and delta (Δ) connections.

This is the synchronous speed of the motor, the speed at which the motor would operate if there were no slippage. The full-load speed of the motor is 3600 rpm minus approximately 150 rpm of slippage, for an operating speed of 3450 rpm.

Four-pole motor operation is

$$rpm = \frac{2F \times 60}{N}$$

$$rpm = \frac{2 \times 60 \times 60}{4} = \frac{7200}{4}$$

$$rpm = 1800$$

Considering slippage of approximately 75 rpm, the motor speed is 1725 rpm.

SLIPPAGE

Slippage is the difference between the synchronous speed, 3600 rpm for a two-pole motor or 1800 rpm for a four-pole motor, and the actual operating speed. For example, in a four-pole motor the magnetic field rotates around the stator at a synchronous speed of 1800 rpm. To develop a magnetic field in the rotor of the motor, the short-circuited copper bars of the rotor must cut through the magnetic field of the stator. With the four-pole stator, the field rotates at 1800 rpm. If the rotor turned at 1800 rpm, the copper bars of the rotor would not cut through the magnetic field of the stator, but would be rotating at exactly the same speed. This situation cannot exist! Since there would be no current in the rotor's copper bars, there would be no rotor magnetic field. There would be no power to cause the rotor to move. The rotor would slow down.

When the rotor turns at a speed less than the synchronous speed, the rotor cuts through the magnetic field of the stator. Current flows in the rotor bars, and a magnetic field is developed in the rotor. The reactions of the stator and rotor magnetic fields cause the rotor to turn. When a motor is operated at no load, the slippage will be small. The rotor cuts through a limited amount of the stator magnetic field. When a load is put on the motor, the motor slows down (more slippage). The rotor cuts through a larger portion of the stator magnetic field. Higher current flows in the rotor, thus providing the power to turn the load.

(**NOTE:** The motor is essentially a transformer. The stator is the primary and the rotor the secondary. When current increases in the secondary [rotor], the current will increase in the primary [stator]. When more power is needed, more power is drawn.)

Induction motors turn at a speed below synchronous speed; they have slippage. A type of motor that runs without slippage is called a *synchronous motor*. A synchronous motor has a different rotor than does an induction motor. Usually, a separate winding is placed on the rotor. This winding is fed DC current through **slip rings** to produce a magnetic field. (No slippage is needed.) Originally, the synchronous motor found little application in air conditioning. However, some of the new variable-speed, three-phase compressors are using permanent magnets in the rotor of the compressor motor. This type of motor would be considered a synchronous motor. The motor would turn at or close to synchronous speed.

TWO-SPEED COMPRESSOR MOTORS

The increase in energy costs has necessitated the development of increased efficiency air-conditioning systems. One development that has provided for increased efficiency is the two-speed compressor. When the demand is high, the compressor operates at high speed; when the demand is low, the compressor is switched to low speed. Demand control may be obtained through the use of a standard two-stage thermostat similar to the one shown in Figure 19–29 on page 232.

Speed control of the compressor motor is obtained by changing the winding connections from two-pole operation at high speed to four-pole operation at low speed. It has already been shown that at 60 hertz a two-pole motor operates at a speed of slightly less than 3600 rpm. A four-pole motor operates at slightly less than 1800 rpm.

Low-Speed Operation

Figure 14–23 shows the motor winding connections for a single-phase, four-pole, low-speed compressor. Consider an instant in time during a cycle of the input voltage. Line L1 is negative, and line L2 is positive. Using the left-hand rule (see Figure 2–16 on page 16) and the direction of electron flow, the polarity of the motor coils may be determined. The direction of electron flow is indicated by the arrows in Figure 14–23. The polarity is shown in the diagram. The motor is four pole and will run at about 1725 rpm.

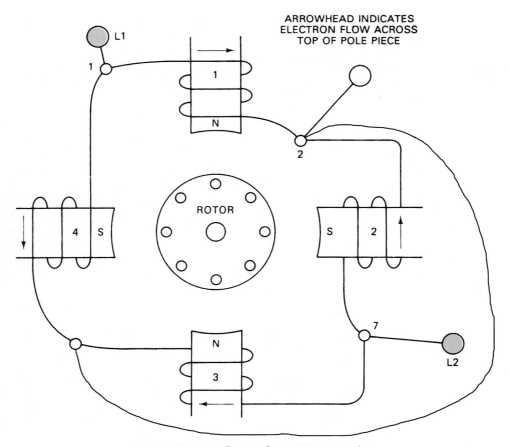

FIGURE 14–23 Four-pole motor connection.

High-Speed Single-Phase Operation

When high-speed operation is called for, the connection of line L2 is changed from motor terminal 7 to motor terminal 2. At the same time, motor terminal 7 is connected to motor terminal 1. The connections are shown in Figure 14–24. Again, consider the instant in time when L1 is negative and L2 is positive. The direction of electron flow is indicated by arrows in Figure 14–24. Poles 1 and 2 are north poles, and poles 3 and 4 are south poles. The motor effectively becomes a two-pole motor, with the two poles as shown in Figure 14–25. The motor speed is about 3450 rpm.

Start Windings

Separate start windings are used for low-speed, four-pole and high-speed, two-pole operation of the two-speed single-phase compressor. Selection of either the low-speed start winding or the high-speed start winding is made by the speed-control relays. The same start and run capacitors are used for either low- or high-speed operation.

Single-Phase Wiring Diagrams

The control circuit connections for low-speed operation of the compressor are shown in Figure 14–26. The heavy lines indicate power connections through closed relay contacts. The low-speed contactor is energized during low-speed operation. The control circuit connections for high-speed operation are shown in Figure 14–27.

THREE-PHASE COMPRESSOR

The three-phase two-speed compressor operates in a manner similar to the single-phase two-speed compressor. The motor windings are switched from four-pole at low speed to two-pole at high speed. Again, a simple two-stage thermostat could be the control input.

Low-Speed Three-Phase Compressor

For low-speed operation, the stator windings of the compressor motor are connected in series as shown

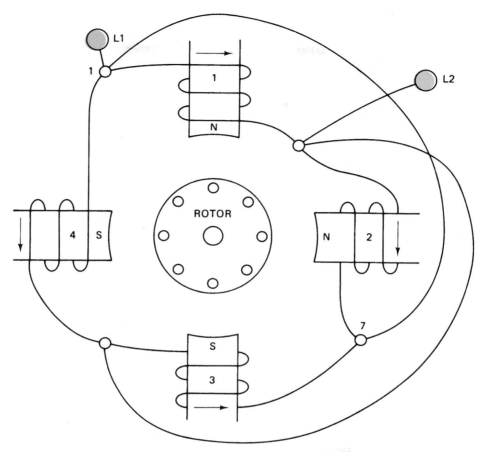

FIGURE 14–24   Four-pole motor connected for two-pole operation.

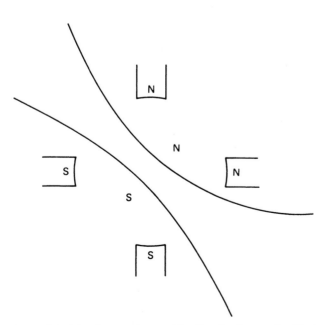

FIGURE 14–25   Four-pole combination for two-pole effect.

in Figure 14–28. Only half the windings are shown for the sake of simplicity. The second half of the motor would have the same winding connections as the first half. Note the different connection of the individual pole pieces 2 and 5. This is the end of the phase 3 winding, which would have started on the other side of the motor at pole pieces 8 and 11 if the complete circuit were shown.

The diagram in Figure 14–28 shows the beginning of phase 1 and phase 2 connection to pole pieces 1 and 4 and 2 and 6, respectively. The ends of these phases are at the other sides of poles 4 and 10 and 9 and 12, respectively.

Consider a point in time, A, on the three-phase waveform (Figure 14–29). Here line L3 is at maximum negative. Line L1 is positive and moving in a negative direction. Line L2 is positive and moving in a positive direction. The small arrows in Figure 14–28 indicate the direction of electron flow across the top of the pole. The polarity of the pole is found using the left-hand rule, shown in Figure 2–16 on page 16. Pole 1 is a maximum north pole and pole 2 is a maximum south pole. The connection of the windings form a four-pole motor.

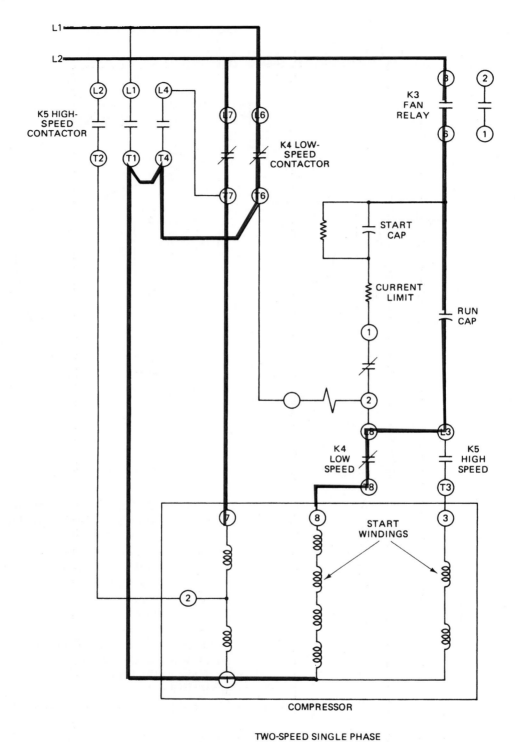

TWO-SPEED SINGLE PHASE

FIGURE 14–26 Low-speed wiring diagram.

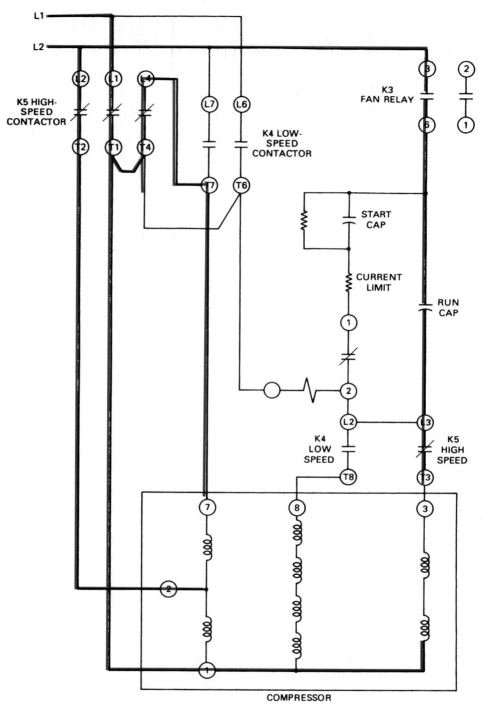

TWO-SPEED, SINGLE-PHASE

FIGURE 14–27 High-speed wiring diagram.

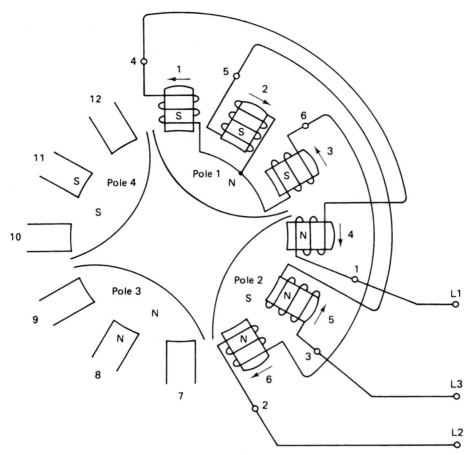

FIGURE 14–28   Three-phase low speed.

In the pole pieces, pole 1 is a weak south pole and becoming weaker. Pole 2 is a maximum strength south pole and pole 3 is a weak south pole and becoming stronger. Pole 4 is a weak north pole and becoming weaker. Pole 5 is a maximum north pole

and pole 6 is a weak north pole and becoming stronger. (See Figure 14–30.)

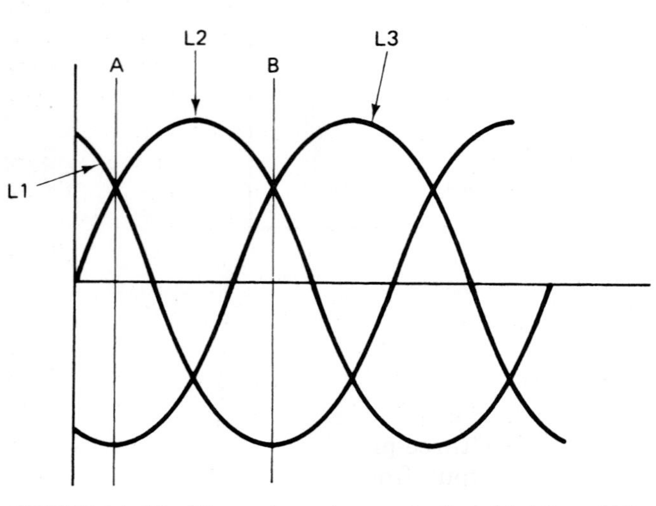

The direction of rotation of the magnetic field is clockwise, as will be the direction of motor rotation. This is a four-pole motor. The motor will run at about 1750 rpm. In Figure 14–31, a schematic diagram of the wiring is shown. In Figure 14–32, a wiring diagram is shown. Note the indication of closed contacts on the low compressor contactor, K4. The windings are in series. The direction of current flow is shown on the three sets of windings, Figure 14–31.

The wiring diagram of a three-phase motor connected for high-speed operation is shown in Figure 14–33. The high compressor contactor, K5, contacts are shown closed here. The direction of current flow through the 1, 2, and 3 terminals is reversed with this connection.

Reversing the direction of current in poles 4, 5, and 6 changes them to south poles at time point A in Figure 14–29. One-half of the motor becomes a north pole, while the other half becomes a south pole. This is effectively now a two-pole motor, as shown in the schematic diagram in Figure 14–34.

FIGURE 14–29   Three-phase sine waves: lines L1, L2, and L3.

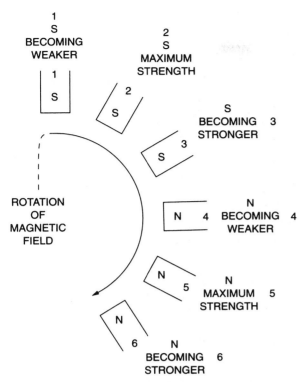

FIGURE 14–30 Magnetic field strength at time point A in Figure 14–29.

Compare the direction of current flow in the windings. Figure 14–31 shows low-speed operation. Figure 14–35 shows high-speed operation.

When the low compressor relay K4 is de-energized and the high compressor relay K5 is energized,

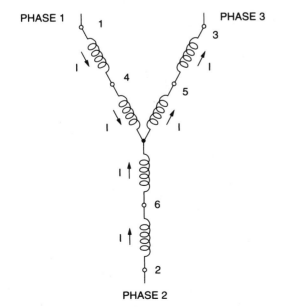

FIGURE 14–31 Schematic three-phase, low-speed with current direction (point A, Figure 14–29).

two actions take place. First, the direction of current flow through three of the coils is reversed. Second, the coils are connected in parallel rather than in series.

Compare the direction of current flow in the coils of Figure 14–31 versus Figure 14–35. The arrows indicate current direction change in terminals 1, 2 and 3. If the direction of current flow through the coil changes, the magnetic polarity of the coil changes.

> **WARNING:** *Do not confuse the series and parallel connection of windings in the two-speed compressor for speed control with series and parallel connection of windings in standard motors for low and high voltage. The number of poles in a standard motor does not change when the winding connections are changed from series to parallel. The speed remains the same; only the voltage requirement changes.*

VARIABLE-SPEED THREE-PHASE COMPRESSOR

Another system of speed control for three-phase is rapidly being developed. The continuously variable speed control for three-phase compressor motors is now becoming available. The following is a general outline of a system that may be followed in the block diagram in Figure 14–36.

Digital logic circuitry is used to develop the three-phase signals. The outputs of the three-phase logic block are three square waves of voltage that are 120° out of phase with each other, just as standard three-phase voltages are 120° out of phase. These waveforms are shown on the block diagram. The input control to the three-phase logic block is a variable-frequency pulse from a frequency generator.

A control input signal determines the frequency of the output. The control input could be a voltage from an electronic thermostat setting the compressor speed requirement. The level of the DC voltage from the thermostat indicates how far the temperature is from the selected temperature. The greater the deviation, the higher the output is. The higher the output is, the higher the frequency output from the frequency generator. The higher the frequency input to the three-phase logic is, the higher the frequency output from the logic circuit. In other words, three-phase voltages of a variable frequency

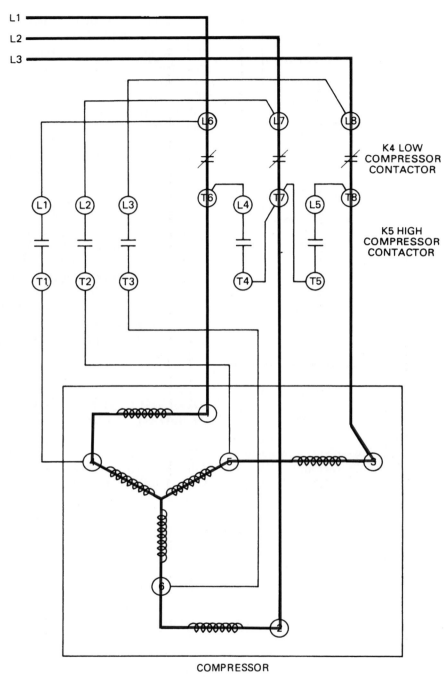

FIGURE 14–32   Three-phase, low-speed wiring diagram.

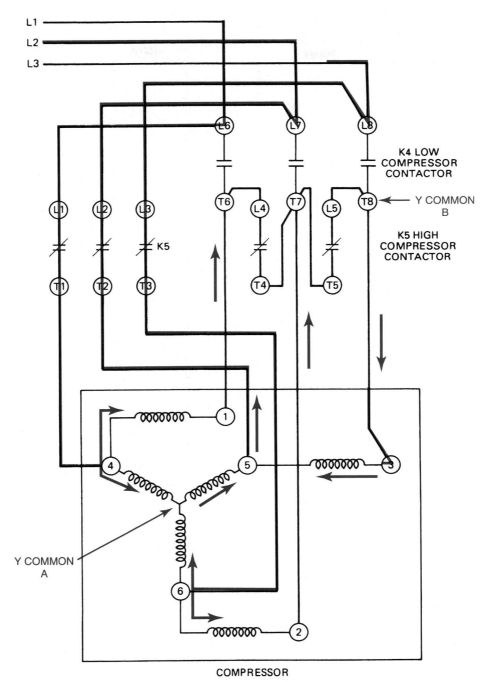

FIGURE 14–33 Three-phase, high-speed wiring diagram.

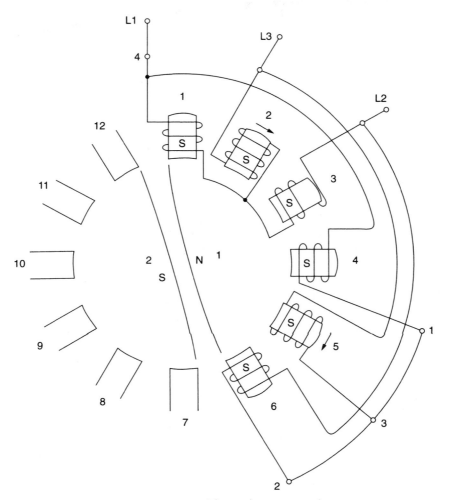

FIGURE 14–34 Three-phase, two-pole.

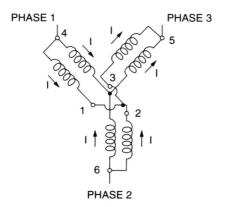

FIGURE 14–35 Schematic three-phase, high-speed with current direction (point A, Figure 14–29).

are available to drive the three-phase compressor motor. The speed of the compressor motor is determined by the frequency of the supply voltage.

$$rpm = \frac{2F \times 60}{N}$$

The three-phase logic circuit feeds the individual-phase wave shaper. The wave shaper provides pulses of current to the three-phase compressor motor. The motor is supplied with three-phase voltage and current and will run at the speed determined by the frequency. The higher the frequency is, the higher the speed.

When the compressor is operated at higher speeds, it can produce higher pressure. In a properly

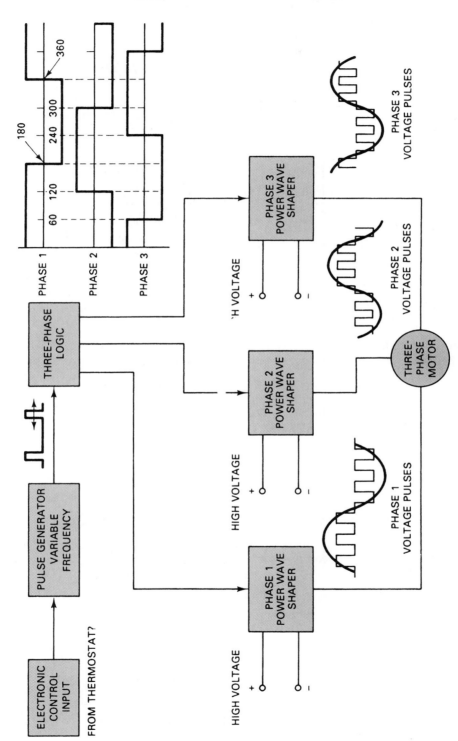

FIGURE 14–36 Variable-speed, three-phase motor diagram.

designed system, this will provide for more rapid cooling. If little cooling is required, the thermostat through the system will cause the compressor to operate at a lower speed, using less power.

SUMMARY

- Single-phase induction motors require a starting system.
- Shaded-pole motors rotate in the direction of the shaded pole.
- Split-phase motors have two sets of windings.
- Capacitor motors provide higher torque.
- Polyphase motors provide their own rotating magnetic fields.
- Variable-speed motors provide for improved efficiency of air-conditioning systems.

PRACTICAL EXPERIENCE

Required equipment   Shaded-pole motor, compressor motor, ohmmeter

Procedure

1. Observe the face of a shaded-pole motor.
2. Locate the shaded pole from one side of the motor.
3. The rotation of the motor is from the unshaded section to the shaded section.
4. Determine the direction of motor rotation, cw or ccw. _____
5. Observe the opposite side of the motor.
6. What is the rotation from this side?
7. Observe the three terminals of a compressor motor.
8. Mark the terminals 1, 2, and 3.
9. Measure the resistance between terminals 1 and 2. _____ ohms
10. Measure the resistance between terminals 2 and 3. _____ ohms
11. Measure the resistance between terminals 3 and 1. _____ ohms

12. The lowest resistance in steps 9, 10, and 11 indicates the run winding.
13. The second-to-lowest resistance is the start winding.

REVIEW QUESTIONS

1. Direct current (DC) motors are the most common type used to drive air-conditioning compressors.
   T _____   F _____
2. In a shaded-pole motor, the shaft always rotates in
   T _____   F _____
3. The direction of rotation of a split-phase motor may be changed by interchanging the connections to the start winding.
   T _____   F _____
4. Centrifugal switches are common in refrigerator sealed compressors.
   T _____   F _____
5. Shaded-pole motors are more efficient than permanent split-capacitor motors.
   T _____   F _____
6. Capacitor-start motors have high starting torque.
   T _____   F _____
7. The run capacitor used with capacitor-start, capacitor-run motors is of a larger value than the start capacitor.
   T _____   F _____
8. Three-phase motors use capacitors in their start circuit.
   T _____   F _____
9. To reverse the direction of a three-phase motor, all three input wires must be changed.
   T _____   F _____
10. Three-phase motors are sometimes started in the wye configuration and then switched to the delta configuration after the motor comes up to speed.
    T _____   F _____

UNIT 15
MOTOR APPLICATIONS

OBJECTIVES

After completion and study of this unit, you should be familiar with the starting devices, starting systems, and voltage control of electric motors. Devices covered are:

- Series start (current) relay.
- Potential relay.
- Push-button.
- Relay-control, three-phase.

Different types of electric motors used in air-conditioning and refrigeration systems were covered in Unit 14. Some of these motors require special control for starting, whereas others are started simply by closing a switch. The air-conditioning and refrigeration technician should become familiar with the different types of motors and the circuits associated with them.

MOTOR SUPPLY VOLTAGE

All motors have a nameplate voltage rating, such as 115 V, 230 V, or 460 V. In most cases, the motor will operate satisfactorily so long as the voltage supply is within plus or minus 10% of the rated voltage. This means that the acceptable supply voltage to a 115 V motor is 103.5 to 126.5 V, and the acceptable supply voltage to a 230 V motor is 207 V to 253 V. If the supply voltage is above or below these limits, excessive heat may be generated in the motor windings, resulting in premature motor failure.

For three-phase motors, there is an additional consideration. The voltage imbalance between the three phases should never exceed 2%. Voltage imbalance is calculated as follows:

1. Take voltage readings between each pair of pins (three readings).
2. Calculate the average of the three readings.
3. Find the maximum difference between one of the voltage readings and the average of the three readings.
4. Divide this difference by the average voltage. If it exceeds 0.02, you will need an electrician to determine whether there is a problem with the utility company supply voltage or the building's power transformer.

EXAMPLE 1

You have taken power supply readings of 235 V, 231 V, and 222 V. The average of the three readings is

$$V_{\text{avg}} = \frac{235 + 231 + 221}{3} = 229\,\text{V}$$

129

The individual deviations from the average for each phase is

$$235 - 229 = 6\,V$$
$$231 - 229 = 2\,V$$
$$229 - 221 = 8\,V$$

Dividing the highest difference by the average, the voltage imbalance is

$$\frac{8\,V}{229\,V} = 0.0349.\ \text{or } 3.49\% \text{ imbalance (unacceptable)} \quad\blacksquare$$

SHADED-POLE MOTORS

Shaded-pole motors are generally used in air-conditioning and refrigeration systems as small fan motors and as timing motors in large residential and commercial units. The shaded-pole motor does not require a special starting control. When power is supplied to the motor winding, the motor rotates. When power is removed from the motor winding, the motor stops but is ready to start again with the application of power. The starting feature of the motor, the shaded poles, is always in the circuit, ready to produce the required starting action.

PERMANENT SPLIT-CAPACITOR MOTORS

The permanent split-capacitor motor, like the shaded-pole motor, does not require special starting considerations. A simple switch device, either manual or automatic, such as a thermostat, can be used to start this motor. When power is applied, the motor runs.

Split-Phase Motors

Split-phase motors include many different types of motors, but with one common characteristic. The split-phase motor, unlike the shaded-pole motor, has two windings inside. One is called the **run winding**, and the other is called the **start winding**. They are connected together at one end, and three wires are connected to the two windings, as shown in Figure 15–1. The wire that is connected to the point where the two windings connect to each other is called the common wire. The wire connected to the other side of the run winding is called the run wire, and the wire connected to the other side of the start

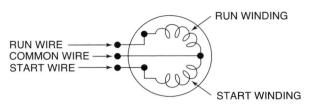

FIGURE 15–1 Wiring for a split-phase motor.

winding is called the start wire. Split-phase motors are used in all single-phase compressors. The three wires penetrate the shell by means of pins as shown in Figure 15–2. They are called the start pin, the run pin, and the common pin. The ceramic insert that the pins pass through in the compressor shell is called a fusite connector.

The wiring connectors that attach to the compressor pins are specially made for this purpose. They are of two types (Figure 15–3). Part a shows the connector used for a simple straight pin on the compressor. Part b shows the connector used for a pin attached to a flat plate. This connector may look like a simple spade connector, but it is actually of different dimensions. A simple spade connector will not fit onto the compressor pin.

START RELAYS

The split-phase motors (except for the PSC motors) require a means to automatically disconnect the start-winding from the circuit after the motor-gets up to speed. One way that start windings can be taken out of the circuit is by means of a centrifugal switch (Figure 15–4). It is a switch that senses the centrifugal force being produced by the rotation of a

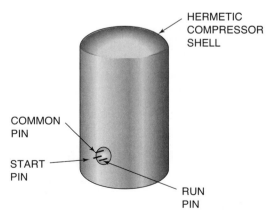

FIGURE 15–2 Electrical connections for a hermetic compressor.

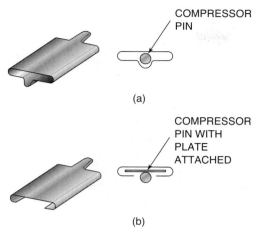

FIGURE 15–3 Compressor terminal connectors. (a) Compressor pin. (b) Compressor pin with plate attached.

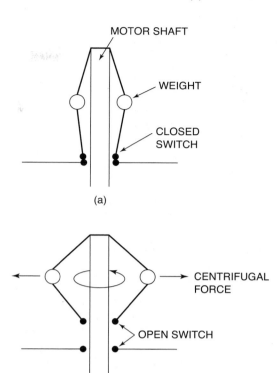

FIGURE 15–4 Centrifugal switch. (a) Centrifugal switch at rest. (b) Centrifugal switch spinning.

motor. When the motor gets up to almost full speed, the centrifugal switch will open, taking out the start winding. When the motor is de-energized, it will slow, and the centrifugal switch will close. The centrifugal switch can be used in applications where the switch is not exposed to refrigerant. It cannot be used for a hermetic compressor motor, which is located inside the refrigeration system, and it is surrounded with the same refrigerant that circulates through the system. If a centrifugal switch were mounted on the end of a hermetic compressor motor, each time the switch opened it would produce a spark, breaking down a small quantity of refrigerant and producing an acid. Over the long term, the acid would attack the motor winding insulation, causing the motor to fail.

In order to avoid the sparking inside the refrigerant system, split-phase compressors use a **start relay** to take the start winding out of the circuit. Three types of start relays will be described:

1. Current relay
2. Potential relay
3. Solid state relay

SPLIT-PHASE MOTORS (CURRENT RELAY)

The split-phase motor is started by the use of a centrifugal switch. The centrifugal switch removes the start (auxiliary) winding from the circuit when the motor shaft approaches operating speed. This is a practical system for starting the motor when it is used in open compressor-type applications. Centrifugal switch operation was covered in Unit 14.

One of the most common uses of a split-phase motor is to "drive" a compressor in small-capacity

ADVANCED CONCEPTS

The reason that there is a spark each time the switch opens is because when the switch contacts separate, there is an instant where the contact is broken, but the contacts are sufficiently close together so that the electrons can jump from the contact that is negative to the contact that becomes positive as the contacts open. This is especially true in circuits that contain inductance (motors, relays) where higher voltages may be induced.

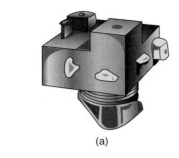

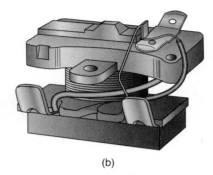

FIGURE 15–5   Current relays. (a) Push-on current relay. (b) Remote mounted current relay.

refrigerators and freezers. The split-phase motor in this application is hermetically sealed along with the compressor as a combined unit.

The most common type of starting control used with small-capacity, hermetically sealed compressors is the series start relay. This type of relay is known as a current relay (Figure 15–5). The coil of the current relay consists of a few turns of relatively large wire connected in series with the run winding of the motor. The contacts of the current relay are normally open and are connected in series with the start winding of the motor. A circuit diagram of a split-phase motor with a current relay is shown in Figure 15–6.

The relay coil is connected in series with the run winding of the motor. The open contacts of the relay are connected in series with the start winding of the motor. When the control switch is closed, 120 volts are applied to the run winding of the motor through the few turns of the relay coil. The start winding is not connected to the power source because the relay contacts are open. A high current flows in the run winding since the motor has not yet started. This same current flows through the relay coil (Figure 15–7a). With a high current flow in the coil, the relay energizes; the normally open (NO) contacts of the relay close. A circuit is completed through the relay contacts, and 120 volts is applied to the start winding of the motor. The motor starts to rotate and gains speed. When the motor approaches operating speed, the current flow in the run winding decreases toward the normal operating current (Figure 15–7b). This reduction causes the relay to de-energize; the relay contacts then return to their normally open position, again removing the start winding from the circuit.

Current relays are commonly used with compressors up to ¾ horsepower capacity. Larger compressors commonly use the potential relay for starting control.

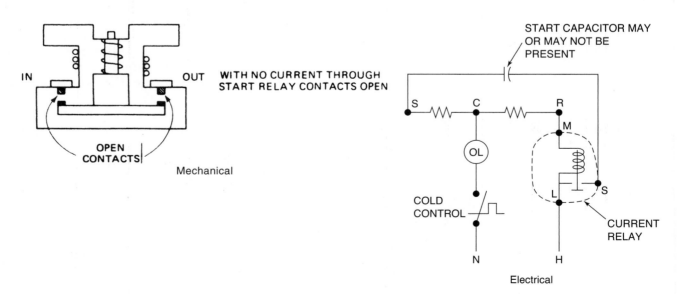

FIGURE 15–6   Series start relay with no current through start relay, contacts are open.

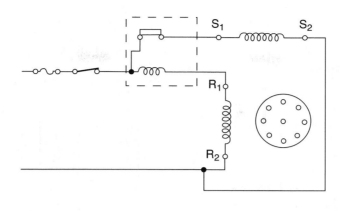

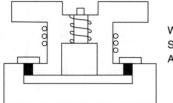

WITH HIGH CURRENT THROUGH
START RELAY COIL CONTACTS
ARE CLOSED

FIGURE 15–7(a) Series start relay with high current through start relay coil, contacts are closed.

Many current-type starting relays operate with the weight of the relay core as a factor in opening the relay contacts. This type of relay must be mounted in the position prescribed in the manufacturer's specifications.

Current relays are designed to pick up (energize) and to drop out (de-energize) at specific current levels. The current relay used in a compressor motor circuit must be matched to the compressor motor. When it is necessary to replace

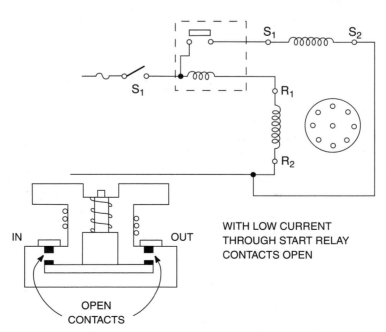

WITH LOW CURRENT
THROUGH START RELAY
CONTACTS OPEN

OPEN
CONTACTS

FIGURE 15–7(b) As motor comes up to speed, current draw decreases and the spring pressure plus gravity opens the contact.

a current relay, an exact replacement must be used.

Current Relay

Two types of current relays are shown in Figure 15–5. The characteristics that are common to all the current relays are

1. There is a coil of large-diameter wire that has few turns and a very low resistance.
2. There is a normally open switch that closes due to the magnetic field produced by the coil, but only if the coil is carrying sufficient current.

Most current relays have three terminals, labelled L, M, and S. This type of current relay is wired into a circuit as shown in Figure 15–6. When power is first applied, there is a circuit from H, through the current relay coil, then through the run winding, and then back to N. At the moment that power is applied, the in-rush current flow through the run winding is very high. This is called the locked rotor amps (LRA). Locked rotor amps are four to six times higher than the running load amps (RLA), sometimes called full-load amps (FLA). Running load amps or full-load amps are the maximum currents that the motor will draw when it is running normally at its design load and speed.

Some current relays (push-on type) simply push on to the start and run pins of the compressor (Figure 15–5). Others are mounted remotely from the compressor and are connected to the start and run pins by wiring.

The locked rotor amps through the run winding are high enough to cause enough magnetic field in the current relay coil to cause the switch between L and S to close. This completes a parallel circuit through the start winding a fraction of a second after the run winding is energized. With both windings energized, the motor starts. As it comes up to speed, the current draw diminishes. When the motor reaches approximately 75% of its rated rpm, the current flow through the current relay coil diminishes to a point where the magnetic field produced by the coil is no longer sufficient to hold the switch closed, and it drops open due to the force of gravity. However, at this point, the motor is rotating fast enough that the run winding alone will be able to bring it up to rated speed. The closing and then opening of the current relay switch all happens within a second or two after voltage is applied to the motor circuit.

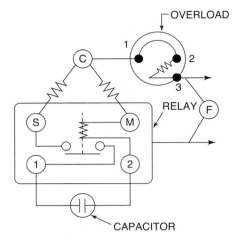

FIGURE 15–8 Four-wire current relay.

The current relay switch is opened by the force of gravity. Therefore, it is important that the current relay be mounted with the shaft that moves this switch perfectly vertical. If it is mounted out-of-vertical, the switch may not open after the motor achieves sufficient speed.

Some current relays have four terminals. Figure 15–8 shows how the four-wire current relay is wired, providing terminals for convenient attachment to the start capacitor. Note that in this wiring diagram, the relay switch is located between the start capacitor and the start pin on the compressor. This is different from the prior diagram, in which the capacitor is located between the switch and the start pin. The operation of the circuit is not changed by this difference.

CAPACITOR-START MOTORS (CURRENT RELAY)

Capacitor-start motors used in refrigeration and air-conditioning applications often use the same type of current relay that is used with a split-phase

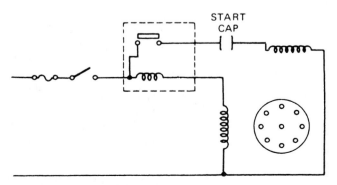

FIGURE 15–9 Series start relay connection, capacitor-start relay.

FIGURE 15–10 Potential relay.

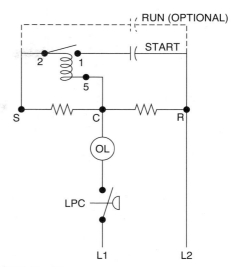

FIGURE 15–11 Wiring schematic—potential relay.

motor. A diagram of a capacitor-start motor circuit using a current relay is shown in Figure 15–9. When the switch is closed, high current flows through the coil of the current relay and the run winding of the compressor motor. The relay contacts close, connecting the start winding of the motor to the supply voltage through the start capacitor.

As the motor approaches operating speed, the run winding current decreases. When the current reaches a predetermined level, the relay de-energizes, removing the capacitor and the start winding from the circuit. It is important to remember that the electrolytic capacitor used with capacitor-start motors is for intermittent use only. It is only in the circuit for a few seconds during the motor start period.

Potential Relay

The potential relay (Figure 15–10) is used with a start capacitor and sometimes a run capacitor as well. The wiring for each is shown in Figure 15–11. The coil of the potential relay is wired in parallel with the start winding. The potential relay switch is wired to take out the start capacitor, but unlike the switch in the current relay, it is normally closed. The potential relay coil will cause the switch to open when the

voltage across it is slightly higher than the applied voltage. This is called the pick-up voltage. When power is applied to the motor circuit, both the start and run windings will be energized, and the motor will start. When the motor gets up to 75%–80% of its rated speed, the voltage across the start winding actually exceeds the applied voltage, and the potential relay coil produces sufficient magnetic force to pull the switch open. With the switch open and the start winding out of the circuit, the start winding acts like a transformer. It is a coil of wire located close to a rotating magnetic field. The start winding produces a voltage in the same way as a transformer secondary winding. The voltage produced by the start winding is sufficient to produce enough current through the potential relay coil to hold the potential relay switch open. The potential relay coil consists of many turns of very thin wire. It can produce a significant magnetic field with a very low current flow.

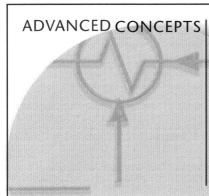

ADVANCED CONCEPTS

Run capacitors come with one terminal identified with a red dot, an M, or a (−) sign. This identifies the terminal of the capacitor plate that is internally located closer to the metal shell of the capacitor. If the capacitor shorts out to the casing, it is more likely this terminal will be the one that gets grounded. The **identified terminal** should be the one that is connected directly to the incoming power source. In this way, if the run capacitor becomes grounded, current will flow directly from the power source to ground, opening the circuit breaker. If the run capacitor is wired the opposite way, it will work fine, but in the event of a grounded capacitor, it can cause damage to motor and/or start relay by allowing the current to flow through the motor start winding to ground (Figure 15–12).

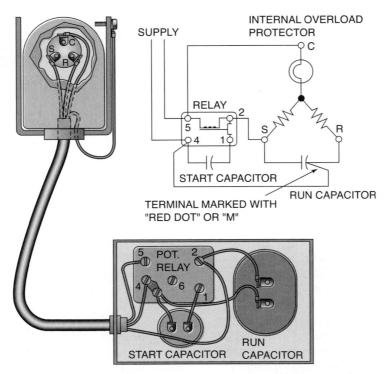

FIGURE 15–12 Physical wiring of potential relay.

Figure 15–12 shows the physical wiring diagram for the potential relay. Potential relays always have terminals numbered 1, 2, and 5, as shown in the previous schematic. Some potential relays also have a terminal 4. This is merely a convenience terminal. A convenience terminal is one that is provided only for the convenient connection of wires. On the potential relay, terminal 4 is not connected to anything inside the relay. It is only for the convenient connecting of three separate wires that would otherwise have to be placed in a wire nut, or all on the same terminal of the start capacitor. An easy way to remember how the wiring of the potential relay is done is to first bring the power into the run and common pins. This will be the running circuit. Then, add the potential relay by memorizing a simple poem, "5, 2, 1, common, start, run." However, you must also remember that I goes to the run pin, but through the start capacitor. Also, if a run capacitor is used, it is wired between the start and run pins of the compressor.

Solid State Relay ("Universal" Replacement Start Relays)

Because current relays and potential relays must be matched to the compressor, it presents a problem to the service technician. You must either stock a lot of different models on your truck, or you must make a trip to the wholesaler to purchase the correct replacement relay each time you need one. In recent years, manufacturers have introduced electronic "universal" start relays. Figure 15–13 shows one type of universal relay for use on compressors up to 1/3 or 1/2 hp (depending on the brand of universal relay). It is simply a two-wire device that is wired between the start pin and the run pin on the compressor. If there is a start capacitor, it is wired in series with the relay. It doesn't matter if the capacitor is wired between the relay and the start pin or between the relay and the run pin as shown. If there is also a run capacitor, it is wired between the start pin and the run pin directly.

This type of relay device does the function of both sensing motor speed and switching out the start winding (and start capacitor). The relay consists of simply a piece of semiconductor material. This material has a very low resistance when it is at room temperature. When power is applied to the common and run pins, current flows through the run winding and the start winding (through the semiconductor material). Because the semiconductor has such low resistance upon start-up, it acts like a closed switch. However, after several seconds of carrying current to the start winding, the semiconductor material heats up and becomes a very high

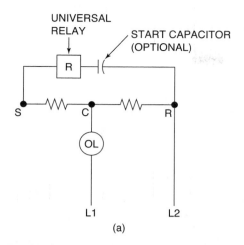

UNIVERSAL RELAY

START CAPACITOR (OPTIONAL)

(a)

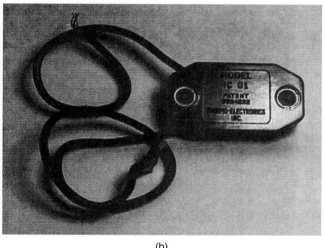

(b)

FIGURE 15–13   Universal start relay 1/3 to l/2 hp.

resistance. At that point, it acts like an open switch, and very little current is allowed to pass through the start winding.

There is not uniform agreement among technicians about how well this relay works. It has the following disadvantages.

1. It doesn't really sense motor speed. It really is only a time delay and therefore is not as accurate as the original relay in taking out the start winding at the ideal instant.

2. The semiconductor must have sufficient time during the off cycle to cool, so that it will once again have low resistance at start-up.

Figure 15–14 shows another type of universal start relay. This can be used on compressors up to 5 hp. It is truly an electronic circuit inside and may

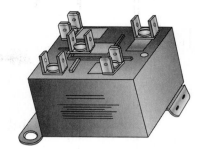

FIGURE 15–14   PR-90 universal relay.

be wired to function as a current relay or a potential relay. The wiring methods are different, and the service technician must carefully follow the wiring diagrams supplied with the relay.

Another universal type of replacement relay is the adjustable potential relay shown in Figure 15–15. The pick-up voltage of this relay is set to match the pick-up voltage of the failed potential relay. When the motor gets up to the appropriate speed so that the start winding generates the required pick-up voltage, the relay contacts will open, taking the start winding and start capacitor out of the circuit. If you don't know what the pick-up voltage is for the failed potential relay, you can probably set the adjustable potential relay for 190 volts for

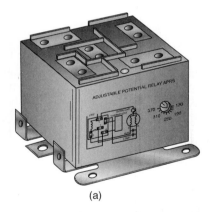

(a)

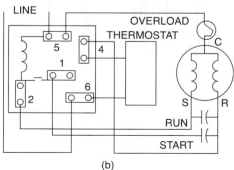

(b)

FIGURE 15–15   Adjustable potential relay.

115 V compressors, and 370 volts for 208/230 V compressors. If those settings are too low, the compressor may have difficulty starting, and you will have to increase the adjustment slightly. If these settings are too high, the internal circuitry of the adjustable relay will open the contacts 1 or 2 seconds after voltage is applied to the compressor. This prevents the start winding or start capacitor from failing due to remaining in the circuit too long.

PSC MOTORS

PSC (permanent split-capacitor) motors are split-phase motors that usually have no start relay. The wiring of a PSC motor is shown in Figure 15–16. The PSC motor is always used with a run capacitor to shift the phase of the voltage supplied to the start winding.

The permanent split-capacitor motor is very efficient electrically, but it has very limited starting torque. It is most commonly used on residential air-conditioning systems that use a capillary tube as the metering device. The capillary lube allows the high-side pressure and the low-side pressure to equalize during the off cycle, so the compressor doesn't have to start against a pressure difference. PSC motors are also commonly used as condenser fan motors.

Sometimes the PSC motor (Figure 15–17) is furnished with two wires for the applied voltage (usually black and white) and two wires for the capacitor (brown and brown). Other times, there are only three wires: black, white, and brown. Figure 15–17 shows how the two configurations are really the same. If you are replacing a failed three-wire motor with a four-wire model, or a four-wire motor with a three-wire model, Figure 15–18 shows how you would modify the wiring.

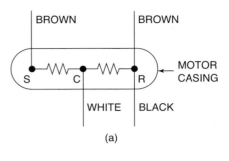

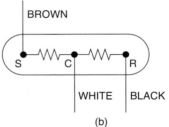
(a)

(b)

FIGURE 15–17   Comparison of 3-wire and 4-wire PSC motor. (a) 4-wire. (b) 3-wire.

PTC Device

Figure 15–19 shows a solid state **PTC device**. PTC stands for "positive temperature coefficient," and it means that as the temperature of the device increases, its resistance also increases. At normal

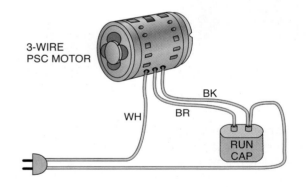

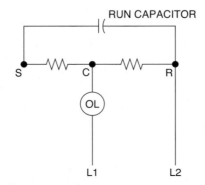
FIGURE 15–16   Wiring for a PSC motor.

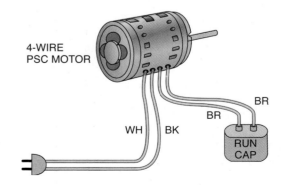
FIGURE 15–18   Wiring of 3-wire and 4-wire PSC motors.

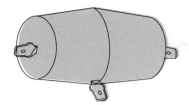

FIGURE 15–19 PTC device.

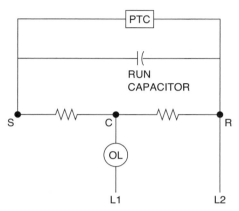

FIGURE 15–20 PTC device in circuit.

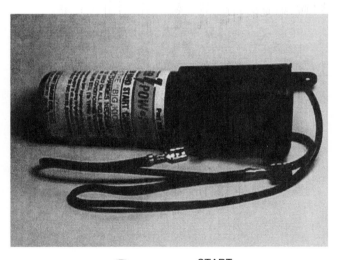

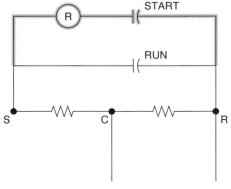

FIGURE 15–21 Hard-start kit.

ambient temperature, the resistance of the PTC device is very low, like a closed switch. The PTC is wired in parallel with the run capacitor Figure 15–20 so that on start-up, the capacitor is bypassed, and the start winding gets full power. The starting current passes through the PTC, causing the PTC to heat up. The PTC resistance increases and has no effect on the circuit after start-up. The run capacitor then takes over as if the PTC were not even there.

Hard-Start Kit

Sometimes, even on new installations using a PSC compressor motor, the compressor will experience difficulty starting because there is insufficient starting torque to overcome the resistance in the compressor between the pistons and the cylinder walls. A **hard-start kit** is a solid-state relay, prewired in series with a start capacitor in a single casing (Figure 15–21). When wired across the terminals of the run capacitor (in parallel with the run capacitor), the start capacitor will add extra starting torque, and then the solid state relay will take the start capacitor out of the circuit.

For new installations, some technicians carry a large start capacitor with wires and insulated alligator clips attached to the terminals. With the disconnect off, the technician attaches the alligator clips to the run capacitor. Power is applied to the unit. When the compressor starts, the start capacitor is manually disconnected. After the compressor is allowed to run for a few minutes, the piston and cylinder will "work in" so that the next time, the unit will start without the start capacitor. If not, then the hard-start kit may be permanently wired into the system.

For old installations that have developed starting problems, the hard-start kit may be a temporary fix, but not a cure. Whatever caused the starting problem to develop (lubrication problem, turn-to-turn short in the start winding, etc.) will probably continue to deteriorate, and a compressor failure may result. Sometimes, if the compressor fails shortly after you have installed the hard-start kit, the customer will blame you for causing the failure. For this reason, use of the hard-start kit on old installations should be approached with caution.

MOTOR ENCLOSURES

Motor enclosures are designed for various environments where the motor may be used. Figure 15–22 shows an open motor with open-ventilation openings

3-SPEEDS
+1 COMMON

FIGURE 15–22 Multispeed motor.

	115 VOLTS	230 VOLTS
COMMON	WHITE (GROUNDED)	PURPLE (UNGROUNDED)
BLACK	HIGH	HIGH
YELLOW	MED-HIGH	MED-HIGH
BLUE	MED	MED
ORANGE	MED-LOW	MED-LOW
RED	LOW	LOW
CAPACITOR	BROWN OR BROWN W/WHITE TRACER	BROWN OR BROWN W/WHITE TRACER

FIGURE 15–23 NEMA lead color codes.

in the shell and end shields. The holes permit passage of external air over the windings. An open drip-proof motor has ventilation openings that are placed so that droplets of liquid falling within an angle of 15 degrees from vertical will not affect the motor performance.

For applications where the motor windings must be protected from the atmosphere, a **totally enclosed motor** is used. It has no ventilation openings in the motor housing (but it is not airtight). The totally enclosed motor may be fan cooled (totally enclosed fan cooled, **TEFC**), nonventilated (totally enclosed nonventilated, **TENV**) or it may use an external source of airflow for cooling (totally enclosed air over, **TEAO**). The TEFC motor includes an external fan in a protective shroud to blow cooling air over the motor. The TENV motor operates without air cooling. It simply operates at a higher temperature, and heat is radiated from the enclosure. The TEAO motor is used to drive a fan. Whenever the motor is energized, the operation of the fan it is driving provides cooling for the motor, which is located in the moving airstream. A broken belt on a TEAO motor can cause the motor to overheat.

Explosion-proof motors are used in environments where the motor may be exposed to explosive atmospheres (flammable vapors or dust). They are designed to be airtight and to withstand an internal explosion without allowing it to propagate to the atmosphere. Explosion proof motors are a rarity because they are so expensive.

MULTISPEED MOTORS

Sometimes a motor will be supplied with multiple power leads (Figure 15–22) to allow running the motor at different speeds. This is done for several reasons:

1. On heating-cooling systems, the fan sometimes runs at a higher speed on cooling and a lower speed on heating.

2. On single-speed systems, the manufacturer uses one-size multispeed motor for a range of cooling unit sizes, and then uses an appropriate speed to match the evaporator airflow to the unit size.

Multispeed motors may be two, three, or four speeds. At most, only two speeds will be wired into the circuit. The NEMA standards for wire colors for the various speeds are given in Figure 15–23 for 115 V and 230 V applications.

Sometimes the winding for one speed of a multispeed motor will burn out, while another unused speed remains functional. It is usually acceptable to simply substitute the still-functioning motor speed instead of replacing the entire motor. However, there are two potential problems. If your substitute speed is slower than the original speed, the air-conditioning system may tend to freeze up the evaporator coil due to insufficient air flow. Check your low-side pressure after you make the change. If it is above 60 psi (for an R-22 system), you should be alright. If the substitute speed is higher than the original speed, the increased air noise may be unacceptable to the occupant. Check with them on the noise level before you leave the job.

MOTOR INSULATION

The wire that forms the stator winding in a motor appears to be a bare copper wire, but it is not. Before the wire is wound into a winding, it gets a

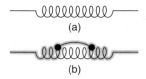

FIGURE 15–24 Turn-to-turn short due to failed winding insulation. (a) Normal winding. (b) Turn-to-turn short.

lacquer or enamel coating that acts as insulation. This keeps each individual turn of the winding electrically isolated from the neighboring turn. If this thin layer of insulation wears out, allowing contact between the turns of the winding, the motor has failed. This is called a turn-to-turn short in the motor. It has the effect of shortening the winding where the turn-to-turn short has occurred (Figure 15–24). A motor in which the motor insulation has failed in this fashion is sometimes difficult to detect. Most times, you don't know what the actual resistance of the motor windings should be, unless you refer to the manufacturer's information (which may not be available to you while you are on the job). The symptoms can be difficulty in starting the motor if the start winding has a turn-to-turn short, or a high amperage draw if the run winding has a turn-to-turn short.

The insulation used on the motor winding is rated according to the maximum allowable operating temperature. The motor nameplate might tell you that it has class, for example, class B insulation. The classifications for motor insulation are shown in Figure 15–25. The maximum temperature referenced in Figure 15–25 is the motor temperature rise plus the maximum ambient temperature where the motor is located. Most common fractional horsepower motors are A and B rated. Generally, you should replace a motor with one of equal or higher temperature class. Replacement with one of lower temperature rating could result in nuisance tripping of the motor overload. Each 18°F rise above the temperature rating of a motor can reduce the motor life by one-half. A rise of 10°C is equivalent to a rise of 18°F.

CLASS	MAXIMUM TEMP
A	105°C (221° F)
B	130°C (266° F)
F	155°C (311° F)
H	180°C (356° F)

FIGURE 15–25 Motor insulation classification.

THREE-PHASE MOTORS

Three-phase motors are not started through the use of phase-shifting devices. A rotating magnetic field is present around the motor stator whenever three-phase voltage is applied to the windings. With larger units, a starting sequence is sometimes used to reduce initial starting current. During starting, the motor is electrically connected in the wye configuration; it is electrically reconnected in the delta configuration when operating speed is approached. Through the use of such connections, the voltage across the motor windings is reduced during the starting sequence but it is at full level during operating speeds.

In Figure 15–26a, the three-phase motor connection for start-up is in the wye configuration. The voltage across each winding is 254 volts. The source voltage is 440 volts. Full-coil voltage operation with the delta connection is shown in Figure 15–26b. The operation of the contactor or relay used to switch from wye to delta operation may be controlled by a time-delay mechanism.

When power is first applied to the motor, the relay is de-energized and the circuit is wye connected. After a fixed time delay during which the motor picks up speed, the relay energizes, changing the motor to a delta connection. The voltage across the motor windings is 440 volts.

Single Phasing of Three-Phase Motors

When a three-phase motor is connected to a three-phase power source, it starts and eventually reaches normal operating speed. If, after the motor is started, one of the three input power connections becomes open, single-phase power is supplied to the motor. The motor continues to operate as a single-phase motor; however, excessively high current is drawn and the motor overheats. If the motor is allowed to continue to operate in single phase, permanent damage to the motor may result. Thermal protective devices are connected in most three-phase motor circuits to electrically disconnect the motor if it becomes overheated.

PUSH-BUTTON, ON-OFF MOTOR CONTROL

A common control used in the start operation of motors is the push-button magnetic-type starter switch. One excellent feature of this system is that the control may be located away from the motor. A diagram of a push-button system is shown in Figure 15–27.

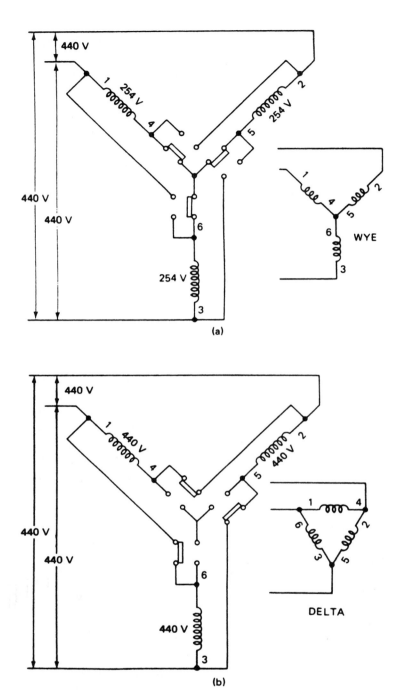

FIGURE 15–26   Wye start, delta run.

The motor is controlled by a magnetic contactor, K-1. A thermal overload protector is included in each line of the motor control. Should excessive current be drawn in either line, the thermal overload will overheat, causing the associated set of contacts to open, interrupting the circuit.

The normally open start button is pressed to start the motor. The current path is from line L1, through the start button, now closed, through the normally closed stop button, through both thermal overload contacts, to the bottom of the contactor coil. The top of the contactor coil is connected directly to line L2. The contactor energizes, supplying power to the motor. The contacts between L1 and T1, when closed, complete the circuit, bypassing the start button. The contactor coil remains energized. The

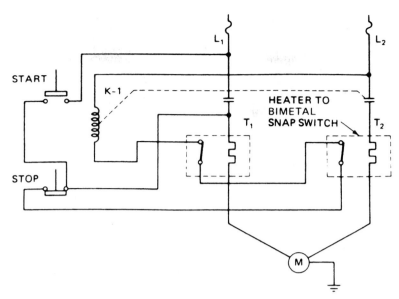

FIGURE 15–27   Push-button motor starter.

contactor coil de-energizes if the contacts of a thermal overload open or if the stop button is pressed.

LOW/HIGH-VOLTAGE OPERATION

Many motors are wound with coils in pairs. With such motors, it is possible to connect them electrically to operate on either high (240) or low (120) voltage. In Figure 15–28a, each coil is designed to function with 120 volts applied. In Figure 15–28b, the coils are connected in parallel for operation from a 120-volt source. So connected, each coil has 120 volts applied to it.

In Figure 15–28c, the coils are connected in series for operation from a 240-volts source. As shown, each coil has 110 volts across it. For a better understanding of this arrangement, review Units 4 and 5.

Single-Phase Motors: Low and High Voltage

Many dual-voltage, single-phase motors have two coils for the run winding and two coils for the start winding. Winding connections for low- and high-voltage operation are shown in Figure 15–29a and b.

Other dual-voltage, single-phase motors have two coils for the run winding and a single coil for the start winding. Winding connections for low- and high-voltage operation are shown in Figure 15–30a and b. When connected for low-voltage operation, the two coils of the run windings (T1-T2 and T3-T4) are connected in parallel across the 120-volt power source. The start winding T5-T6 is also connected directly across the 120-volt power source.

For operation at higher voltage, the run windings are connected in series. The start winding is connected to the junction of T2 and T3. Slightly less than one-half of the source voltage appears across the start winding during motor start, as shown in Figure 15–31a. After the motor comes up to operating speed, the switch device removes the start winding from the circuit. A schematic diagram of the circuit is shown in Figure 15–31b.

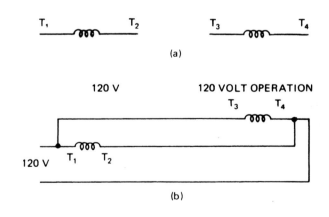

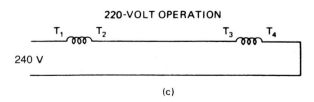

FIGURE 15–28   Examples of 120 V/240 V operation.

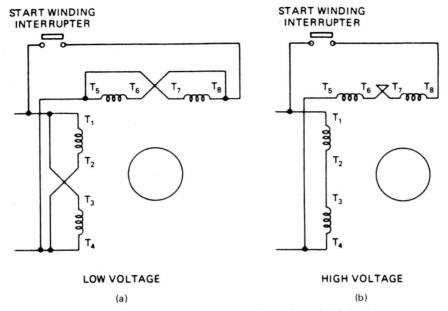

FIGURE 15–29 Standard dual-voltage motor winding connections.

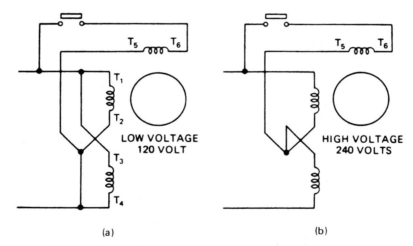

FIGURE 15–30 Single-start winding, dual-voltage operation.

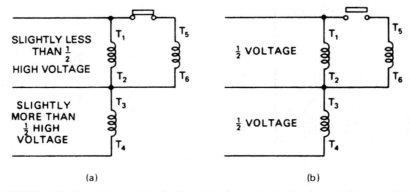

FIGURE 15–31 Voltage distribution (single-start winding), start (a), run (b).

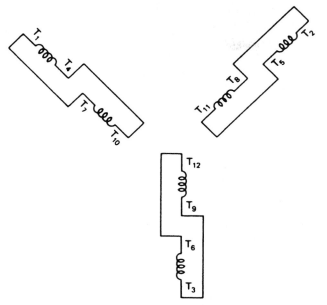

FIGURE 15–32   Three-phase motor, low-voltage parallel connection.

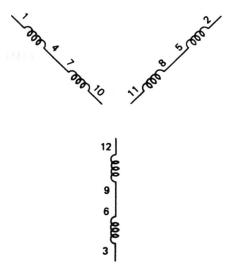

FIGURE 15–33   Three-phase motor, high-voltage series connection.

Three-Phase Motors: Low and High Voltage

The connection of windings for operation on high or low voltage with three-phase motors is a series or parallel consideration. In Figure 15–32, the motor is connected for operation at the low voltage. The winding pairs of each phase are connected in parallel. For high-voltage operation, the winding pairs for each phase are connected in series. This configuration is shown in Figure 15–33.

Direction of Rotation (Phase Reversal)   The direction of rotation of a three-phase motor is dependent on the connection of the power to the motor terminals. Many compressors, such as scroll, rotary, screw, and centrifugal, must turn in the correct direction to function properly. When installing systems containing three-phase components, be certain to follow the manufacturer's instructions for proper phase connection.

DUAL-VOLTAGE MOTORS

Three-phase motors are commonly available with dual-voltage ratings. A single dual-voltage motor may be used on one of two different supply voltages.

Some common dual-voltage ratings are 115/230 V for single-phase motors or 230/460 V for three-phase motors. For single-phase dual-voltage motors, one pair of wires emerging from the motor is used for the lower voltage, and a different pair of wires is used if the motor is to be connected to the higher voltage. For three-phase motors, the dual-voltage motor is different from a single-voltage motor in the following ways:

1. There are six different windings inside the motor instead of just three.
2. Nine wires emerge from the casing instead of just three.

The nine wires are attached to the motor windings as shown in Figure 15–34. Depending on which

---

ADVANCED CONCEPTS

The reason that dual-voltage motors are manufactured is so that a single motor part can be stocked for two different applications. The added cost of making a motor capable of handling two different voltages is more than offset by the reduction in storage costs of stocking two different motors.

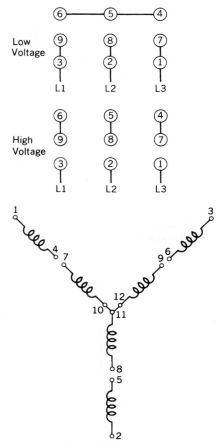

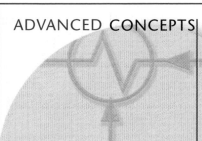

FIGURE 15–34   Dual-voltage motor.

voltage is being used, you will connect the nine wires to each other and to the incoming three-phase power supply as shown. These same instructions will appear on the motor.

HORSEPOWER RATING

The **horsepower** rating of a motor is a measure of how much work it can do each second, minute, or hour. Some small fan motors will be rated in watts,

rather than horsepower. (746 watts is equivalent to 1 hp). The horsepower requirement for a motor is determined by the amount of work it must perform. For example, if a motor is driving a pump that moves 100 gpm of water against a pressure difference of 30 psi, the pump will require approximately 2.5 hp of input to its shaft. The amount of work required to be done on a device like a pump, a fan, or a compressor is often called **brake horsepower (bhp)**. For the pump described earlier, you might find a 3-hp motor, because motors are not commonly manufactured in a 2.5 hp size. If the 3-hp motor were to fail, you would have to replace it with another motor no smaller than 3 hp. If you tried to use a 2-hp motor, the pump would still try to draw 2.5 bhp, and the motor would overload. If you used a 5-hp motor, you would be wasting money, and the motor efficiency would be reduced slightly, but the pump would still only draw 2.5 hp, and the pump performance would not change simply because you installed a higher hp motor (assuming the same pump/motor rpm). The horsepower rating of a motor is usually found on the motor nameplate.

Some motors are not rated in horsepower. For example, hermetic compressor motors will not show a horsepower rating on the nameplate. Other times, you will not be able to read the horsepower rating of the failed motor. In these cases, you can purchase a replacement motor that has a similar amp draw.

Motors can be rated for both their full-load amps (FLA) and their locked rotor amps (LRA). Sometimes full-load amps are called running load amps (RLA). Full-load amperage is the current that the motor will draw when it is operating at its rated horsepower output. The chart in Figure 15–35 shows what some average values are for the full-load amperage of various motors. At part load, the motor amps will be roughly proportional to the load.

Locked rotor amps is the current that the motor draws on start-up. This starting current is usually four to six times higher than the full-load amps, but

ADVANCED CONCEPTS

You will notice that the different wiring schemes for the two different voltages each use a pair of windings to form a single winding connected to a single leg of the three-phase power. For the low-voltage wiring, the pair of windings is wired in parallel. For the high-voltage wiring, the pair of windings is wired in series. When the higher voltage is used, the motor winding resistance is higher because the loads forming each leg are in series. The motor produces the same horsepower as the lower voltage arrangement, but it consumes approximately half as much amperage. The operating cost is roughly the same, regardless of which voltage supply is being used.

		Motor HP	150	125	100	75	60	50	40	30	25	20	15	10	7½	5	3	2	1½	1	¾	½
Single Phase	115 V	Full Load Current												100	80	56	34	24	20	16	13.8	9.8
		Power Factor %												89	88.5	87.5	86	84	83	80	77	73
		Starting Current												575	460	322	195	138	115	92	80	56
	230 V	Full Load Current												50	40	28	17	12	10	8	6.9	4.9
		Power Factor %												89	88.5	87.5	86	84	83	80	77	73
		Starting Current												288	230	161	98	69	58	46	40	28
Three Phase	220 V	Full Load Current	353	293	223	180	144	120	103	75	64	52	40	27	22	15	9	6.5	5	3.5	2.8	2
		Power Factor %	91.5	91.4	91.2	91	90.8	90.6	90.4	90.2	90.1	90	89.5	89	88.5	87.5	86	84	83	80	77	73
		Starting Current	2118	1758	1338	1080	864	720	618	450	384	312	240	162	132	90	54	39	30	21	16.8	12
	440 V	Full Load Current	172	144	117	90	72	60	52	38	32	26	20	14	11	7.5	4.5	3.3	2.5	1.8	1.4	1
		Power Factor %	91.5	91.4	91.2	91	90.8	90.6	90.4	90.2	90.1	90	89.5	89	88.5	87.5	86	84	83	80	77	73
		Starting Current	1032	864	702	540	432	360	312	228	192	156	120	84	66	45	27	19.8	15	10.8	8.4	6
	550 V	Full Load Current	138	117	94	72	58	48	41	30	26	21	16	11	9	6	4	2.6	2	1.4	1.1	.8
		Power Factor %	91.5	91.4	91.2	91	90.8	90.6	90.4	90.2	90.1	90	89.5	89	88.5	87.5	86	84	83	80	77	73
		Starting Current	828	702	564	432	348	288	246	180	156	126	96	66	54	36	24	15.6	12	8.4	6.6	4.8

FIGURE 15–35 Average motor current ratings.

it only lasts for a few seconds. Once the motor gets up to speed, the amp draw diminishes.

☞ If the nameplate is completely missing on the failed motor, you will not know the horsepower rating or the amp rating. In this case, you must look at the device that it is driving and determine how much horsepower it requires. For example, say you have discovered a condenser fan motor that has failed. It drives a propeller-type fan blade. You can go to the wholesaler and look up the performance of a propeller type fan with the same number of blades, the same diameter, the same blade pitch, and the same rpms. Chances are that the motor requirements of two different fans with all these same characteristics will have a similar bhp requirement.

SERVICE FACTOR

Some motor nameplates will show a **service factor (SF)**. It may be between 1.00 and 1.35. This number shows the percentage of full load that the motor can run at for short periods of time, without damaging the motor. For example, if a motor rated for 22 FLA had a service factor of 1.15, it would be acceptable if the load, on occasion, loaded the motor to 22 × 1.15 = 25.3 amps. When replacing a motor with an identical horsepower motor, the service factor of the new motor should be the same or higher than the failed motor.

FRAME SIZE

The frame size for a motor refers to its outside dimensions that have been standardized by the **National Electrical Manufacturers Association**

(NEMA). It is fortunate that NEMA has set standards for motor sizes, because if your replacement motor is the same frame size as the one it replaces, you know it will be the same physical dimensions. This includes the mounting holes, the diameter, the shaft size, the distance of the shaft from the mounting feet, motor length, and all those dimensions that are important if the new motor is to fit into the same space as the failed motor.

Simply matching the horsepower rating of the failed motor does not ensure that it will be the same frame size. Refer to Figure 15–36, which shows the various horsepower ratings that are available in each frame size. A 1/6-hp motor is available in either a 42 frame size or a 48 frame. Similarly, a 1/2-hp motor is available in either a 48 frame size or a 56 frame.

SHAFT SIZE

Matching the shaft size of the failed motor is important because it must mate properly with the pulley or fan hub that the shaft fits into. Even though you match the frame size, subcategories of the frame size may allow for different shaft diameters or shaft lengths. If the replacement motor has a shaft of a smaller diameter than the failed motor, you can use a ring between the shaft and the hub to effectively

NEMA Frame	Shaft Diameter	hp Range
42	⅜	1/20 1/15 1/12 1/10 1/8 1/6 1/5
48	½	1/6 1/4 1/3 1/2 ¾
56	⅝, ⅞	1/12 1/8 1/6 1/4 1/3 1/2 ¾ 1 1½ 2
66	¾	1 1½
143T	⅞	1 1½
145T	⅞	1 1½ 2 3

FIGURE 15–36 Frame sizes.

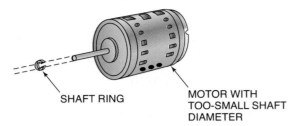

FIGURE 15–37   Shaft ring to increase shaft size.

make the shaft size larger (Figure 15–37). The shaft ring is open in one place to allow a spot to fasten the hub to the shaft with a setscrew. If the shaft of the new motor is too long, it can be cut to length with a hacksaw. Shafts are made of soft material, and cutting them are not difficult.

Figure 15–38 shows the "flat" on a shaft. A fan or pulley may be fastened to the shaft by means of a setscrew. It is important when tightening a setscrew onto the shaft, that it bears down onto this flat. Otherwise, the device may quickly work its way loose after you leave the job. Some larger motors have two flats, 90 degrees apart, so that two setscrews may be used for added strength.

MOTOR ROTATION

A replacement motor must be selected to rotate in the same direction as the failed motor. Single-phase motors may be designated as clockwise, counterclockwise, or reversible. The motor will rotate in the designated direction, even if the power leads are switched. Usually (but not always), the direction of motor rotation is designated as the direction you would see when viewing the motor from the shaft end (Figure 15–39). However, not all manufacturers follow this convention. When purchasing a replacement motor, you must confirm which convention that manufacturer uses. Some will designate the rotation

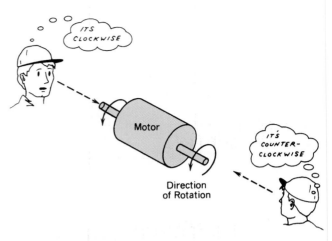

FIGURE 15–39   Motor rotation direction.

direction with an arrow on the motor body. Others will say on the box something like CCWLE (counterclockwise rotation when viewed from the lead end).

Some single-phase motors are reversible (Figure 15–40). The reversing plug shown may be connected so that it connects purple-purple and yellow-yellow to give one direction of rotation. Or, it may be connected so that it connects purple-yellow and purple-yellow to give the other direction of rotation.

A reversible motor is a good choice to carry on your truck if you are going to carry any commonly used motors. It costs slightly more than a uni-direction motor, but you only need to stock one motor of a given horsepower for both clockwise and counterclockwise applications.

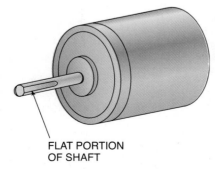

FIGURE 15–38   Flat portion of motor shaft.

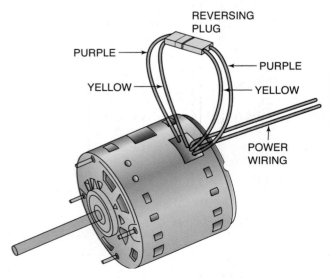

FIGURE 15–40   Single-phase motor with reversing plug.

DIRECTION OF
MAGNETIC FIELD
ROTATION

MOTOR ROTATION IS CCW, VIEWED FROM SHAFT END.

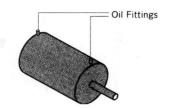

FIGURE 15–42   Motor oil fittings.

REASSEMBLED MOTOR. MAGNETIC FIELD
IS UNCHANGED, BUT MOTOR ROTATION IS
CW, VIEWED FROM SHAFT END.

FIGURE 15–41   Mechanically reversing rotation.

Some motors, even though not reversible electrically, may be reversible mechanically. If you can disassemble the motor and reassemble it so that the shaft emerges from the opposite end of the stator housing, the stator field direction will be unchanged, but the direction of the shaft will be reversed (Figure 15–41).

For three-phase motors, the direction of the new motor will not be specified. That is because three-phase motors will operate in either direction. When you connect a three-phase motor to the three power leads, you will not know which direction it is going to rotate. If, after you start it up, it rotates in the wrong direction, switching any two power leads will reverse the direction of rotation.

BEARINGS

The rotating portion of the motor is supported at each end by a bearing or a bushing. A bearing is similar to the ball bearings that separate a rotating wheel from a fixed axle on a car or a skateboard. A bushing is simply a smooth brass sleeve that supports a rotating shaft. Bearings are better than bushings from the standpoint of lower friction and therefore better motor efficiency. Bushings are superior to bearings from the standpoint of not needing as much (or any) lubrication. Motors with bushings are also quieter and less expensive than motors with bearings.

The bearing (or bushing) at the end of the motor where the shaft emerges is called the inboard bearing. The other end where the wiring is attached is called the outboard bearing.

When lubrication ports are provided, the motor should be lubricated with a few drops of 30 W oil once a year for most applications. When installing a new motor with oil ports, make sure that the motor is oriented so that they are above the elevation of the shaft (Figure 15–42). Otherwise, it will be very difficult to lubricate the motor because the oil would have to defy the laws of gravity and flow uphill.

The primary causes of bearing failure in motors are lack of lubrication, overloading, and overheating. Overloading of a bearing is most commonly

ADVANCED **CONCEPTS**

The direction of rotation of any split-phase motor may be reversed by switching the leads on either the start winding or the run winding (but not both). However, unless the ends of one of the windings is brought outside the motor enclosure, you are unable to access them to reverse the direction of rotation.

caused on belt-drive applications where an over-zealous technician makes the belt too tight. The sideways pull on the bearing will result in premature failure. On direct-drive fans, an out-of-balance fan wheel can cause the same result. Overheating can be prevented by maintaining proper motor ventilation by keeping the ventilation ports clean.

MOTOR SPEED

The rotation speed of a motor is determined by the number of poles in the stator winding. Figure 15–43 shows a four-pole motor and a six-pole motor. On a four-pole motor, the magnetic field rotates at 1800 revolutions per minute. This is called the **synchronous speed**. However, the motors used in heating, air conditioning and refrigeration work generally run at a speed slightly slower than synchronous speed. For example, a four-pole motor is usually rated at 1725 to 1750 rpm. A six-pole motor has a synchronous speed of 1200 rpm, but is rated at 1050 to 1075 rpm. The difference between the synchronous speed and the actual speed is called **slip**.

Sometimes, you find a failed motor, and you can't determine from the nameplate what the rated speed is. You can determine the correct speed for the replacement motor by disassembling the failed motor and counting the poles. If there are six poles, buy a 1075-rpm motor. If there are four poles, replace it with a 1725-rpm motor.

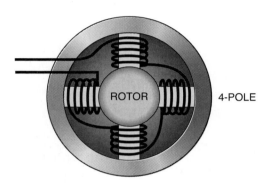

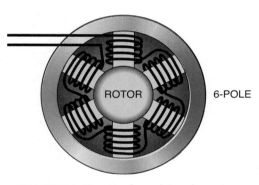

FIGURE 15–43 4-pole and 6-pole motors.

AUTO RESET VS. MANUAL RESET OVERLOAD

Some motors have built-in overload protection. Others rely on some external method of sensing and preventing overload. Most motors that have internal overload protection use some sort of thermostat

ADVANCED **CONCEPTS**

The solid-state speed controller uses a device called a **triac.** The triac operates to "chop off" a portion of the voltage signal coming to the motor. As the knob on the triac is adjusted, it chops off more or less of the voltage signal. As more of the input voltage signal is dropped, the amperage through the motor winding is decreased, reducing motor speed. The triac does *not* act like a resistor and does not generate any significant amount of heat. The adjustment of the triac may be accomplished from some mechanical sensing device, such as a bellows that senses head pressure. The condenser fan motor speed can then be adjusted automatically in response to the sensed head pressure.

MORE ADVANCED **CONCEPTS**

A true variable frequency drive varies both the voltage and the frequency of the output voltage supplied to the motor. These devices are more expensive, but result in very precise control of motor speed. They also optimize power consumption and motor efficiency. The lower-cost motor speed controls generally found in A/C equipment usually vary the line voltage to the motor but not the frequency. This results in somewhat lower overall motor efficiency at reduced loads, but this is of little concern on small-fan motor applications.

imbedded in the stator windings. When the motor draws more than its rated current, the motor windings get warmer than normal and the thermostat opens, shutting down the motor. When the motor then cools, the thermostat recloses, and the motor can attempt to start again. This is called **automatic reset.**

Other motors, particularly those used to drive the auger on an ice flaker, are **manual reset.** When these overloads sense an abnormal condition, they operate a switch located on the motor shell. The switch will stay open until it is manually pressed to reset it. When you push on this reset button, if the motor was actually out on overload, you will feel a slight "click."

VARIABLE-SPEED MOTORS

Small variable-speed motors are used in some high-efficiency furnaces (blowers) and condensing units (condenser fans) to match the fan performance with the load requirements. They are usually either shaded-pole motors or PSC motors, and they use a separate speed controller.

One method of controlling the speed is to insert a coil of wire in series with the motor winding. The coil does not have a high resistance, so it does not generate much heat or waste energy. But the back EMF that it creates reduces the current flow through the motor winding, therefore reducing its torque and increasing its slip. Figure 15–44 shows how a rotary-speed switch can be used to insert varying amounts of coil in series with the motor coil. As the selector switch moves to lower speed settings, the voltage actually applied to the motor decreases. The same effect can also be achieved with a device called an **autotransformer**. It has multiple taps that provide different voltages to the motor.

Infinitely variable speed can be achieved by using a solid-state speed controller.

TRICKLE HEAT CIRCUIT

Figure 15–45 shows a circuit that is sometimes used on compressor motors to take the place of a crankcase heater. There is a **trickle heat** capacitor that allows a small amount of current to pass through the motor windings, even when the contactor is open.

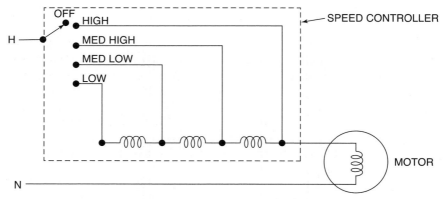

FIGURE 15–44 Motor speed selector.

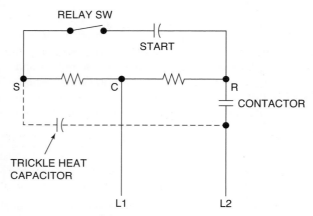

FIGURE 15–45 Trickle heat circuit.

This produces a small amount of heat to prevent refrigerant vapor from condensing in the compressor and diluting the oil.

When the contactor switch closes the trickle heat circuit has no effect.

TROUBLESHOOTING A MOTOR CIRCUIT

In the circuit of Figure 15–46, a motor is controlled through the use of a 24-volt contactor. When 24 volts are applied to the contactor coil (C and D), the contactor should energize. The contacts close, and 220 volts is applied to the motor. The motor should be running.

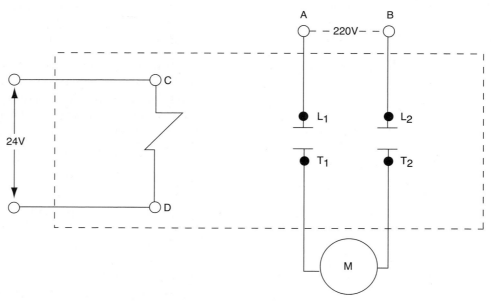

FIGURE 15–46 Motor control circuit.

If the motor is not running, a malfunction has occurred. The situation calls for troubleshooting.

Troubleshooting Procedure

The standard procedure in troubleshooting is to localize, isolate, and locate the problem or source of trouble. In this situation, we have already localized the problem, the motor circuit. To isolate and locate the problem, complete the following steps.

1. Measure the voltage of the contactor coil at C and D. If zero voltage is available, check the source prior to the contactor coil. If 24 volts is available, proceed to step 2.

2. Measure the voltage at L1 and L2 of the contactor. If 220 volts are not available, check the power source. If 220 volts are available at L1 and L2, proceed to step 3.

3. Measure the voltage at T1 and T2 of the contactor. If 220 volts are not available, the contactor is bad. If 220 volts are available and the motor is not running, check the motor.

SUMMARY

- This chapter provided some deeper insight into the actual electromagnetic principles underlying common HVAC motors.
- Most motor failures are relatively easy to diagnose and correct, occasionally a particularly troublesome repeat failure condition occurs.
- Often these can be traced to low line voltage or mechanical overload conditions, but on occasion an intermittent start component or improperly sized run capacitor is found to be the cause.
- Technicians armed with a good understanding of the role of capacitors and start relays in single-phase induction motors are far less likely to overlook or misdiagnose the problems that these devices sometimes cause.
- Those with the curiosity to explore further may find stepper motors, reluctance motors, and direct-current motors particularly interesting.

PRACTICAL EXPERIENCE

Required equipment   Ohmmeter, series start relay (current relay), potential relay

Procedure

1. Connect the ohmmeter across the contact terminals of the series start relay.

2. Does the meter indicate an open or short terminal?

3. With the meter connected to the relay, turn the relay over.

4. Did the circuit through the relay change from open to short or vice versa?

5. Connect the meter across terminals 2 and 5 of the potential relay.

6. Measure and record the resistance of the relay coil. _____ ohms

7. Connect the ohmmeter across terminals 1 and 2 of the relay.

8. Are the contacts open or shorted?

Required equipment   Dual-voltage single-phase or three-phase motors

Procedure

1. Observe the nameplate on the motor.

2. Locate wiring information for low-voltage operation.

3. Properly connect the motor for low-voltage operation. Are the coil sets connected in series or parallel for low-voltage operation?

4. Have the motor checked by supervisory personnel for proper connection.

5. Connect the motor to the proper voltage through suitable control devices.

6. Turn the motor on and observe the motor rotation.

7. Disconnect the motor from power.

8. Make motor winding connections required to reverse the direction of the motor.

9. Repeat steps 5, 6, and 7.

10. If time permits, repeat this experiment with the motor connected for high-voltage operation.

REVIEW QUESTIONS

1. The shaded-pole motor is used in low-power applications.

   T _____   F _____

2. A permanent split-capacitor motor requires a special starting device called a centrifugal switch.

   T _____   F _____

3. The starting device used in split-phase motors in hermetically sealed compressors is the centrifugal switch.

    T _____   F _____

4. The potential relay coil is connected in series with the run winding of the motor.

    T _____   F _____

5. The current relay has contacts that are open when the relay is de-energized.

    T _____   F _____

6. The potential relay has contacts that open when the motor approaches operating speed.

    T _____   F _____

7. It is not good practice to make adjustments or repairs on relays.

    T _____   F _____

8. All current relays may be operated in any position.

    T _____   F _____

9. Many high-horsepower motors in air-conditioning systems use potential relays.

    T _____   F _____

10. Current or series relays are used in high-horsepower applications.

    T _____   F _____

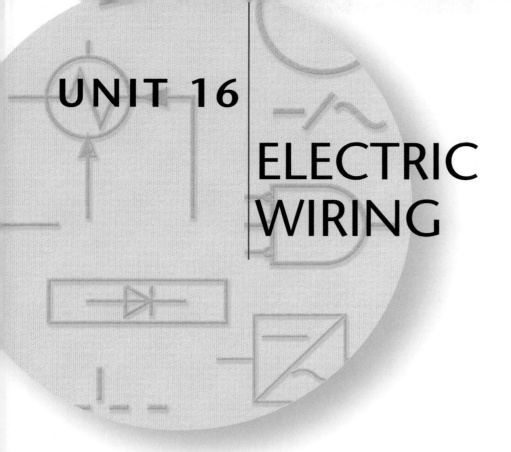

UNIT 16

ELECTRIC WIRING

OBJECTIVES

After completion and study of this unit, you should be familiar with:

- Power wiring as related to heating refrigeration and air conditioning.
- The components used to control that power, including sizing conductors, aluminum wiring, insulation, conduit, disconnect devices, solderless connectors, plugs, strain reliefs, thermostats, and conduit bending.

UTILITY POWER DISTRIBUTION

The utility company distributes electricity from the power-generating station at a very high voltage of 120,000 V or higher. The reason for such a high transmission voltage is to reduce power loss and to minimize the required transmission line sizes. Both the wire size and the power loss ($I^2 \times R$) vary with the amperage. Therefore, the power company distributes power at high voltage and low amperage.

The high voltage power is supplied to a **substation** that contains transformers that reduce the power down to a usable voltage. For residential use, additional transformers are supplied by the power company to provide single-phase power.

Three wires are supplied to the individual residence. After passing through the meter, the power comes into a circuit breaker box. Individual circuits may use either hot leg with a neutral leg to provide 115 V, or an individual circuit may use two different (out-of-phase) 115 V legs through a pair of circuit breakers ganged together to provide 230 V (see Figure 16–1). Each 115 V circuit has a neutral wire that comes back to the circuit breaker box ground to complete the circuit. The 230 V circuit is supplied from two different circuit breakers whose trip levers are mechanically connected together. If either circuit breaker trips, it will also pull the other breaker switch open.

SIZING CONDUCTORS

Power wiring is the business of a qualified electrician. Heating, refrigeration, and air conditioning service technicians will only do power wiring that is incidental to the installation of equipment. The HVAC/R technician must also be able to recognize when a problem rests with the power wiring and not with the HVAC/R equipment. Because power wiring can carry such large amounts of current, the laws that govern its installation are primarily concerned with two potential hazards:

1. The size of the conductor must be large enough in cross section to carry the required amount of current.

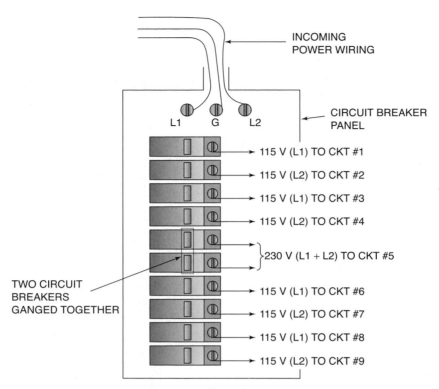

INCOMING
POWER WIRING

CIRCUIT BREAKER
PANEL

L1   G   L2

→ 115 V (L1) TO CKT #1
→ 115 V (L2) TO CKT #2
→ 115 V (L1) TO CKT #3
→ 115 V (L2) TO CKT #4
} 230 V (L1 + L2) TO CKT #5

TWO CIRCUIT
BREAKERS
GANGED TOGETHER

→ 115 V (L1) TO CKT #6
→ 115 V (L2) TO CKT #7
→ 115 V (L1) TO CKT #8
→ 115 V (L2) TO CKT #9

FIGURE 16–1   Circuit breaker panel.

2. The physical installation of the power wiring must be mechanically sound to prevent power wiring from shorting together.

The ability of a wire to carry power is analogous to the ability of a pipe to carry water. If you try to push too much water through a pipe, its velocity will be very high, and there will be a high friction rate between the water and the inside of the pipe wall. This will cause a high pressure drop in the pipe. Similarly, when electrical wiring is too small to carry the required load, there is a voltage drop in the wire.

Wire size is defined according to its **AWG** (American Wire Gauge). The largest wire is 0000 (4/0). As the numbers get larger, it denotes a smaller wire diameter. The smallest standard wire is number 50. Figure 16–2 gives information for some of the more commonly used wire sizes. The wiring used to supply power to a unit must have an **ampacity** (amp carrying capacity) greater than the current draw of the equipment being supplied. Where exceptionally long runs of power wiring are required, larger wire sizes may be required. The resistance values shown in Figure 16–2 can be used to determine how much voltage drop will occur in a wire.

AWG	DIA (IN.)	OHMS/000 FT		AMPACITY	
		COPPER	ALUMINUM	COPPER	ALUMINUM
0000	.460	.049	.080	195	150
4	.204	.248	.408	70	55
8	.128	.628	1.03	40	30
10	.102	1.00	1.64	30	25
12	.081	1.59	2.61	20	15
14	.064	2.52	4.14	15	—

FIGURE 16–2   Amp carrying capacities of wire.

EXAMPLE 1

Select the proper wire size and determine the voltage drop for a copper wire to supply a 35-amp load at 440 V. The length of the power wiring is 60 ft.

Solution

A number 10 wire can only carry 30 amps; therefore we would choose a number 8 wire with an ampacity of 40 amps. The resistance of this wire is .628 ohms per 1000 ft. Therefore, the resistance of a 60-ft. length would be

$$0.628 \, \text{ohms} \times 60/1000 = 0.038 \, \text{ohms}.$$

The voltage drop in the wire is

$$\text{volts} = \text{amps} \times \text{resistance}$$
$$= 35 \, \text{amps} \times 0.038 \, \text{ohms}$$
$$= 1.33 \, \text{V}. \quad \blacksquare$$

Copper is the most popularly used conductor. Aluminum is sometimes used due to its low cost. Aluminum wiring cannot carry as much amperage as copper for a given AWG size. Note also that the ampacity values in Figure 16–2 are based on a 140°F rating on the wire insulation. For higher temperature ratings, the ampacity would be higher.

Oversizing wire diameter causes no operational problems. It is just a waste of money. However, undersizing wire can cause overheating, voltage drop, and motor burnout due to low voltage supply.

Diagnosing Undersized Power Wiring

On construction trailers and other temporary structures, it is not uncommon to find an air conditioner that blows fuses from time to time, for no apparent reason. One possibility is that the temporary power wiring to the air conditioner is undersized. This makes it difficult for the compressor to start, and on start-up, the compressor will draw higher than FLA for a longer period. Sometimes this will be long enough to blow the fuse.

Start your diagnosis with the disconnect to the air conditioner open. Measure the voltage available at the unit. This will be equal to the voltage being supplied from the other end of the power wiring because there is no current flow and therefore no voltage drop through the wire. If the available voltage is within the acceptable voltage rating of the air conditioner (nameplate voltage plus or minus 10%), continue. Using an analog voltmeter, read the voltage at the disconnect when you close it and the compressor motor tries to start. Immediately, it will drop because of the LRA being supplied through the power wiring and the voltage drop that is caused by this high draw. After the compressor starts, the amp draw diminishes, and the voltage available at the disconnect should once again rise to almost the same voltage that you measured with the unit off. If the voltage available at the air conditioner remains significantly lower during normal operation than it was with the unit off, it indicates either undersized power wiring or a poor connection in the power wiring that is causing a voltage drop.

☞ If you are using a digital voltmeter, you will not be able to read how low the voltage drops as the unit is starting. This is because the digital meter takes a few seconds to "zero in" on the correct reading. If the voltage being measured is changing, the digital meter cannot read it accurately as it is changing.

ALUMINUM WIRING

In the 1970s many cities changed their building codes to allow the use of aluminum wiring due to its lower cost. However, two characteristics of aluminum wiring caused some unexpected and undesirable results. Those properties are

1. The oxide coating that forms on the surface of aluminum wiring does not conduct electricity (unlike copper oxide which is a fine conductor).

2. The coefficient of thermal expansion for aluminum tubing is quite high. As its temperature increases, its size changes considerably.

These two characteristics combined to cause a number of fires. Here's how. When the aluminum wiring was tightened under a connector lug, the copper oxide coating between the wire and the connector would present a slight resistance to the flow of electricity. This caused the generation of a slight amount of heat (due to the $I \times R$ drop of the aluminum oxide coating) and a slight amount of expansion of the aluminum wire. Over a period of years, the continuing cycle of expansion and contraction would cause the mechanical connection to loosen slightly. This, in turn, allowed air to the aluminum, caused more aluminum oxide to form, creating more electrical resistance and more heat to be generated, more expansion, more loosening of the connection, and an ever-accelerating march toward a failure of the connection due to heat.

Most cities outlawed the use of aluminum wire in buildings when the failure mode was discovered, but there are still millions of feet of aluminum wire out there, lurking in wait to cause you problems. The most common problem you will encounter is a failure of the connection of aluminum wire to a circuit breaker. When you find a loose connection, turn off the power and remove the wire. Clean the oxidation from the surface of both the wire and the connector lug. You can apply a special compound (you'll probably have to go to an electrical wholesaler) to the wire to inhibit the reformation of the oxide coating. Then the connection can be remade. Make sure that you get it *really tight*.

WIRE INSULATION

In the previous section, you saw that wire size was determined only by the current-carrying requirement of the wire. It had nothing at all to do with the voltage in the conductor. Selection of the wiring insulation is just the opposite. It depends only on the voltage inside and the potential physical abuse to which the insulation is likely to be exposed. Some of the different wire insulation designations are

1. Type RH is a heat-resistant rubber insulation used in dry applications where the temperature will not exceed 167°F.

2. Type RHH is similar, except that it is rated up to 194°F.

3. Type RHW is a heat-and-moisture-resistant rubber used in dry and wet locations at temperatures not to exceed 167°F.

4. Type TW insulation is a moisture-resistant thermoplastic for wet and dry locations up to 140°F.

5. Type THHN is a heat-resistant thermoplastic for dry locations up to 194°F. It has an outer covering consisting of a nylon jacket.

6. Type THW is a moisture- and heat-resistant insulation for wet and dry applications up to 167°F.

CONDUIT

To protect power wiring from mechanical damage, it is run inside a protective pipe called **conduit**. Conduit can be threaded like a water pipe or it may be thin-wall tubing using special connectors. The thin-wall conduit is the most popular due to its low

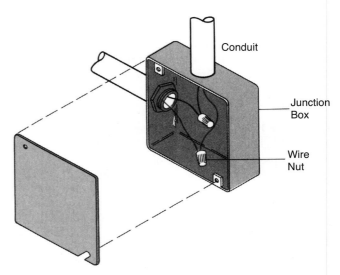

FIGURE 16–3   Junction box.

cost. It is referred to as **EMT (electro-mechanical tubing)**.

Wires are pulled through the conduit from the power source to the load. Any wire connections that are required are made with wire nuts inside a junction box (Figure 16–3). You will hear technicians refer to this as a J-box.

Conduit may be rated for indoor or for outdoor service. Outdoor conduit connections are gasketed to keep rain from getting inside to the wiring. Conduit supplying power to walk-in freezer boxes can present a unique challenge. The box can cool the conduit enough so that the moisture inside the conduit condenses, resulting in an ice-ball forming on the junction box.

Conduits are available in sizes starting at 1/2 inch diameter. The size of the conduit required is determined by the National Electric Code, according to the number and diameter of the wires. The maximum number of conductors allowed in conduit or EMT is given in Figure 16–4. The maximum number of conductors allowed in an electrical box is given in Figure 16–5.

Figures 16–6 shows flexible conduit as opposed to **rigid conduit**. This may also be rated for indoor or outdoor service. Codes govern the maximum allowable lengths of flexible conduit that may be used. Rigid conduit is used for the long, straight runs. At the end near the air-conditioning or refrigeration unit, flexible conduit is used to simplify the installation. Indoor flexible conduit is commonly called **armored cable** or **BX cable**. Outdoor flexible conduit is commonly called **SealTite**, although this brand name is only one of the many brands available.

TYPE OF WIRE	CONDUCTOR SIZE AWG KCMIL	CONDUIT SIZE (INCHES)					
		$\frac{1}{2}$	$\frac{3}{4}$	1	$1\frac{1}{4}$	$1\frac{1}{2}$	2
TW, XHHW (14 THRU 8)	14	9	15	25	44	60	99
	12	7	12	19	35	47	78
	10	5	9	15	26	36	60
	8	2	4	7	12	17	28
RHW AND RHH (WITHOUT OUTER COVERING), THW	14	6	10	16	29	40	65
	12	4	8	13	24	32	53
	10	4	6	11	19	26	43
	8	1	3	5	10	13	22
TW, THW, FEPB (6 THRU2), RHW AND RHH (WITHOUT OUTER COVERING)	6	1	2	4	7	10	16
	4	1	1	3	5	7	12
	3	1	1	2	4	6	10
	2	1	1	2	4	5	9
	1		1	1	3	4	6
	$\frac{1}{0}$		1	1	2	3	5
	$\frac{2}{0}$		1	1	1	3	5
	$\frac{3}{0}$		1	1	1	2	4
	$\frac{4}{0}$			1	1	1	3
	250			1	1	1	2
	300			1	1	1	2
	350				1	1	1
	400				1	1	1
	500				1	1	1
	600					1	1
	700					1	1
	750					1	1
THWN, THHN, FEP (14 THRU 2) FEPB (14 THRU 8) XHHW (4 THRU 500 KCMIL)	14	13	24	39	69	94	154
	12	10	18	29	51	70	114
	10	6	11	18	32	44	73
	8	3	5	9	16	22	36
	6	1	4	6	11	15	26
	4	1	2	4	7	9	16
	3	1	1	3	6	8	13
	2	1	1	3	5	7	11
	1		1	1	3	5	8
	$\frac{1}{0}$		1	1	3	4	7
	$\frac{2}{0}$		1	1	2	3	6
	$\frac{3}{0}$		1	1	1	3	5
	$\frac{4}{0}$		1	1	1	2	4
	250			1	1	1	3
	300			1	1	1	3
	350			1	1	1	2
	400				1	1	1
	500				1	1	1
	600				1	1	1
	700				1	1	1
	750					1	1
XHHW	6	1	3	5	9	13	21
	600				1	1	1
	700					1	1
	750					1	1

FIGURE 16–4   Maximum number of conductors in conduit.

BOX DIMENSIONS, INCHES TRADE SIZE OR TYPE	MAXIMUM NUMBER OF CONDUCTORS			
	14 AWG	12 AWG	10 AWG	8 AWG
$4 \times 1\frac{1}{4}$ RND. OR OCTAG.	6	5	5	4
$4 \times 1\frac{1}{2}$ RND. OR OCTAG.	7	6	6	5
$4 \times 2\frac{1}{8}$ RND. OR OCTAG.	10	9	8	7
$4 \times 1\frac{1}{4}$ SQUARE	9	8	7	6
$4 \times 1\frac{1}{2}$ SQUARE	10	9	8	7
$4 \times 2\frac{1}{8}$ SQUARE	15	13	12	10
$4\frac{11}{16} \times 1\frac{1}{4}$ SQUARE	12	11	10	8
$4\frac{11}{16} \times 1\frac{1}{2}$ SQUARE	14	13	11	9
$4\frac{11}{16} \times 2\frac{1}{8}$ SQUARE	21	18	16	14
$3 \times 2 \times 1\frac{1}{2}$ DEVICE	3	3	3	2
$3 \times 2 \times 2$ DEVICE	5	4	4	3
$3 \times 2 \times 2\frac{1}{4}$ DEVICE	5	4	4	3
$3 \times 2 \times 2\frac{1}{2}$ DEVICE	6	5	5	4
$3 \times 2 \times 2\frac{3}{4}$ DEVICE	7	6	5	4
$3 \times 2 \times 3\frac{1}{2}$ DEVICE	9	8	7	6
$4 \times 2\frac{1}{8} \times 1\frac{1}{2}$ DEVICE	5	4	4	3
$4 \times 2\frac{1}{8} \times 1\frac{7}{8}$ DEVICE	6	5	5	4
$4 \times 2\frac{1}{8} \times 2\frac{1}{8}$ DEVICE	7	6	5	4
$3\frac{3}{4} \times 2 \times 2\frac{1}{2}$ MASONRY	7	6	5	4
$3\frac{3}{4} \times 2 \times 3\frac{1}{2}$ MASONRY	10	9	8	7

FIGURE 16–5   Number of conductors in an electrical box.

FIGURE 16–6   Flexible conduit. (Courtesy of Anamet Electrical, Inc.)

DISCONNECT

Figure 16–7 shows a disconnect switch. It is usually located on or near the equipment that it controls. It is basically a set of knife switches that open all the legs (two or three, depending on whether the incoming power is single-phase or three-phase) of the incoming power supply to a load. When the handle is in the down position, the knife switches are open (Off), and when the handle is in the up position, the switches are closed (On). Disconnect boxes are provided with holes to accommodate a padlock, to allow locking the handle in either the On position or the Off position. A service technician might lock the disconnect in the Off position when s/he is working on a unit. An owner might lock the disconnect in the On position to prevent it from being turned off by mischievous children.

Many local codes require the installation of a disconnect switch within a line of sight from the unit it controls. This is a safety precaution, so that if the service technician shuts down the unit, s/he can make sure that the unit will not be turned back on by another person. (Of course, the safest way is to lock the disconnect switch open.)

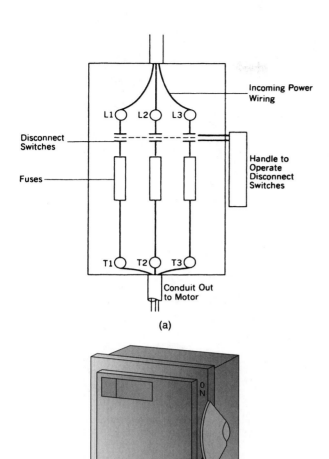

(a)

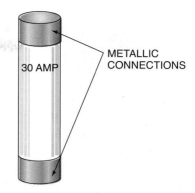

FIGURE 16–8   Cartridge fuse.

(b)

FIGURE 16–7   Disconnect switch.

On construction jobs, **locking out** the disconnect is common, where there may be several tradespeople working at the same time. Following are some important safety considerations about lockouts.

1. If you lock out a unit, have a tag on your lock that clearly identifies you as the owner of the lock. The tag must bear sufficient identification so that anyone who sees the tag will be able to locate you.

2. If there is already another person's lock attached to the disconnect, do not assume that this is sufficient protection for you. Add your own lock.

3. Never take it upon yourself to cut off another person's lock, even If you are sure that they have mistakenly left it on after going home for the day. Advise the owner's repre-

sentative, foreman, or other responsible person in charge. There may be a reason for the switch to remain locked.

The disconnect switch can be either fused or nonfused. Cartridge-type fuses (Figure 16–8) are used in the fused disconnect switch. One size disconnect switch may accommodate several different size fuses. For example, the physical size is the same for I5-, 20-, 25-, and 30-amp fuses. A different size is used for 40-, 45-, 50-, and 60-amp fuses. Where the fuses are the same physical dimensions, you can replace one fuse with a fuse of a different amp rating. But be sure that if you are replacing a fuse with a higher-rated fuse, that you do not exceed the amp rating given on the nameplate of the unit. The standard ratings for the disconnects are 30, 60, 100, and 200 amps (and higher).

EXAMPLE 2

You have arrived at a job site and found that the 45-amp fuse has blown. You can find no short or other apparent cause. Can you replace it with a 50-amp fuse to prevent another nuisance trip-out?

Solution

Physically, the 45-amp and the 50-amp fuses are the same dimensions. So long as the "Maximum Fuse Size" rating on the nameplate is 50 amps or higher, you can use the 50-amp fuse.   ■

The maximum fuse size on the unit nameplate will sometimes be identified as the HAC or HACR rating. *Do not* use Locked Rotor Amps as an indicator of the maximum allowable fuse size. The wiring inside the unit can handle the LRA for only a few seconds. The maximum permissible fuse rating will be significantly lower than the LRA rating.

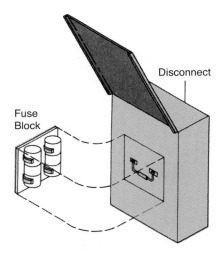

FIGURE 16–9 Pull box.

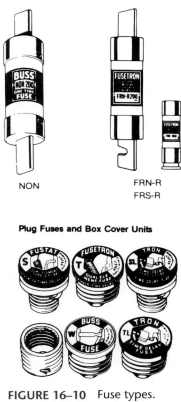

NON

FRN-R
FRS-R

Plug Fuses and Box Cover Units

FIGURE 16–10 Fuse types.
(Courtesy of Cooper Industries.)

Figure 16–9 shows another type of fused disconnect. It is probably a safer design, because the fuses can be removed by simply pulling on the handle and removing the entire fuse block. The fuse block may be returned upside down to the receptacle, and this will provide the same protection as having turned the handle of the disconnect in Figure 16–9 to the Off position.

FUSES

Figure 16–10 shows various types of **fuses**. A fuse is a thin wire that carries current to a load. The wire inside can only carry a limited voltage without overheating and breaking. In this way, the fuse senses the amount of current that the wire is carrying and acts like a switch that opens if the current rating is exceeded.

A common misconception is that fuses are used to protect a load such as a motor. This is not true. A fuse is sized to protect the downstream wiring. For example, suppose you have a motor rated to run at 26 amps. The wiring that supplies the motor can handle 30 amps, and a 30-amp fuse is provided. Suppose the motor starts drawing 28 amps. Eventually, the motor will fail because of the overload, but the fuse will never detect it until after the motor failure results in a short circuit.

Fuses that protect motor circuits have an application problem that is not faced by fuses that protect circuits with resistive loads. The motor will draw locked rotor amps on start-up, but will drop down to its normal operating amperage within a few seconds. The wiring inside the unit is sized to supply the running load amps. In order to prevent blowing the fuse during the few seconds of locked rotor amps, we use a dual-element or **slow blow** or **time-delay** fuse. This type of fuse will allow the momentary surge of power at start-up (4 to 6 times the running load amps) without blowing. If this current persists for more than a few seconds, the fuse will blow. In the event of a short-circuit condition, there will be no time delay in the operation of the fuse.

When changing a fuse, the power supply should first be turned off. In the case of cartridge fuses, they can be pulled using a plastic fuse puller as shown in Figure 16–11.

Fuse Pullers

FIGURE 16–11 Plastic fuse pullers. (Courtesy of Cooper Industries.)

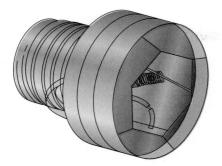

FIGURE 16–12 Plug-type fuse.

Technicians have been killed (electrocuted) using screwdrivers, pliers, fingers, and other inappropriate tools for removing cartridge fuses. Do not do this. This is no idle safety warning.

For circuits of less than 30 amps, sometimes plug-type fuses are used (Figure 16–12). They are sometimes called Edison-base fuses, because years ago they all had base sizes that are the same as an ordinary socket. However, newer plug type fuses each have a different size base for the different amp ratings.

CIRCUIT BREAKERS

Circuit breakers (Figure 16–13) serve the same purpose as fuses. They protect the downstream wiring. They have the advantage of being able to be reset after they have tripped. Besides the convenience, the

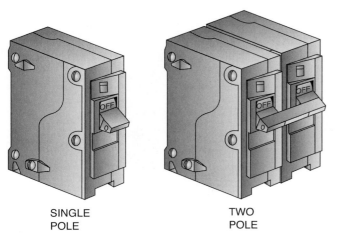

SINGLE
POLE

TWO
POLE

FIGURE 16–13 Circuit breakers.

possibility of personal injury that is present when changing a fuse is avoided. However, they probably provide a slightly lower level of protection for the following reasons:

1. When a circuit breaker switch has not been operated for a long period of time (one year or more), the contacts tend to become somewhat fused together. It will still provide adequate short-circuit protection, but for a gradually added overload, the circuit breaker may be able to carry far more than its rated amperage.

2. The current interrupting capacity of a circuit breaker is approximately 10,000 amps, compared to a fuse short-circuit rating of 100,000 amps.

POWER CORDS

A power cord for an appliance sometimes is not supplied with the appliance (such as a residential furnace). It must be connected by the service technician. Use a 14–3 cord of the required length. 14–3 means that the cord contains three 14 ga wires. At one end, you must attach the plug. A 115 V plug is shown in Figure 16–16b. The U-shaped connection is the ground. It is longer than the other connectors so that it will make contact first when the plug is inserted into an outlet (and, when removing the plug from the outlet, the ground wire breaks contact last).

The terminals on the attachment plug are

1. Brass for the Hot blade
2. Silver for the Neutral blade
3. Green hex screw for the ground

The power cord has a black, a white, and a green wire inside the sheath (Figure 16–16a). Remove to 1 inch of the outer sheath. Strip 1/2 inch of insulation from each conductor. Insert the conductor through the center hole of the plug. Twist the ends of each conductor (to prevent them from spreading), and bend each around the appropriate terminal (black on brass, white on silver, green on green). It is important that you bend the wires around the terminals in a clockwise direction. Tighten the terminals, taking care to keep all strands of the conductor under the screw head. Inspect for loose strands, which can cause shorts.

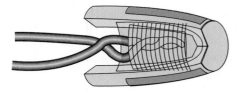

FIGURE 16–14   Wire nut.

SOLDERLESS CONNECTORS

The electrical connections made by the service technician will almost always be nonsoldered mechanical connections. The most common type of connector is the wire nut (Figure 16–14). The wire nut has a coil of wire inside. When bare conductors are placed inside the wire nut, the wire nut may be "screwed" onto the wires, connecting the wires together and to the internal coil of wire. If the conductor has been stripped of insulation to the correct length, the wire nut will extend past the end of the insulation so that no bare conductor is visible.

After the connection has been made, pull hard on each individual wire. If the connection has been properly made, it will not come loose.

Do not twist the wires together prior to inserting them into the wire nut. They will become twisted by the wire nut. If you are connecting a solid wire to a stranded wire (Figure 16–15), leave the stranded wire slightly longer than the solid wire. Otherwise, the wire nut may grab securely

only onto the solid wire, with the stranded wire wrapped around the solid wire. Over time, this may cause the connection between the two wires to deteriorate, creating a resistance, then heat (due to the $I \times R$ drop through the resistance), and eventually a burned connection that fails.

After the wires have been attached to the plug, place the insulating disc over the blades, and tighten the clamp around the cord where the cord enters the plug. If the plug does not have a clamp to relieve strain on the terminals when the cord is pulled, you will have to make your own as shown in Figure 16–16c. The black-and-white wire can be tied as shown prior to making the connection to the terminals. Then the knot may be pulled into the plug body. It will take the strain off the terminals.

At the other end of the cord, connect the conductors to the unit wiring using wire nuts or a terminal strip if provided (black to black, white to white, and green to green or if no green is provided, to the casing).

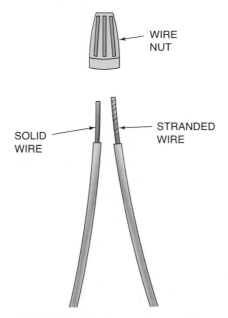

FIGURE 16–15   Connecting solid to stranded.

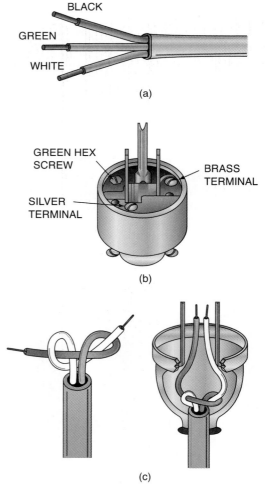

FIGURE 16–16   Wiring a plug.

TERMINAL STRIPS

Sometimes, a manufacturer provides terminal strips for the wiring connections made by the manufacturer and for the field wiring. It serves two purposes:

1. It eliminates the need for wire nuts, making for a neater installation.
2. The terminals can be labeled to correspond with terminal identifications on a wiring diagram. This makes it easy for the technician to troubleshoot. All you need to do is find the terminal numbers on the wiring diagram where you want to measure voltage, and go to the appropriate terminal numbers on the strip.

STRAIN RELIEF

Figure 16–17 shows several types of strain relief that are used where a power cord or a thermostat wire passes through a hole in the metal casing of a unit. The strain relief serves two functions:

1. It prevents the insulation form being cut where it rubs on the sharp metal edge of the unit.

2. It prevents strain on the wiring connections when the cord is pulled.

A wire should never be routed through a metal hole without using a strain relief.

WORKING WITH THERMOSTAT WIRE

Connecting thermostat wire to a room thermostat subbase is one of the most-watched tasks you will perform. A customer's opinion of your competence will very likely be affected by how professionally you can do this task. That means it must be done quickly, and when you're done, it should look good.

Start by stripping 2–3 in. of sheathing from the thermostat wires. Manipulate the wire to flatten the thermostat wires inside, so they are laying side by side. This will allow you to insert the blade of a knife between the thermostat wires, without cutting into the individual wire insulation. Figure 16–18 shows a technique that your mother warned you against, but

FIGURE 16–17   Strain relief.

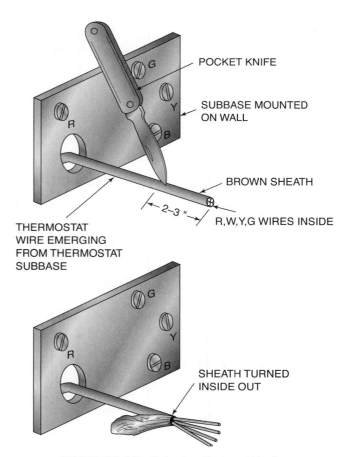

FIGURE 16–18   Stripping thermostat wire.

it is very effective if you are very careful. Push the blade through one side of the sheath, and then cut the sheath out to the end (make sure you don't put your fingers between the blade and the handle if you are using a folding pocketknife). Then you can pull the cut portion of sheath back, turning it inside out and cutting off the excess. If there are more wires inside the sheath than you plan to use, don't cut them off. Instead, fold them back and wrap them around the remaining sheath. That way, they will be available for future use, if needed.

Push the sheathing and any excess thermostat wire back into the wall. Lay out the individual wires to the individual terminals on the subbase, and strip the insulation as shown in Figure 16–19. Note that you are making each wire a different length to match the distance to the terminal. Loosen the terminal screw. Grab the end of the exposed thermostat wire with a needle-nose pliers, and pull it around the terminal in a clockwise direction. Tighten the screw terminal to secure the wire. Then, still holding the end of the wire in the pliers, rock the end of the wire back and forth until it breaks off close to the terminal. Repeat for each terminal. The result will be a connection where the insulation goes all the way up to the screw terminal, the wire is the exact right length, and the tag end of the wire is as short as possible.

When you "rock" the tag end of the wire back and forth, you must roll your wrist so that there is no bending of the wire at the pliers. If you do it wrong, the wire will break at the pliers. If you do it right, the wire will break at the terminal. With practice, you can make it break at the terminal every time.

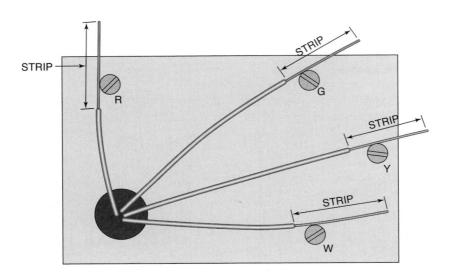

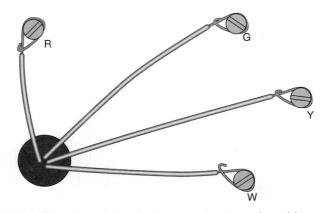

FIGURE 16–19 Connecting the thermostat wire to the subbase.

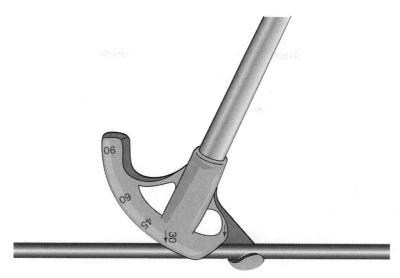

FIGURE 16–20   EMT bender.

CONDUIT BENDING

Installation of conduit is usually done by the electrician. But there may be times when the air-conditioning technician may have to install incidental runs of conduit. EMT (thinwall conduit) can be bent using an EMT bender, as shown in Figure 16–20. They come in 1/2″, 3/4″, and 1″ sizes. The difficulty in using a conduit bender is in predicting the length of the finished bend. Each bender (depending on the manufacturer) has a certain amount of take-up (Figure 16–21). For example, if the bender you are using has a 6-in. take-up, and you want to make a bend that will produce a stub that is 16″ from the outside of the main leg, begin by subtracting the take-up dimension from the desired 16″ dimension (10″). Place the bender with mark "B" positioned 10″ from the end of the tube (Figure 16–22). With the tube on the floor, pull on the tube bender until the stub rises to vertical. The resulting stub will terminate 16″ from the floor.

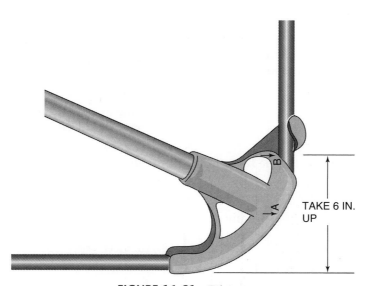

FIGURE 16–21   Take-up.

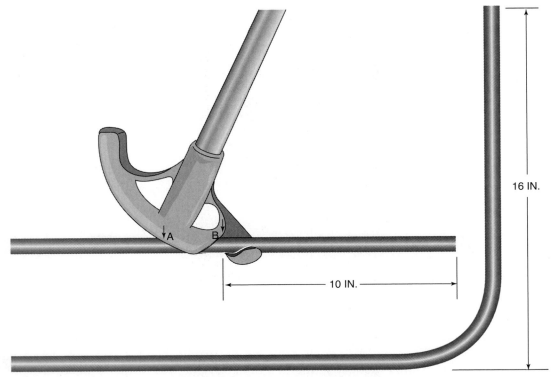

16 IN.

10 IN.

FIGURE 16–22   Positioning the bender.

SUMMARY

- Air-conditioning, refrigeration, and heating technicians must be familiar with most wiring techniques used by electricians.

PRACTICAL EXPERIENCE

1. Using a thermostat subbase mounted on a board, practice the given technique until you can do it quickly and easily.

REVIEW QUESTIONS

1. Edison-base fuses are replacing S-type fuses in industry.
   T _____   F _____

2. Thermal circuit breakers may be reset immediately.
   T _____   F _____

3. According to the chart Figure 16–2, Amp carrying capacity of wire, what would be the drop in voltage in 200 feet of #12 copper wire carrying 16 amperes? _____ volts

4. Is 16 amperes within the current capacity of #12 copper wire? _____.

5. What is the current capacity of #8 copper wire? _____

6. According to the Maximum number of conductors in a conduit chart in Figure 16–4, how many #12 wires with TW insulation may be placed in 3/4-inch conduit? _____

7. How many #8 wires with THHN insulation may be placed in 1-inch conduit? _____

8. According to the Number of conductors in an electrical box chart in Figure 16–5, how many # 12 wires may be in a 4 × 2 1/8 box? _____

9. How many wires, size 14, in a 4 11/16 × 2 1/8 square box? _____

10. How many wires, size 8, in a 3 × 2 × 3 1/2 device? _____

UNIT 17

SEMICONDUCTOR DEVICES

OBJECTIVES

After completion and review of this unit, you should be able to understand:

- What semiconductor materials are.
- The effect of the addition of impurities to making usable semiconductors.
- The action inside a semiconductor diode.
- The electrical characteristics of a diode.
- What zener voltage regulator diodes are.
- How transistors are constructed and how they operate.
- Special semiconductor devices.

There has been an increase in the use of semiconductor devices in the control of air-conditioning and refrigeration systems. An example of such a device, the anti-short-cycle timer, is shown in Figure 17–1. The purpose of the device is to keep power from being immediately reapplied to the compressor motor if power is interrupted for any reason. The power is kept from the compressor motor until such time as high head pressure is allowed to bleed down through the evaporator. The device contains transistors, diodes, resistors, capacitors, and a relay. In most cases, this type of device is not repaired in the field.

Service technicians should be familiar with the electronic devices associated with air-conditioning and refrigeration control devices to be able to determine whether the device is functioning properly.

DEFINITION

A **semiconductor** is a material that falls somewhere between the good conductors (most metals) and the pour conductors or **insulators** (such as glass or wax). Semiconductors function in electronic circuits because of their crystalline structure and because they

FIGURE 17–1   Anti-short-cycle timer. (Courtesy of BET Inc.)

169

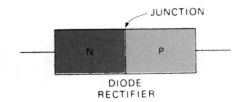

FIGURE 17–2   Junction diode.

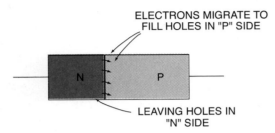

FIGURE 17–3   Semiconductor diode.

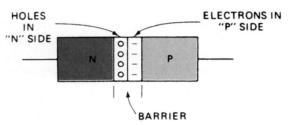

FIGURE 17–4   Potential hill barrier.

are doped with materials called *impurities*. The impurities provide characteristics to the semiconductor material that make the material useful in rectifiers and amplifiers.

The basic semiconductor materials are germanium and silicon. **Silicon** is the most popular and is covered here. When the pure silicon material is doped with an impurity, it becomes either P-type silicon or N-type silicon, depending on the material added as the impurity. The *P* stands for positive and the *N* for negative. The impurity combines with the silicon crystals and modifies the crystalline structure.

If P-type impurity is added, the crystals of silicon form improperly, leaving a space for an electron in the structure; this space is called a *hole,* and it is movable in the silicon. If N-type impurity is added, the crystals of silicon form with an extra electron available; this extra electron is movable.

DIODES

A **diode** is a device that allows electrons to flow through it in only one direction. Diodes are useful in converting alternating current into direct current. In the construction of a diode, P-type silicon and N-type silicon are joined together (Figure 17–2). Each is electrically neutral before they are joined. The excess electron in the N-type material and the hole (the space for an electron) in the P-type material relate only to the crystalline structure. When the two materials are joined, some of the ex-

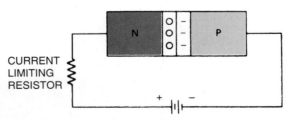

FIGURE 17–5   Reverse bias.

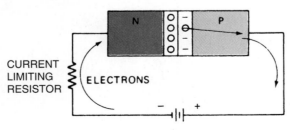

FIGURE 17–6   Forward bias.

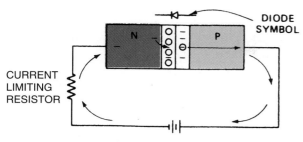

FIGURE 17–7 Conducting diodes.

cess electrons in the N type migrate over and fill the holes in the P material (Figure 17–3).

When the electrons from the N side move over to the P side, a potential is developed across the junction. The potential is called a *barrier.* The N material loses some electrons and is positively charged; the P material gains electrons and is negatively charged (Figure 17–4). No further action takes place unless an outside potential is applied to the diode.

If a voltage is applied to the diode, as shown in Figure 17–5, electron current does not flow. Electrons cannot flow because the application of the external potential increases the width of the barrier. The positive terminal of the battery pulls a few more electrons from the N side, increasing the number of holes and widening the barrier. The negative terminal of the battery adds few electrons to the

P side, widening the barrier here. The application of the battery potential to the diode increases the resistance of the diode. This is called *reverse bias.*

If the battery connection is changed, a completely different situation exists (Figure 17–6). The application of the positive to the P material allows for the movement of an electron from the barrier area toward the positive potential. An electron from the N side may move across the junction to fill the hole left by the moved electron in the P side. An electron may travel through the battery and out the negative terminal to the N side of the diode to replace the electron that moved across the junction. This is called *forward bias.*

The procedure is continuous, and electron current flows through the diode (Figure 17–7). The resistance of the diode is reduced to a low level. The symbol for a diode is shown in Figure 17–7. Electrons flow against the arrowhead.

If an alternating voltage is applied to the diode, the diode conducts every other half-cycle. In the circuit in Figure 17–8, current flows through the diode and R_1 every other half-cycle. The input voltage to the resistor diode combination is represented as E_{in}; the output voltage across R_1 is represented as E_{out}. The circuit is called a *half-wave rectifier.* A capacitor may be added to the circuit to provide filtering. In Figure 17–9, the connection of the capacitor and the effects on the output waveform are shown.

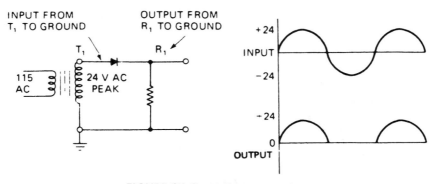

FIGURE 17–8 Half-wave rectifier.

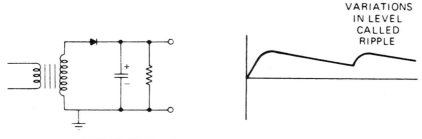

FIGURE 17–9 Filter capacitor maintains voltage.

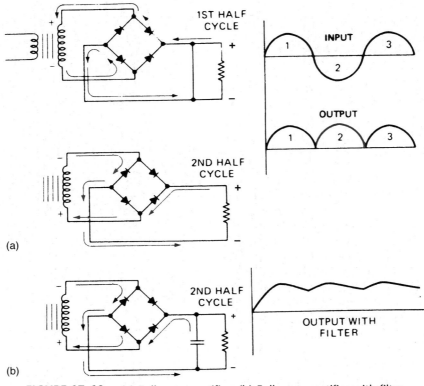

FIGURE 17–10 (a) Full-wave rectifier; (b) Full-wave rectifier with filter.

FULL-WAVE BRIDGES

The full-wave bridge circuit is used to provide an output voltage with less ripple. The circuit is shown in Figure 17–10. Figure 17–10a shows the path for electron flow during the first half-cycle. Figure 17–10b shows the path for electron flow during the second half-cycle. Note that the ripple amplitude has decreased in the filtered output.

BREAKDOWN VOLTAGE

When a diode is conducting in a circuit, a low-level voltage will appear across the terminals of the diode. This is called the diode's *forward voltage*. If the voltage across the diode is of reverse polarity, the diode will not conduct. There is a maximum re-verse voltage that the diode can withstand. If this voltage is exceeded, the diode will break down and conduction will occur. The diode may or may not be damaged, depending on the amount of current flow.

ZENER DIODES

The breakdown voltage across a diode remains fairly constant and is set by the doping ratios. This characteristic provides a means of developing voltage regulators. Voltage-regulating diodes, or **zener diodes**, are diodes developed to operate in circuits in which the applied voltage exceeds the diode's breakdown voltage. The diode operates in the breakdown region and has a relatively constant voltage across it. The symbol for a zener diode is shown in Figure 17–11.

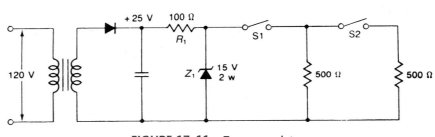

FIGURE 17–11 Zener regulator.

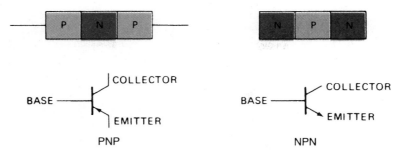

FIGURE 17–12   Transistors.

Consider the circuit in Figure 17–11, in which a 25-volt DC power supply is connected through R_1, a 100-ohm resistor, to Z_1, a 15-volt, 2-watt zener diode. The switches S1 and S2 control the connection of load resistors R_1, 500 ohms, and R_2, 500 ohms. With switches S1 and S2 open, the +25 volts from the supply exceeds the breakdown voltage of the zener diode. Current will flow up through the zener and the resistor R_1 to the source voltage. The current will rise to 0.1 (100 mA) ampere, at which point 10 volts will appear across the resistor R_1, and 15 volts will be developed across the zener diode. The zener diode has 15 volts across it and 100 mA through it. According to Ohm's law, its resistance is 150 ohms.

When switch S1 is closed, current will flow through resistor R_2. This would tend to increase the voltage across R_1 and thereby decrease the voltage across the zener. When the voltage across the zener decreases, the zener resistance increases. The current through the zener decreases from 100 mA to 70 mA. With 70 mA through the zener diode and 30 mA through the load resistor R_2, the current through R_1 remains at 100 mA. The 100 mA through R_1 continues to provide a 10-volt drop across R_1, and the output voltage remains 15 volts.

Closing switch S2 increases the load current by another 30 mA to a total of 60 mA. The current through the zener drops to 40 mA. The current through R_1

remains at 100 mA, and the output voltage remains at 15 volts. Actually, the voltage across the zener will decrease slightly as the current drawn by the load is increased. It is the slight decrease in voltage across the zener that provides for the change in zener resistance.

TRANSISTORS

Transistors are three-terminal semiconductor devices that can be used to amplify signals. Two types of transistors are available, the PNP transistor and the NPN transistor (Figure 17–12). The N and P refer to the type of impurity used in the silicon.

The main characteristic of transistors is that a small current flow in the emitter-base circuit controls a larger current flow in the emitter-collector circuit. Transistors, like diodes, require the proper polarity of applied voltage for proper operation. The emitter-base junction must have voltage applied in the forward direction for conduction. The applied voltage must oppose conduction in the base-collector junction. An example of a transistor amplifier is shown in Figure 17–13. Note the reverse in polarity of applied voltages to the NPN and PNP circuits.

In Figure 17–14, the 100-K resistor provides for a small emitter-base current. A larger emitter-collector

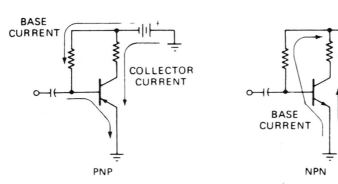

FIGURE 17–13   Transistor bias.

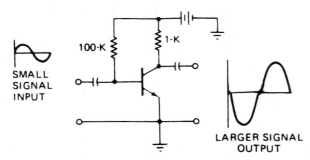

FIGURE 17–14 Transistor amplifier.

FIGURE 17–15 Integrated circuit. (Courtesy of BET Inc.)

current also flows. An input signal varies the emitter-base current, and an amplified version of the current flows in the emitter-collector circuit. The varying collector current produces a varying voltage across the 1-K collector resistor.

A single transistor can provide a signal gain of more than 100. A combination of three transistors in sequence could provide a signal gain of more than 1,000,000.

INTEGRATED CIRCUITS

Miniaturization processes have provided for the manufacture of complete amplifier circuits in a single, small package. Examples of amplifiers in the form of integrated circuits are shown in Figure 17–15. There are more components in the integrated circuit than in the anti-short-cycle unit that was shown in Figure 17–1 on page 169.

Operational Amplifiers (OP AMP)

A common component found in practical use in refrigeration control circuits is the operational amplifier. The device comes in many shapes and sizes, with one to four individual amplifiers in a single unit.

An integrated circuit such as shown in Figure 17–15 could contain one or two operational amplifiers.

The symbol for an operational amplifier is given in Figure 17–16.

The gain of an amplifier is in the range of 200,000; that is, a small signal at the input is amplified 200,000 times at the output.

Control of the output is usually obtained using *feedback resistors*. An example is shown in Figure 17–17. Power supply connections are not shown. A positive 1-volt signal is fed into the amplifier through a 1-K resistor to the inverting input of the amplifier. The output of the amplifier, a negative 10 volts, is fed back to the inverting input through a 10-K resistor.

Consider just the resistor combination. There are 11 volts across the two resistors, as shown in Figure 17–18, and essentially 0 volts at the junction and at the amplifier inverting (−). Actually, there would be +0.00005 volts at the amplifier inverting input. This voltage, amplified by a factor of −200,000, produces the −10 volts at the output.

SILICON-CONTROL RECTIFIERS AND TRIACS

A silicon-control rectifier (SCR) is a three-terminal device that has two states of operation, off and on. The symbol of a silicon-control rectifier, a circuit example, and the voltage waveforms of input, the gate, and the output are shown in Figure 17–19.

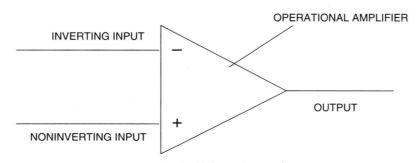

FIGURE 17–16 Operational amplifier symbol.

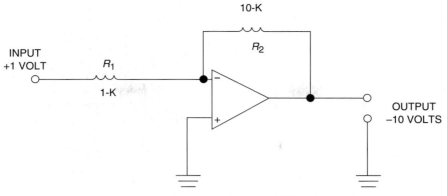

FIGURE 17–17 Operational amplifier circuit.

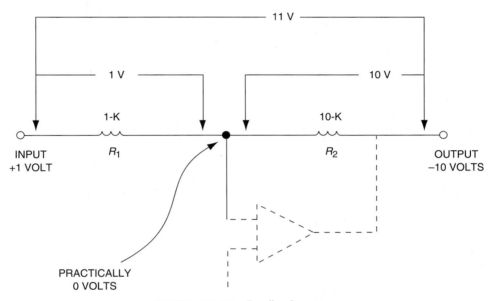

FIGURE 17–18 Feedback system.

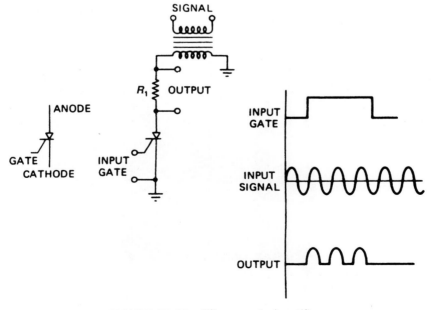

FIGURE 17–19 Silicon-control rectifier.

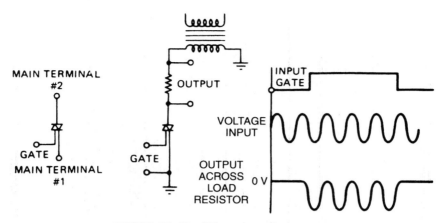

FIGURE 17–20   Triac, controlled output.

The signal that allows conduction is a positive reference between the gate and cathodes. When there is a positive reference on the gate, the SCR conducts whenever the anode is positive. Note that the positive half-cycle of the input appears in the output only during the positive gate period.

Another three-terminal semiconductor device is the *triac* (Figure 17–20). The triac allows for conduction in both directions when the gate signal is positive with reference to input main terminal 1.

The triac circuit shown in Figure 17–20 is used in many modern control devices. One such device is the dual time delay shown in Figure 17–21. The dual time delay provides time delay in turning a circuit on and time delay in turning the circuit off. Time delay in control of the inside blower motor is useful in air-conditioning and heating systems.

It is practical to delay turning on the inside blower motor until after a heating system has had time to heat the plenum area of the furnace. Otherwise, cold air in the plenum will be moved into the room to be heated. It is also proper to continue to move the heat in the plenum into the space to be heated after the heating device has been turned off.

The circuit in Figure 17–22 shows how this might be accomplished in a gas furnace system. The dual time delay is set for 15-second delay on and 25-second delay off. (The required delay depends on the particular system.) When the thermostat calls for heat, the gas solenoid valve is actuated, turning the heater on. Approximately 15 seconds later, after the plenum has had time to heat, the time-delay relay energizes the inside blower relay. The contacts of the inside blower relay feed power to the blower motor. The

FIGURE 17–21   Dual time-delay relay. (Courtesy of BET Inc.)

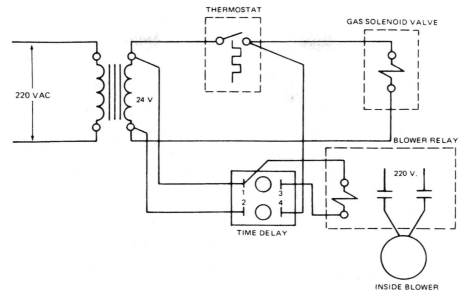

FIGURE 17–22 Connection of time delay.

operating blower moves warm air from the plenum to the space to be heated.

When the room reaches the selected temperature, the thermostat opens. The gas solenoid valve de-energizes, turning the furnace heating unit off. The time delay keeps the blower relay energized. The blower continues to move the heated air from the plenum to the area to be heated. The temperature of the plenum reduces. After 25 seconds, the time delay causes the blower relay to de-energize, and the inside blower motor stops. A considerable amount of energy may be saved using a time-delay system.

TEMPERATURE SENSING

The materials that are used to make up resistors are usually temperature sensitive. Most materials increase in resistance as temperature increases. There are, though, special resistors that are designed to have a relatively large variation in resistance as temperature changes. They are called *temperature-sensitive resistors*.

Temperature-Sensing Element

A special temperature-sensitive semiconductor device has recently been developed. The device produces a linear variation in output voltage with temperature. Connected as in the circuit in Figure 17–23, a linear output voltage of +10 mV/°F can be obtained for temperatures between +5°F and +300°F.

Simple Thermometer for Temperature Control (Evaporator Defrost)

The LM34 shown in Figure 17–23 could be combined with the operational amplifier in Figure 17–16 to provide temperature control.

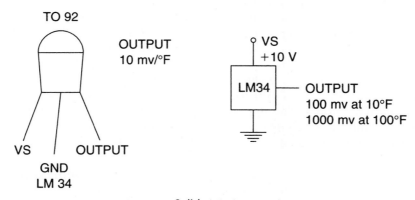

FIGURE 17–23 Solid-state temperature sensor.

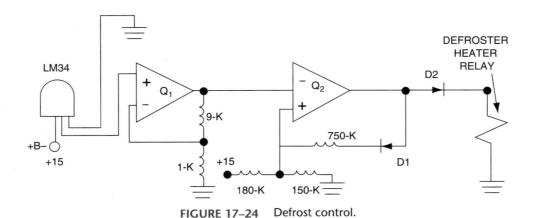

FIGURE 17–24 Defrost control.

In Figure 17–24, temperature sensor LM34 is shown connected to a two-amplifier circuit. The output of the second amplifier could be used to control a heater, turning it on and off by the temperature demand designed into the circuit.

If the temperature requirement control is to turn the heater off at 75°F and on at 68°F, a circuit as shown in Figure 17–24 will provide such control. The resistors associated with Q2 provide the on/off control of an icebox defrost heater.

The output of the LM34 is 0.01 volts per degree Fahrenheit, or 0.75 volts at 75°F. Amplifier Q1 provides a gain of 10 and 7.5 volts at the input of Q2 at 75°F.

Amplifier Q2 with a positive output has a feedback network that provides +7.5 volts at the junction of the feedback resistors and at the (+) input to the amplifier. When the amplifier output is negative, the effect of the output is removed from the feedback by the diode D1. The voltage at the (+) input of Q2 is reduced to +6.8 volts.

Defrost control in normal operation is provided as follows: After a fixed period of time, the defrost system is allowed to operate. Since the box had been cooling, the temperature was well below 75°F.

The +7.5 volts of the (+) input of Q2 provides a positive output and the defrost heaters turned on. The defrost heater causes the temperature to rise until the LM34 reaches 75°F. The output of the LM34 is now 0.75 volts. Amplifier Q1 provides a gain of 10 and an input of +7.5 volts at the (−) input of Q2. As this input rises to something just over +7.5 volts, the output switches to a negative voltage. The defrost heater relay is de-energized, and the heater shuts off.

The voltage at the (+) amplifier input is now +6.8 volts since the (+) voltage feedback through diode D1 and the 750-K resistor is cut off. The evaporator coil cools off. When the temperature reaches 68°F, (+6.8 volts at the [−] Q2 input), the output of

Q2 will switch to a positive voltage. The evaporator heater will switch on, causing the temperature to rise again.

COMPRESSOR MOTOR PROTECTORS

Compressor motors are designed to operate within a fixed temperature range. They are also designed to operate within a fixed current range. (See Figures 15–8 and 15–12 for examples of thermal overloads and their connection.)

Under special conditions, the thermal overload can become an active element in compressor failure. This is not to suggest that the overload should be removed. The compressor could burn out very rapidly without the overload. It is just that the overload is active in the operation.

Low Voltage (Brownout)

Low-voltage effect on compressors depends on how close to its rated load the compressor is operating.

Underloaded Compressor   Consider an underloaded compressor. This compressor is drawing less than full-load current during normal operation. As line voltage lowers, the current drawn by the compressor motor will decrease. The decreasing current flow with lowering voltage will continue until a point is reached at which required power is greater than available power. The current will then increase rapidly with lowering voltage.

Fully Loaded Compressor   With a fully loaded compressor, current flow will immediately start to increase with a lowering of line voltage.

Effect of Increasing Current During Brownout
Increasing current flow in the compressor causes increased heat. The increased heat, along with the increased current flow through the thermal overload, causes the overload to open.

After a period of time, the overload cools and the contacts close. The compressor tries to start. If the voltage is too low, the compressor cannot start. High current is drawn, and the overload again opens. This process may continue as follows:

1. Overload opens.
2. Overload cools.
3. Overload closes.
4. Compressor tries to start.
5. High current is drawn.
6. Overload opens.
7. Repeat 2 through 6.

If this process is allowed to continue, the compressor will finally burn out.

Low-Voltage–Anti-Short-Cycle Cutout

Recent developments include the low-voltage–short-cycle cutout. This device is designed to provide a delay in the application of power to the compressor after it becomes available to the system. The device also removes power from the compressor and keeps it from the compressor if the voltage supplied is 10% below the specified voltage. Most compressor motor specifications allow for 10% voltage decrease if a minimum operating voltage is given.

The circuit for connection of the low-voltage–anti-short-cycle device is given in Figure 17–25. A photograph of a low-voltage–anti-short-cycle cutout is shown in Figure 17–26.

There is a delay of approximately four minutes between the application of power to the device and

FIGURE 17–26   Low-voltage–anti-short-cycle cutout. (Courtesy of BET Inc.)

power to the compressor. If at any time the line voltage decreases below 105 V, power will be removed from the compressor. Power will not be returned to the compressor until four minutes after voltage above 105 V is supplied to the system.

LIGHT-EMITTING DIODES (LEDS)

All semiconductor diodes produce light (photons) when conducting. The light is not seen because the material silicon (Si) or germanium (Ge) is opaque. The photons are produced but do not escape the device.

Light-emitting diodes (LEDs) are constructed of materials that are translucent to light. The light can escape and therefore can be seen. One of the earliest LEDs available produced the color red. Shortly after the red LED came green LEDs. LEDs are now available in white, red, orange, yellow, green, and the latest addition, blue. Light-emitting diodes (LEDs) are used as indicating devices. They are often used to indicate the on-off condition. They are small, bright, and produce little heat. Figure 17–27 shows an example of a light-emitting diode.

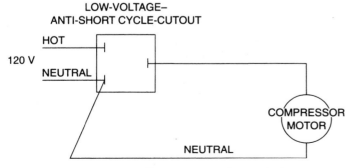

FIGURE 17–25   Compressor motor protector.

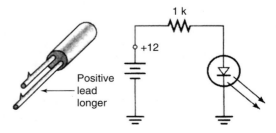

FIGURE 17–27   Light-emitting diode.

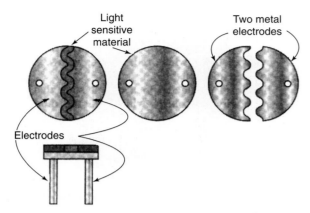

Light sensitive material

Two metal electrodes

Electrodes

FIGURE 17–28 Photoconductive cell.

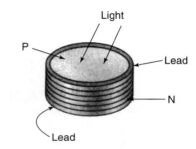

FIGURE 17–29 Photoconductive cell symbols.

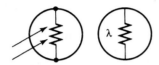

PHOTOCONDUCTIVE CELLS

Photoconductive cells or **photocells** are usually made using cadmium sulfide (CdS) or cadmium selenium (CdSe) as the cell's active material. The light-sensitive material is deposited on a glass surface. Two electrodes are fixed over the glass substrate and provide a controlled path for light to act on the active surface.

The resistance of a photocell is high in darkness and low when light strikes its surface. A photoconductive cell is shown in Figure 17–28. The symbols used for photoconductive cells are shown in Figure 17–29. The arrows in Figure 17–29 indicate the light that the device is sensitive to. The second symbol uses the Greek letter lambda (λ) to represent the wavelength of light.

Light

P

Lead

N

Lead

FIGURE 17–30 Photovoltaic cell.

PHOTOVOLTAIC CELLS

A photovoltaic cell (Figure 17–30) converts light energy directly into electrical energy. It is a junction device made from a P-type layer and an N-type layer. When light strikes one of the junction surfaces while the other is isolated from the light, a voltage is produced across the junction.

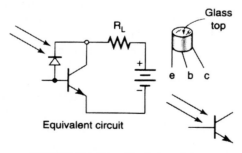

R_L

Glass top

e b c

Equivalent circuit

FIGURE 17–31 Phototransistor.

PHOTOTRANSISTORS

By incorporating a photovoltaic cell in the same body as a transistor, a light-sensitive transistor can be produced. The equivalent circuit is shown in Figure 17–31.

When light strikes the surface of the photodiode within the transistor, it produces a voltage that turns the transistor on. The greater the amount of light striking the phototransistor, the greater the collector current.

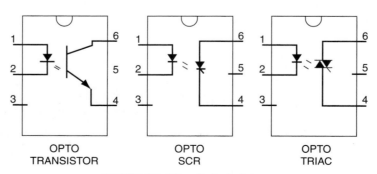

OPTO TRANSISTOR

OPTO SCR

OPTO TRIAC

FIGURE 17–32 Opto isolators.

TABLE 17–1   Thermistor Resistance

°C	Resistance
125	2 K
100	5 K
75	12 K
50	33 K
25	100 K
20	127 K
15	162 K
10	209 K
5	272 K
0	355 K
−10	469 K
−15	625 K

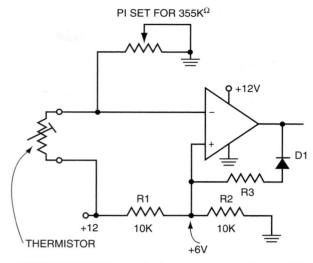

FIGURE 17–34   Electronic thermostat set to detect 0°C.

OPTO ISOLATORS

An opto isolator is a device used to isolate two voltage levels with light as the information-transmitting medium. They are usually 4- or 6-pin devices with an output of a transistor, silicon-controlled rectifier, or triac. The input is a light-emitting diode that, when activated, turns on the output device. Figure 17–32 shows the diagram of three devices.

When current is passed through the diode, the diode emits light and turns the output device on.

THERMISTORS

A thermistor is a temperature variable resistor. Thermistors are available with a positive or negative temperature coefficient, that is, the resistance increases with increased temperature (positive) or the resistance decreases with increased temperature (negative). An example of a useable thermistor has a temperature variation according to Table 17–1.

COMPARATOR SERVICE

An op amp may be used to compare two signals to determine which is the larger. In Figure 17–33, there is a fixed voltage of plus one volt connected to the noninverting input. Whenever the voltage applied to the inverting input is above plus one (1), the output will be zero volts. If the inverting input is below plus one (1) volt, the output will be positive.

ELECTRONIC THERMOSTAT

Using the information on thermistors and an op amp in comparator service. It is relatively simple to put together an electronic thermostat such as shown in Figure 17–34.

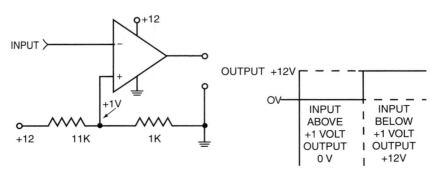

FIGURE 17–33   Comparator service.

The pot in Figure 17–34 is set for 355 K, which is the resistance of the thermistor at 0°C (Table 17–1). Also note that thermistor has a negative temperature coefficient (i.e., as temperature decreases, resistance increases).

With a temperature above 0°C, the thermistor's resistance increases to 355 K at 0°C. The voltage at the inverting input decreases to less than +6 volts, and the output goes to the maximum positive +12 V. The thermostat output will switch from +12 V to 0 V, depending on whether the temperature is below or above 0°C.

As temperature decreases, the thermistor's resistance increases to 355 K ohm at 0°C. The voltage at the inverting input decreases to less than +6 volts, and the output goes to the maximum positive, +12 volts. The thermostat output will switch to +12 volts to 0 volts, depending on whether the temperature is below or above 0°C.

Since P1 is adjustable, different temperatures may be selected for the switch-over point.

The feedback circuit of D1 and R3 provides for a selected switching range. If the thermometer switches output to +12 at 0°C, it will not switch back to 0 volts until +5°C or so, depending on the feedback.

PRACTICAL APPLICATION

Using the thermistor, opto isolator, comparator, and a triac, a system as shown in Figure 17–35 can be put together controlling a compressor motor.

In the compressor control system, Figure 17–35, the temperature selector P1 is adjusted to 625 K ohms or −10°C (14°F) or icebox temperature. Assume the temperature in the box is above −10°C. The compressor is running. The box is cooling down. The thermistor's resistance is less than 625 K ohms, so the voltage at the amplifier inverting input is greater than +6 volts. The amplifier output is zero (0) volts. With 0 volts at its base, Q1 is turned off. The light-emitting diode of X2 is turned on since there is a path for current from ground to the diode and R5 to +12 volts.

With X2 turned on, Q2 is conducting since there is a current path for gate drive through R6 and the triac in X2. When Q2 is conducting, the compressor motor is running. The box temperature is decreasing. When the box temperature reaches −10°C, the thermistor reaches 625 K ohms. The inverting input voltage decreases to +6 volts and X1 is ready to switch to +12 volts output. Q1 conducts at 12 ma dropping the full 12 volts across R3. There is nothing left for the diode of X1. It turns off, shutting off its output. Q2 shuts off stopping the compressor. Cooling stops.

Since the compressor is not running, the box heats up. When it reaches the temperature controlled by the feedback (R2 and D1) about −5°C, the output of X1 switches to zero volts, turning off Q1. X2 turns on along with Q2. The compressor starts and cooling resumes, keeping the box temperature between −10°C and −5°C.

The previous explanation is for information only. The technician would not be expected to trou-

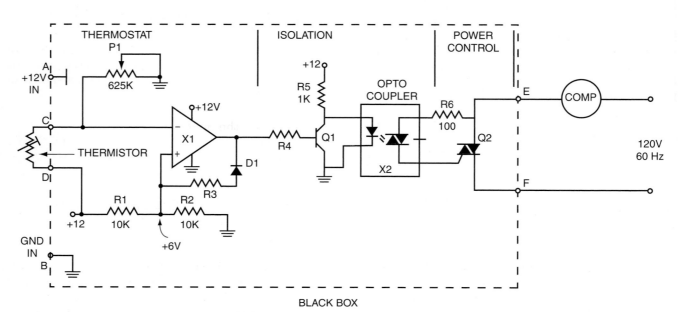

BLACK BOX

FIGURE 17–35 Compressor control system.

bleshoot inside the so-called black box (bb). In this system, if the box were not cooling, the technician should be able to troubleshoot the system by measuring the voltages at the black box terminals.

1. If the DC voltage is supplied to the black box, there should be +12 volts between A and B.

2. The voltage from terminal C to B is temperature dependent. If the box temperature is high, the voltage C to B should be above +6 volts.

3. If the compressor is not running, the voltage from E to F should be 120 volts AC.

In many cases, the black box is not supplied with DC voltage. It is supplied with 24 volts AC. This voltage is rectified and filtered inside the box providing the operating voltage for the electronics.

Do not worry about what is going on inside the black box. Your concerns are the input and output voltages.

SUMMARY

- There are many uses for semiconductors in control devices, and the popularity of semiconductors is increasing.

- The air-conditioning and refrigeration technician should be familiar with semiconductor devices and be able to determine whether they are operating properly or are malfunctioning.

- Whenever a control does not perform the required function, it should be checked for proper inputs. These inputs usually are a supply voltage and the control signals. If these two are present but the black box does not perform according to specifications, it should be replaced.

- Repair of semiconductor devices is seldom practical in the field.

PRACTICAL EXPERIENCE

Required equipment   Analog volt-ohm-milliammeter (VOM), NPN transistor such as 2N3904.

NOTE: Some analog meters provide positive voltage on the black lead in ohmmeter service. In this procedure the lead that is positive lead be connected to the collector of the transistor.

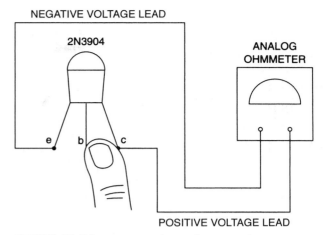

FIGURE 17–36   Finger used as collector, to base resistor.

Procedure

1. Select resistance on the multimeter and select the highest $R\times$ scale, probably $R\times$ 10,000.

2. Connect the black lead (negative) to the transistor emitter.

3. Connect the red lead (positive) to the transistor collector.

4. What does the meter read? ____ (See Note 1.)

5. Wet one of your fingertips.

6. Place your fingertip between the transistor collector and the base lead, as shown in Figure 17–36. (See Note 2.)

7. What does the meter read? ____ (See Note 2.)

8. In step 3, the transistor is turned (off/on).

9. In step 6, the transistor is turned (off/on).

10. Why, in your opinion, is there a difference (if any) in steps 3 and 6? (See Explanation.)

Notes

1. In step 3, the meter should indicate (high resistance), full scale, or close to full scale, since there is no emitter-base current to turn the transistor on.

2. In step 6, the meter should jump up the scale, indicating a lower resistance. The transistor is turned on.

Explanation   When the wet fingertip is placed between the collector and base, current flows through the skin from the positive (+) meter lead through the transistor base. When emitter-base current flows, the transistor turns on, lowering its resistance. The battery in the analog meter provides the necessary power to operate the transistor.

Conclusions

1. Semiconductors have resistance that falls between good conductors and good insulators.
2. "Doped" semiconductor materials are useful in the manufacture of diodes and transistors.
3. Diodes have low resistance in one direction and high resistance in the other direction.
4. In a transistor, a small base current controls a larger collector current.

REVIEW QUESTIONS

1. The two types of material used in semiconductors are cobalt and silicon.

    T_____   F_____

2. Diodes are used to change DC to AC.

    T_____   F_____

3. Diodes normally conduct in both directions.

    T_____   F_____

4. The output of a full-wave rectifier is easier to filter than the output of a half-rectifier.

    T_____   F_____

5. Transistors are two-terminal semiconductors.

    T_____   F_____

6. In a transistor, a small signal current in the collector controls a large signal current in the base.

    T_____   F_____

7. Transistors can be used to make amplifiers.

    T_____   F_____

8. Complete amplifiers can be formed as miniature integrated circuits.

    T_____   F_____

9. Silicon-control rectifiers allow for current flow in both directions.

    T_____   F_____

10. Triacs require a negative gate for turn-on.

    T_____   F_____

UNIT 18

FREEZER CIRCUITS

OBJECTIVES

After completion and review of this unit, you should be able to understand:

- Simple freezer circuit operation.
- Thermostat (cold) control of a freezer.
- Klixon operation.
- Current measurement with an amprobe.
- Defrost timer operation.
- Defrost termination control.
- Head pressure control of multiple fans.
- Modulating fan speed.
- Electronic fan speed control.
- Pumpdown circuits.
- Liquid level controls.

FREEZER CIRCUIT NO. 1

Figure 18–1 is an interconnection diagram for a freezer. It shows the simplest possible refrigeration circuit. It consists of one switch and one load. The thermostat is located inside the freezer, and it senses the freezer temperature. The compressor is located below the freezer. The wiring for the thermostat, the compressor, and the power supply (plug) all come together in a small box called a **junction box**

(sometimes referred to as a J-box). The detailed drawing shows how the wires are connected inside the junction box. Figure 18–2 shows the same wiring, but in a ladder diagram. When the freezer temperature rises above the set point of the thermostat, it will cause the compressor to run. It will continue to run until the thermostat senses that the freezer has cooled to a temperature below the thermostat set point. Then the thermostat will open its contacts, and the compressor will stop running. The thermostat in a small refrigerated box is sometimes called a **cold control**.

Notice the symbol that is used for the thermostat in the ladder diagram. The "squiggle" below the switch tells you that this is an automatic switch that opens and closes its contacts in response to a change in the *temperature* that it senses. Notice also the position of the switch mechanism. It is shown *below* the wire. This tells you the *action* of the thermostat. It closes on a rise in temperature. In other applications (e.g., heating), you will find thermostats that close on a fall in temperature. Figure 18–3 shows the two different symbols for thermostats. The principal of operation of this cold control is that a temperature change can cause the liquid inside a sensing element to expand or contract. Figure 18–4 shows a cold control with a sensing line. The expansion and contraction of this liquid causes changes in the hydraulic pressure inside the sensing line, and this change in pressure can be used to operate a switch using a bellows, as shown in Figure 18–5.

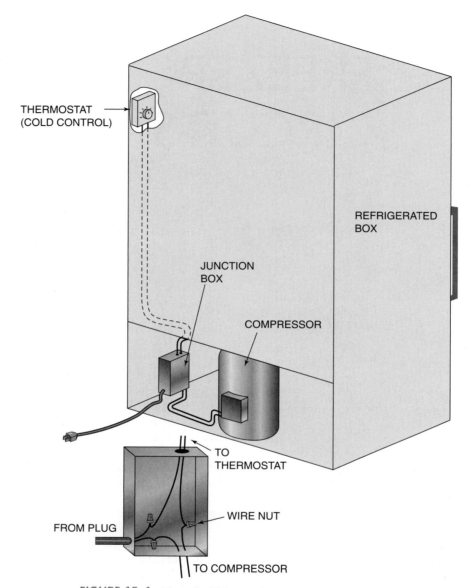

THERMOSTAT
(COLD CONTROL)

REFRIGERATED
BOX

JUNCTION
BOX

COMPRESSOR

TO
THERMOSTAT

WIRE NUT

FROM PLUG

TO COMPRESSOR

FIGURE 18–1   Household freezer interconnection diagram.

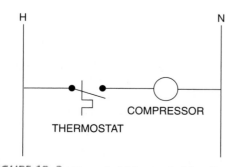

H   N

COMPRESSOR

THERMOSTAT

FIGURE 18–2   Household freezer ladder diagram.

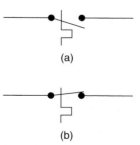

(a)

(b)

FIGURE 18–3   Thermostat symbols. (a) Cooling thermostat—closes on a rise in temperature. (b) Heating thermostat—opens on a rise in temperature.

FIGURE 18–4　Cold control with temperature sensing line.

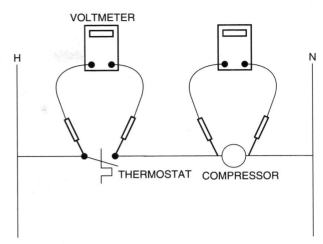

FIGURE 18–6　Troubleshooting freezer Circuit No. 1.

The temperature in the freezer is not maintained at the exact temperature corresponding to the set point of the thermostat. Instead, it is controlled over a *range* of temperatures. For example, even though the thermostat might be set for −10°F, the thermostat contacts might not close until the freezer temperature rises to −8°F and might not open until the refrigerator temperature drops to −12°F. In this example, we would say that the thermostat has a **differential** or **range** of 4°F. In other terminology,

we might say that the switch makes at −8°F, and breaks at −12°F.

Troubleshooting

The only possible customer complaint that could result from this simple circuit is that the compressor does not run. Use the voltmeter to measure voltage across the switch and across the compressor motor (Figure 18–6). There is no right or wrong place to start. You should measure whichever is easier to measure first. If you find that there are 115 V across the thermostat, then it is open and must be replaced. If you find that there are 115 V across the compressor, then it has failed. If you don't find 115 V across either, then you must measure the voltage across Hot and Neutral to determine that you probably do not have voltage available to the refrigerator.

FREEZER CIRCUIT NO. 2

In Figure 18–7 we have added another device called an **overload**. This type of overload is commonly called a "**Klixon**," although this is actually jargon

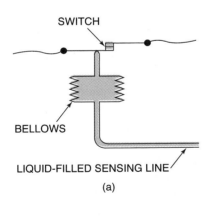

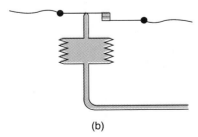

FIGURE 18–5　Switches operated by hydraulic pressure. (a) Closes on a rise in temperature. (b) Opens on a rise in temperature.

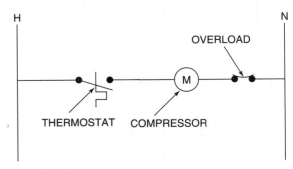

FIGURE 18–7　Freezer circuit with overload.

FIGURE 18–8 Klixon overload.

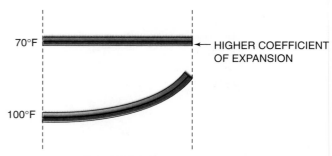

FIGURE 18–9 Bimetal element.

for this type of overload. Klixon is actually a registered brand name that has fallen into common use, like Freon, Kleenex, or Amprobe. A picture of this type of overload is shown in Figure 18–8.

An overload is a switch that senses the current passing through it and opens if the current exceeds a predetermined rating. In this case, the overload is carrying the same current as the compressor. It must be selected so that if it carries a current that exceeds the maximum allowable current through the compressor, it will open, thereby preventing the compressor motor from burning out. Two observations are in order about the overload:

1. It must be matched to the load that it is protecting. Different compressors can be rated for a wide range of maximum allowable currents. Therefore, you cannot arbitrarily replace an overload with a different one. If the replacement you select is rated for a too-low amperage, it will stop the compressor from operating when it is still within its acceptable operating amperage range. If the replacement you select is rated for a too-high amperage, it may allow the compressor to operate at an amperage draw that exceeds its maximum allowable rating.

2. The overload is a **safety control**, as compared to the thermostat, which is an **operating control**. An operating control is a switch that is expected to open and close regularly, to turn the system on and off as required by the area in which temperature is being controlled. However, a safety control is one that will never open unless a malfunction occurs.

The overload relies on two different principles to make it work:

1. The more current that flows through a wire, the more heat will be generated.

2. Different metals expand at different rates when heated. If two metals with different coefficients of expansion are joined together,

it is called a **bimetal**. When a bimetal element is heated, the differential expansion causes the bimetal element to bend. A bimetal element is shown in Figure 18–9.

Figure 18–10 shows the internal construction of the Klixon overload. When it is carrying less than its rated amps, the heater wire produces very little heat. But when the current exceeds the rated amp draw, the heat from the heater wire will be sufficient to cause the bimetal element to warp, opening the switch, and de-energizing the compressor. When the heater wire cools, the bimetal will warp back to its original closed position. Each time the overload opens or closes, there will be an audible "pop," similar to the noise made when you push on the bottom of an oil can.

Troubleshooting

If the compressor is not running, either the thermostat or the overload could be open, or the compressor

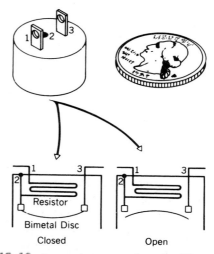

FIGURE 18–10 Internal construction of a Klixon overload.

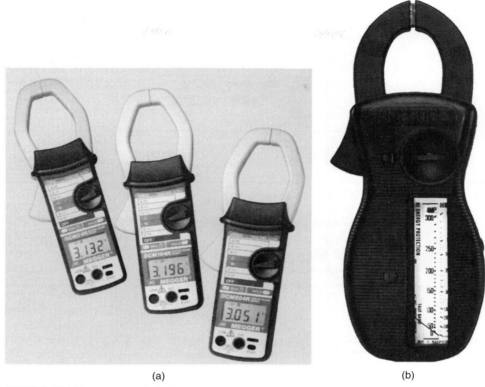

(a)   (b)

FIGURE 18–11 Amp meters. (a) Digital amp meter. (Courtesy of AVO International, manufacturer of Biddle®, Megger®, and Multi-Amp®, products.) (b) Analog amp meter. (Courtesy Amprobe)

may have failed (we will assume that power is available). Using your voltmeter, measure across each of the three components. Whichever has voltage across it is the one that has failed.

If the compressor turns on, runs for a few seconds, turns off, and then repeats the cycle a few minutes later, it is most likely **cycling** on the overload. You will probably hear the overload open and close just before the compressor turns off and on. A reading of 115 V across the overload when the compressor is off, and then a 0 V reading across the overload when the compressor is on will confirm it. In this event, it may be *either* the overload or the compressor that is at fault. There may be something wrong with the compressor that is causing it to draw too many amps, and the overload is working properly. Or, the compressor may be fine, and the overload is opening at a too-low amperage. In order to determine which is the problem, you must measure the amperage when the compressor turns on, and compare it to the amperage rating on the nameplate of the compressor. If your amperage reading exceeds the nameplate rating, the compressor is the problem. However, if the measured amperage is lower than the nameplate

rating, then you must replace the overload. An amp reading may be taken by using an amp meter, shown in Figure 18–11. Many modern VOMs have an accessory amp probe (Figure 18–12) available that reads out amps on the VOM. To use it, you simply clamp the jaws of the meter around the wire whose current you wish to measure and plug the probe wires into the appropriate jacks on the VOM.

FIGURE 18–12 Accessory amp probe. (Courtesy AEMC Instruments.)

ADVANCED CONCEPTS

A clamp-on amp meter works on the principal of sensing the amount of magnetic field that is generated around a current-carrying wire. On a 120-V single-phase circuit, if you clamp around two wires at the same time that are carrying current in opposite directions, their magnetic fields will oppose each other. If the currents are at the same level, the resulting magnetic field will be zero, and the meter will indicate zero amps. See Figure 18–13.

ADVANCED CONCEPTS

If you wind a wire around the jaws of the amp meter, the magnetic fields will add, and the amp reading will equal the actual amps times the number of turns passing through the jaws (Figure 18–14). This is particularly useful when trying to read very low amperages that won't accurately register on the lowest amp scale.

The accessory amp probe has its own batteries and must be turned on when in use. Technicians commonly forget to turn the accessory amp probe back off, leaving them with a dead instrument the next time it is needed. Some accessory amp probes have a "sleep" mode. The probe will automatically turn itself off when not in use for a few minutes. This is an option well worth looking for when you buy this instrument.

This circuit is an example of switches in series. Some circuits have several safety controls wired in series with the load. If any one of the safety controls detects a problem with current draw, temperature, or pressure, that particular switch will open and de-energize the load.

FREEZER CIRCUIT NO. 3

In Figure 18–15, we have added a condenser fan motor in parallel with the compressor. When the thermostat closes, it will energize both loads in parallel. Where there are multiple loads in a unit, they will almost always be wired in parallel, so that each load "sees" the full-circuit voltage.

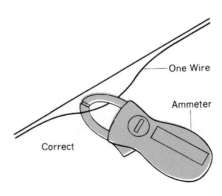

One Wire

Ammeter

Correct

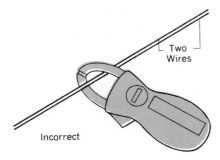

Two Wires

Incorrect

FIGURE 18–13 Using an amp meter.

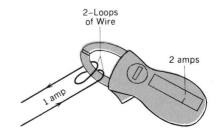

2–Loops of Wire

2 amps

1 amp

FIGURE 18–14 Multiplying the amp reading.

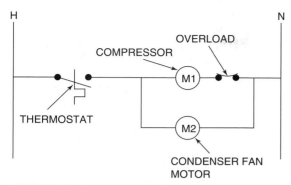

FIGURE 18–15   Freezer with condenser fan motor.

Note that the condenser fan is wired in parallel with both the compressor and overload. In this way, the overload only carries the compressor amps, but not the condenser fan motor amps. If mistakenly wired so that the overload carries the current from both loads, nuisance trips on the overload will result.

Troubleshooting

1. If neither the compressor or the condenser fan is operating, the problem is almost surely a faulty thermostat (assuming that there is power available). Although theoretically the thermostat might be closed, and both the compressor and condenser motors might be burned out, it is so unlikely that we would not investigate that possibility first. Also, note that if the overload was open, it would prevent the compressor from running, but it would not prevent the condenser fan from running. Always direct your initial investigation to a device whose failure could cause all of the symptoms you observe. In the great majority of troubleshooting service calls, you will find that there is only one device that has failed. Go to the thermostat first. If it measures 115 V across it, it is open.

2. If the compressor is running but the condenser fan is not, the problem must be a failed condenser fan motor. Although an open thermostat can prevent the condenser fan from running, that is not the case in this situation. If the thermostat was open, the compressor could not run. But since the compressor *is* running, the thermostat *must* be closed.

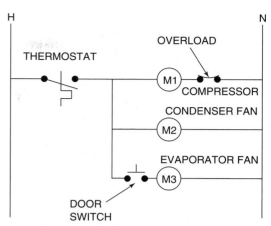

FIGURE 18–16   Freezer with two fan motors.

3. If the condenser fan is running but the compressor is not, the problem must be that either the compressor has failed or the overload is open. Note that an open overload will shut down only the compressor. Measure voltage across the compressor and the overload. Whichever has voltage across it has failed.

FREEZER CIRCUIT NO. 4

In Figure 18–16, we have added two more devices, an evaporator fan and another switch. The switch symbol tells us that it is a **momentary switch** (Figure 18–17). A momentary switch may have contacts that are **normally open** or **normally closed**. When the movable part of the switch (the push button) is pressed, the switch will operate. A normally open switch will close, and a normally closed switch will open. When the movable part of the switch is released, the switch contacts return to their normal position. The symbols for a normally open and a normally closed momentary switch are shown in Figure 18–18. The momentary switch for the evaporator fan is normally open. When the freezer door closes, it presses on the momentary switch, causing it to close and allowing the evaporator fan to run

FIGURE 18–17   Momentary switch.

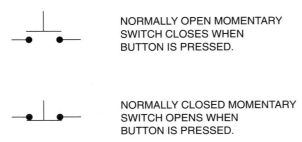

FIGURE 18–18   Symbols for momentary switches.

whenever the cold control closes. A momentary switch that is operated by a door is commonly referred to as a **door switch**.

FREEZER CIRCUIT NO. 5

In this circuit, we have added a cabinet light and another momentary switch. Note that this momentary switch is normally closed. When the door is closed, it will press on the switch, opening its contacts, and turning the light off. Some refrigeration systems have two different switches, each operated by the freezer door. However, in this diagram, note that the two momentary switches are connected by a dotted line. This means that they are ganged together and operate simultaneously. The momentary switch for this circuit looks just like the one in Figure 18–19, but instead of two terminals, it has four. One pair of terminals is the normally open fan switch. The other pair of terminals is the normally closed cabinet light switch.

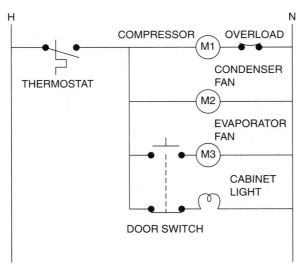

FIGURE 18–19   Freezer Circuit No. 5.

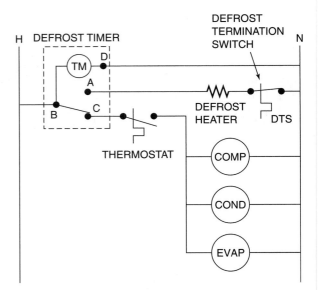

FIGURE 18–20   Freezer with automatic defrost.

FREEZER CIRCUIT NO. 6

A freezer operates with a coil temperature of well below freezing. Moisture that condenses out of the air onto the coil will freeze. If the evaporator is of a finned-tube design, the spaces between the fins will be completely blocked within a few hours unless the coil is periodically defrosted. Figure 18–20 shows the wiring diagram with an automatic defrost system. This system consists of three devices:

1. A defrost timer (Figure 18–21). It has a time clock inside that moves gears and cams that operate an SPDT switch. The switch is in one position for approximately 7½ hours and in the other position for approximately ½ hour. There is a knob that may be rotated manually to advance the timer, causing the switches to operate.

2. A defrost heater. This may look like the heater in an electric oven. It may be a quartz glass rod with an internal heater. It may be an insulated heating element covered with aluminum foil See Figure 18–22. Defrost heaters come in many shapes and sizes.

3. A defrost termination switch (Figure 18–23). It is a thermostat that is fastened to the evaporator coil. It opens when the evaporator is above a temperature of 45°F.

Figure 18–24 shows the arrangement of the electric defrost heaters and defrost termination

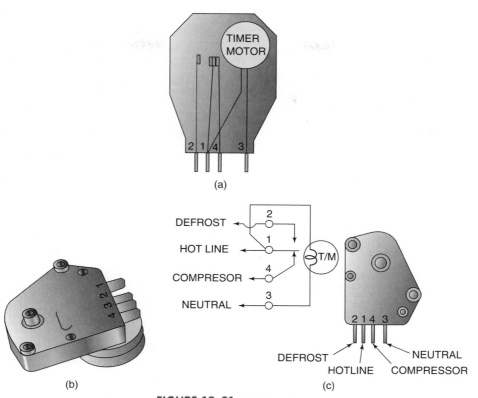

FIGURE 18–21 Defrost timer.

switch mounted on an evaporator coil of the type that would be found in a walk-in freezer.

When the switch is in the 7½-hour position (the "freeze" position), the operation of the compressor, condenser, and evaporator is normal. The timer motor is also energized, turning the gears inside the defrost timer. After the timer motor runs for 7½ hours, the switch moves to the other position (the "defrost" position). It will stay there for 1/2 hour. During that time, the compressor, condenser, and evaporator are de-energized. The heater coil is ener-

gized, and the ice will melt off the evaporator. [NOTE: the 7½-hour and 1/2-hour times are just for example. Timers may operate on a six-, eight-, ten-, or twelve-hour frequency, and the defrost time is also variable, depending on the manufacturer.] When all the ice has melted, the evaporator coil temperature will start to rise above 32°F. When it reaches 45°F, the defrost termination thermostat opens, de-energizing the heater. This prevents the box from overheating when "defrost" has been accomplished, but the timer is still in the "defrost"

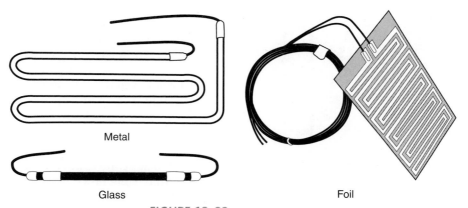

FIGURE 18–22 Defrost heater.

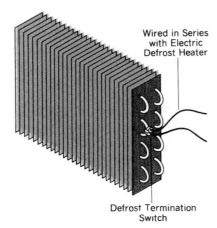

FIGURE 18–23 Defrost termination switch.

position. After the defrost termination thermostat opens, only the timer motor is energized. Neither the heater nor the refrigeration circuit is energized until the defrost timer motor switch returns to the "freeze" position.

Troubleshooting—Box Is at Room Temperature, Compressor Not Running

Check to see that there is power to the unit (the cabinet light comes on). Rotate the defrost time clock

FIGURE 18–24 Defrost heaters. (Courtesy of Witt Division of Ardco.)

manually. If the compressor comes on, it means that the defrost timer was not switching automatically. Measure voltage at the power input terminals of the timer (see following practice pointer) to make sure that 115 V is available to run the timer motor. Then replace the defrost time clock.

☞ The terminal arrangement of most 4-wire defrost timers is the same (Figure 18–21a). There are three terminals together and one by itself. Power comes in the center terminal in the group of three (terminal 1). From there it goes into the timer motor and into the SPDT switch. The current that leaves the timer exits at the lonesome terminal (terminal D), which is connected to the other side of the circuit. The current that enters the SPDT switch is switched out to either the "freeze" circuit or the "defrost" circuit. Usually (but not always), the switched terminal closest to the edge of the timer (terminal 2) goes to the defrost circuit, and the switched terminal closest to the lonesome terminal (terminal 4) goes to the compressor circuit. The numbering of the terminals may differ from timer to timer, but the arrangement of terminals is pretty universal.

Troubleshooting—Compressor Running, Evaporator Coil Airflow Is Blocked with Ice

Manually advance the defrost timer until it goes into "defrost."

☞ You can tell when the timer is in the "defrost" mode by listening as you rotate the manual advance knob. You will be able to rotate the manual advance several turns before you hear the first click. Then, a slight additional manual advance will produce a second click. After the first click, the clock is in the "defrost" mode. After the second click, the clock is in "freeze" mode.

Check to see if the evaporator is defrosting. This can be done three ways:

1. Listen. You will be able to hear a "sizzle" as the heater melts the ice.
2. Look at the drain pan that collects the water that runs down a hose from the evaporator. If water is coming out, the defrost is working.

3. Measure the amps through the defrost heater. If amps are flowing, the heater is working.

If the "defrost" is taking place, it means that the defrost heater and defrost termination thermostat are both okay. Change the defrost timer.

☞ There is a potential trap in the preceding discussion. If the evaporator is concealed, you can be fooled into thinking that the heater is alright because you read amps. Sometimes, several heater elements are wired in parallel. There may be one or more heater elements burned out, but as long as there is at least one functioning heater element left, you will read amps on the defrost cycle. The only way to discover this trap is to inspect the coil as it defrosts to make sure that all of the coil is defrosting.

If, after placing the timer in the "defrost" position, no defrost occurs, there are two likely problems:

1. The defrost timer switch between terminals 2 and 1 is not making contact.

2. The defrost timer switch *is* making contact, but either the heater or the defrost termination switch is open.

NOTE: For the following troubleshooting sequence, it is assumed that Hot is connected to terminal 1 and that Neutral is connected to terminal 3. Even if this is opposite from the actual wiring, the same voltage readings described following would be applicable.

Check voltage between terminals 1 and 3. It should be 115 V. This means that electrical pressure is coming into the SPDT switch at terminal 1. Then check for electrical pressure at terminals 2 and 4, by measuring 2–3 and 4–3. There should be electrical pressure available at the terminal going to the defrost circuit. A 115-V reading between 2 and 3 would indicate that there is electrical pressure available at terminal 2. If 2–3 and 4–3 both read zero volts, then the SPDT switch is not making contact in the defrost position, and the defrost timer must be replaced.

If you get a 115-V reading between 2 and 3, that means that voltage is being supplied to the heater circuit. But we know that the heater circuit is not working. So, either the defrost heater has failed or the defrost termination switch is open. Some disassembly of the unit will be necessary to gain access to the evaporator. Then, measure voltage across the heater

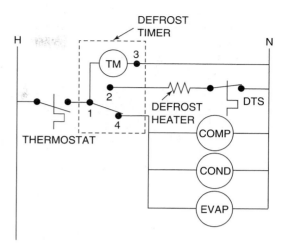

FIGURE 18–25 Freezer Circuit No. 7.

and the defrost termination thermostat. Whichever one shows 115 V has failed.

FREEZER CIRCUIT NO. 7

This circuit (Figure 18–25) is the same as Freezer Circuit No. 6, except that the cold control is in a different place. In this circuit, the timer motor only runs when the cold control is energized. There is an advantage to doing it this way. Instead of going into "defrost" every eight hours, this system will go into "defrost" more frequently during periods of heavy usage and less frequently during periods of light usage. Because this time clock doesn't run all the time, it would probably use a timer that switches the SPDT switch every four or six hours. This would result in a "defrost" taking place probably every eight to ten hours, because it would take that long for the timer motor to get six hours of run time.

The disadvantage to this system is that it will "defrost" at different times each day, compared to regular eight- or twelve-hour intervals, which would start a "defrost" at the same times each day. This could possibly result in a nuisance service call from an owner who notices the refrigeration system not running when the box is above 0°F.

FREEZER CIRCUIT NO. 8

This circuit (Figure 18–26) is the same as Freezer Circuit No. 6, except that the evaporator fan has been wired differently. When Freezer Circuit No. 6 returns to the "freeze" cycle, the evaporator coil is warm (45°F), and the box is also warmer than normal. At

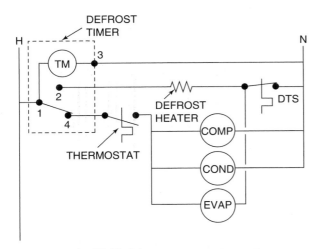

FIGURE 18–26 Freezer Circuit No. 8.

the beginning of the freeze cycle, the load on the compressor is abnormally high because the evaporator coil and the box are both warm. This could cause the compressor to overload. Freezer Circuit No. 8 addresses this problem by providing a time delay for the evaporator fan. When the defrost timer returns to the "freeze" position, the compressor and condenser fan motors are energized. But the evaporator fan motor does not start because the defrost termination switch is still open. When the compressor and condenser fan run without the evaporator fan, the load on the compressor is reduced. The evaporator temperature (and suction pressure) will drop to a more-normal value relatively quickly. When the defrost termination switch senses that the evaporator coil is cold, it will close, turning on the evaporator fan.

👉 Some technicians who are not familiar with the evaporator fan delay circuit have been fooled. A technician may arrive at a walk-in freezer to find that nothing is running, the box is at room temperature, and the product has been removed for storage elsewhere. After discovering the problem and repairing it, the technician starts up the system. While checking the operation the technician notes that the evaporator fan doesn't run. Rather than tearing into the wiring again, leave it alone for a few minutes. The evaporator fan will probably restart all by itself!

👉 Some older refrigeration systems use a five-wire defrost timer to provide an evaporator time-delay. The extra terminal is wired to the evaporator fan motor only. There is an extra cam inside the defrost timer that allows this extra terminal to become energized a few

minutes after the compressor is energized. If you cannot locate a replacement five-wire defrost timer, use a four-wire timer. The wire going to the evaporator fan motor may be placed on the same terminal as the wire going to the compressor.

FREEZER CIRCUIT NO. 9

This circuit (Figure 18–27) is similar to Freezer Circuit No. 6, except that the defrost timer has *three* terminals instead of four. Note how the internal wiring of the timer motor has changed. The timer motor is now wired between terminals 2 and 4, and terminal 3 is missing.

When the SPDT switch is made between 1 and 4, the refrigeration components run, as in the previous circuits. But the circuit through the timer motor is different. Starting from Hot, current travels to terminal 1, then terminal 4, then to the timer motor, then to terminal 2, then to the defrost heater, then to Neutral. Even though the same amps flow through the timer motor and the heater, the timer motor works and the heater does not. Recall that when loads are wired in series, each takes a portion of the total available voltage drop. The voltage drop across

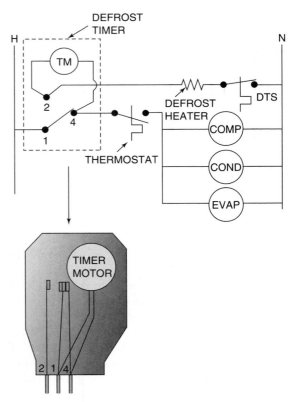

FIGURE 18–27 Freezer Circuit No. 9.

each is proportional to their resistance. The resistance of the timer motor is *much* higher than the resistance of the heater. The timer motor takes maybe 118 V, whereas there is only a 2-V drop across the heater. The current is sufficient to run the timer motor normally (a few tenths of an amp), but it is insufficient to cause the heater to produce any heat. The heater simply acts as a wire, connecting the downstream side of the timer motor to Neutral.

After a while, the timer motor causes the SPDT switch to switch to the 1–2 position. Then, the circuit through the timer motor is from Hot to terminal 1, then terminal 2, then to the timer motor, then to terminal 4, then to the compressor-condenser-evaporator in parallel, then to Neutral. Again, the compressor-condenser-evaporator combination presents very low resistance compared to the timer motor. The timer motor operates, and the refrigeration system, is off. Notice that the timer motor does not care which way the current flows. It always rotates in the same direction, just like any other clock motor.

ELECTRONIC DEFROST TIMER

The defrost timers used in residential and small commercial freezers may all look about the same, but there can be major differences in the internal gears that provide wide variations in the frequency and duration of the "defrost." For example, a system in which the defrost time clock operates only when the compressor operates might use a four-hour frequency, compared with a continuously running time clock defrosting every eight hours. If the compressor runs 50% of the time, each system will "defrost" three times per day. In order to be able to match all of the different frequencies and durations of defrost timers that are used, the service technician would need to carry quite an assortment of defrost timers.

Figure 18–28 shows a new-design replacement defrost timer. Instead of the timing being determined by the internal gears, it is electronic in operation. Two adjustment knobs are provided. One allows setting the defrost duration from 10 to 35 minutes. The other allows setting of the defrost frequency for 4, 6, 8, 10, or 12 hours. With these adjustments, the technician can use this one timer to simulate the operation of any of the common timers.

HEAD PRESSURE CONTROL

Most refrigeration systems and some air-conditioning systems must operate during the winter. If they are air-cooled systems, and they are using outside air as

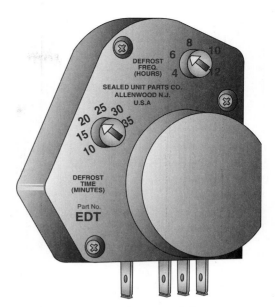

FIGURE 18–28　Electronic defrost timer.

the condensing medium, operational problems can result when the outdoor temperature drops too low. The low temperature increases the capacity of the condenser dramatically, and it lowers the head pressure. If the head pressure drops too low, there is insufficient pressure differential between the high-pressure and low-pressure sides to push sufficient refrigerant through the metering device. **Head pressure control** is any method that is used to reduce the condensing capacity so that the head pressure does not drop too low. One method of head pressure control is called **fan cycling**. Figure 18–29 shows how a pressure switch that senses high side pressure can be used to turn the condenser fan off if the head pressure drops too low. In Figure 18–29a, the condenser fan will not turn on until the head pressure builds to 250 psi. As soon as the condenser fan starts, the head pressure may drop to below 250 psi. In order to prevent the condenser fan from cycling on and off, the cut-out setting is well below the cut-in setting. In Figure 18–29a, once the condenser fan starts, it will not turn off again until the head pressure drops below 210 psi.

 When replacing a head pressure switch, here's how to set the cut-in and cut-out.

1. Estimate the maximum ambient temperature around the condenser coil.
2. Add 30°F to that temperature, and convert the result to the corresponding pressure for the refrigerant in your system. Set the cut-in to that pressure.

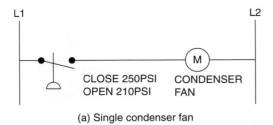

(a) Single condenser fan

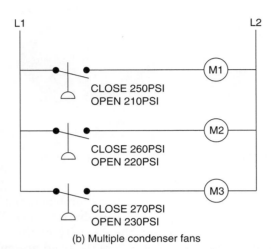

(b) Multiple condenser fans

FIGURE 18–29   Condenser fan cycling. (a) Single condenser fan. (b) Multiple condenser fans.

3. Determine the lowest acceptable condensing temperature from the manufacturer's data, and convert that to a pressure. Set the cut-out to that pressure. If no manufacturer's data is available, set the cut-out to a pressure that corresponds with 65°F.

For multiple condenser fans as shown in Figure 18–29b, the pressure settings are slightly different for each fan motor. When the head pressure rises to 250 psi, M1 is energized. If that is sufficient to keep the head pressure from rising to 260 psi, then M1 will be the only condenser fan motor to run. If the ambient temperature later rises, the head pressure will also rise, bringing on M2 when the head pressure reaches 260 psi, and M3 when (and if) the head pressure reaches 270 psi. When the ambient temperature drops, the head pressure will also fall, turning off the condenser fan motors in the reverse order from which they turned on.

Electronic Head Pressure Control

Head pressure control can also be accomplished by modulating the speed of the condenser fan motor instead of motor cycling. Some electronic fan speed controllers are a combination of a pressure sensor

and electronics. The unit is mounted to the top of the receiver or liquid line after the condenser, where it will sense the high side pressure. Most will mount on a standard 1/4″ SAE flare connector or Shrader valve. As the sensed pressure drops below the set point of the controller, the electronics wired in series with the condenser fan motor will reduce the motor speed. Generally these speed controllers are limited to use with permanent split capacitor and shaded pole fan motors (the different types of motors are described in unit 14 and 15). Figure 18–30a shows how this type of electronic speed control is wired into the circuit. The operation of the electronic head pressure controller is shown in Figure 18–31b.

Figure 18–30 shows another electronic head pressure control. Instead of sensing pressure, it uses a **thermistor** probe between terminals 9 and 10 to sense the liquid line temperature (which rises and falls to correspond with head pressure). A thermistor is simply an electronic device whose resistance varies with the sensed temperature. When 24 V is applied to terminals 5–6, the controller will apply full voltage to the fan motor. After the motor starts (a few seconds), the output from the controller will be determined by the temperature sensed by the thermistor. As the temperature being

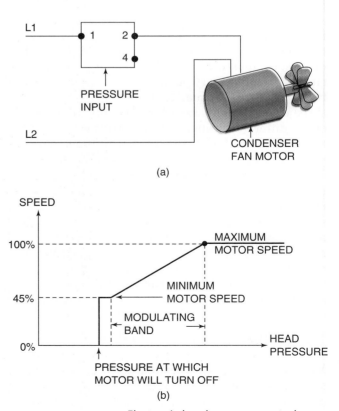

FIGURE 18–30   Electronic head pressure control.

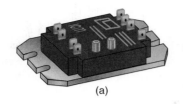

(a)

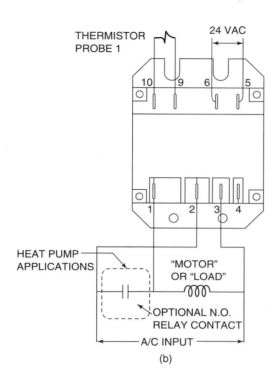

(b)

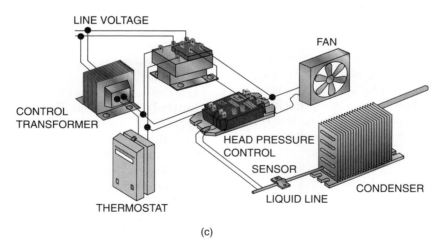

(c)

FIGURE 18–31   Electronic head pressure control.

sensed decreases, the output voltage decreases. A low temperature cut-off setting is provided to allow the technician to set the minimum condenser fan rpm. If the liquid line temperature falls below that point, the condenser fan will cycle off. There are two adjustments on the controller. One sets the initial full-voltage output time. The other sets the cut-out temperature. The physical wiring of another solid-state speed controller is shown in Figure 18–31. Instead of sensing head pressure directly, it uses a thermistor to sense the temperature of the liquid line.

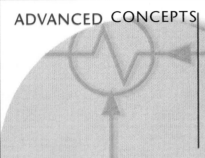

ADVANCED CONCEPTS

The output of the speed controller in Figure 18–31 is not really a reduced output voltage. If the voltage going to the motor were really lower than the rated voltage of the motor, the motor would fail. Actually, the output from this controller is not a reduced voltage, but part of each cycle of output voltage is cut off (see Figure 18–32). This has the effect of reducing the motor speed, and if you measured the voltage output of this controller at part load with a standard volt meter, it would appear to be less than line voltage. But by "chopping off" a portion of each time cycle, overheating of the motor is not a problem.

Troubleshooting

If the condenser fan doesn't run, either the condenser fan has failed or it is not receiving voltage from the speed controller. If correct voltage is available from terminals 1–3, then the motor has failed. If correct voltage is not present from terminals 1–3, it could be because the controller has failed or the thermistor has failed, making the controller think that it is so cold outside that the condenser fan should be cycled off. The thermistor used on this type of controller is a negative temperature coefficient (as sensed temperature increases resistance decreases). Short the thermistor terminals (9–10), simulating a very hot outside air temperature. If the output voltage becomes the same as line voltage, the thermistor has failed.

RELAY/CONTACTOR CIRCUIT NO. 1

Figures 18–33 and 18–34 show different types of relays and contactors. They work on the same principle of operation, but the contactor is used in circuits for larger loads. A schematic is shown in Figure 18–35. The relay in Figure 18–33c is called a plug-in relay. It plugs into a base, and the wiring is connected to the base instead of directly to the relay. You can determine which terminals are which by

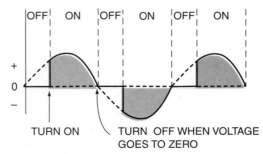

FIGURE 18–32 Output signal from electronic speed control.

looking inside the clear case and following the wires from each switch and the coil until they emerge at the base of the relay. The terminal arrangement shown is popular, but not universal. The principal of operation of a relay or contactor is as follows:

1. A voltage is applied to a coil of wire (the coil), creating a magnetic field.
2. The magnetic field draws a metal bar into the magnetic field, causing a switch (or multiple switches) to operate. The number of switches that operate are called **poles**. Thus, you can have a single-pole, two-pole, or three-pole control relay or contactor.
3. When the metal bar is drawn into the magnetic field, it may cause the switch to close (normally open) or it may cause the switch to open (normally closed). Contactor switches are always normally open.

A circuit using a single-pole contactor to start a motor is shown in Figure 18–36. During the "off" cycle, as the box temperature rises, the corresponding pressure being sensed by the **LPC (Low Pressure Control)** also rises. When the cut-in setting of the LPC is reached, the switch closes, energizing the coil in the compressor contactor. The magnetic field produced by the coil closes the switch in the compressor circuit. There are two aspects of this device that sometimes cause confusion for the student.

1. The circuit through the LPC and the coil is *completely separate* from the circuit through the switch contacts and the compressor. The current flow through the LPC and coil is very low, whereas the current flow through the switch contacts and compressor may be as high as the amp rating of the switch (20, 30, 40 amps or more). There is *no electrical connection* between the coil of a contactor

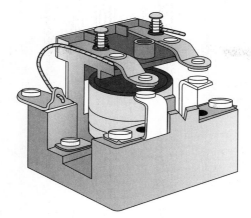

(a) RELAY-OPEN CONTACTS

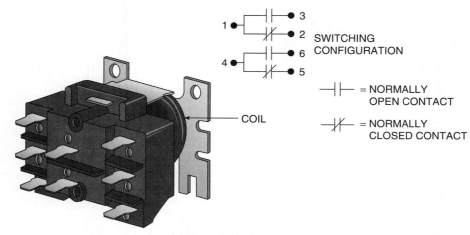

(b) RELAY-ENCLOSED CONTACTS

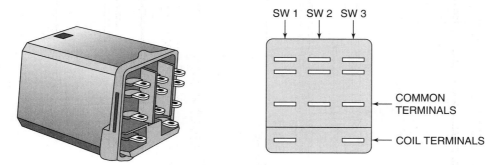

(c) ENCLOSED RELAY

FIGURE 18–33   Control relays.

and the switch that is operated from its magnetic field. Later in the text, you will see contactors where the coil operates in a different voltage circuit than the switch.

2. Even though the ladder wiring diagram shows the switch contacts in a different place than the coil, both are actually located in the same physical device. Note that both the coil and the switch are labeled CC. In circuits with several control relays or contactors, you will be able to determine which switches are operated by which coils, because each switch will be identified the same as the coil that operates it.

Single Pole
(24 & 30 amps)

Single Pole with Bus Bar
(25 & 30 amps)

Two Pole
(20, 25, 30 amps)

Three Pole
(25, 30, 40 amps)

Three Pole
(50, 50, 75 amps)

FIGURE 18–34   Contactors. (Courtesy Joslyn Clark Controls, Inc.)

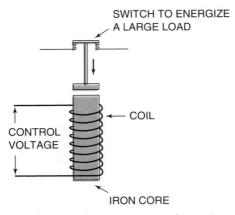

SWITCH TO ENERGIZE
A LARGE LOAD

COIL

CONTROL
VOLTAGE

IRON CORE

FIGURE 18–35   Contactor schematic.

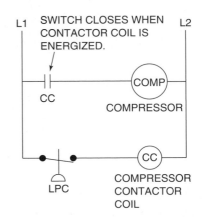

L1   SWITCH CLOSES WHEN   L2
      CONTACTOR COIL IS
      ENERGIZED.

CC

COMP

COMPRESSOR

LPC

CC

COMPRESSOR
CONTACTOR
COIL

FIGURE 18–36   Single-pole contactor circuit.

Note that the LPC controls the operation of the compressor, but only indirectly. The contactor is being used as an intermediary device, so that the LPC does not have to carry the full current draw of the compressor. When a relatively light-duty switch is used to control the operation of a much larger device in the manner shown, the switch is called a **pilot-duty** device. Note that for a smaller compressor that draws far less current, the very same LPC could be used to control the compressor directly. Then, the very same LPC would be called a **line-duty** device.

CONTACTOR CIRCUIT NO. 2

Contactor Circuit No. 1 used a coil to operate a single switch. In 230-V single-phase applications, even though breaking one of the legs is sufficient to stop the operation of the load, many safety codes require breaking both legs (L1 and L2). Figure 18–37 shows how an unsuspecting technician can get shocked from compressor wiring, even if the contactor has been de-energized.

A compressor motor being controlled by a two-pole contactor is shown in Figure 18–38a. Figure 18–38b shows the exact same system, but in a pictorial wiring diagram. It sometimes confuses students because the two switches, which are actually right next to each other, are shown in different places on the ladder diagram. But remember, the ladder diagram makes no attempt to show physical locations of devices. The common designation for terminals on a contactor are L1, L2, and L3 for the power coming into the switches and T1, T2, and T3 for the power going out to the motor.

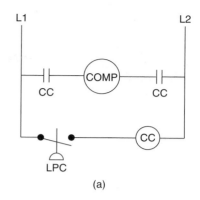

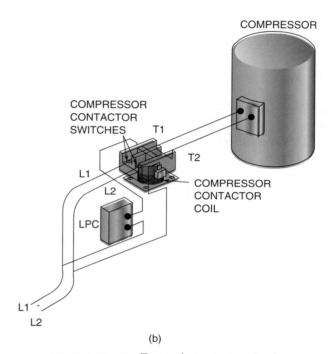

(b)

FIGURE 18–38 Two-pole contactor circuit.

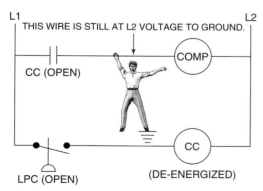

FIGURE 18–37 Unsafe condition caused by a single-pole contactor.

CONTACTOR CIRCUIT NO. 3

Figure 18–39 shows a wiring schematic for a large three-phase cooling system. Note that the contactor has three poles. Some electrical codes permit the use of a two-pole contactor to stop a three-phase load, but using a three-pole contactor is safer for the same reasons that the two-pole contactor is safer on a single-phase 230-V system.

The compressor has a 230-V three-phase motor, but the condenser fan motors are 230-V single-phase. The condenser fan motor and the control circuit each take their power from two of the three legs of the three-phase power supply. The contactor would be wired as in Figure 18–40. Most contactors

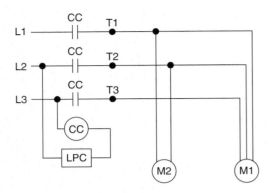

M1 = COMPRESSOR MOTOR
M2 = CONDENSER FAN MOTOR
CC = COMPRESSOR CONTACTOR
LPC = LOW PRESSURE CUT-OUT

FIGURE 18–39   Three-pole contactor circuit.

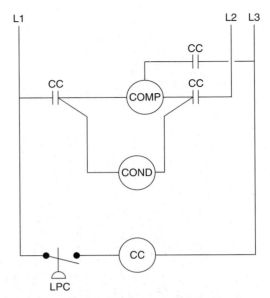

FIGURE 18–41   Three-phase ladder diagram.

have multiple connectors on the T1, T2, and T3 terminals to allow for the connection of multiple loads.

A three-phase system can also be shown in a ladder diagram, as in Figure 18–41.

CONTACTOR CIRCUIT NO. 4

Figure 18–42 shows a circuit for an even larger refrigeration system. Instead of operating the compressor and the condenser fan(s) from a single contactor, each of the condenser fans has its own contactor.

Each condenser fan motor is tied into a different pair of legs of the three-phase. This is done to attempt to balance (as much as possible) the amp draws on the

three legs. If there were a severe load imbalance, it could cause a voltage imbalance on the three wires going to the compressor motor, resulting in motor overheating. Note also that each condenser fan contactor is controlled from a pressure switch called a fan-cycling switch. Although the fan cycling switch senses pressure on the high-pressure side of the refrigeration system, it is different from an HPC. The HPC *opens* when the sensed pressure rises. The fan cycling switch *closes* on a rise in pressure, thereby bringing on the fans one at a time, as needed. The ladder diagram for this circuit is shown in Figure 18–43.

Troubleshooting—Box Warm, Compressor/Condenser Not Running

Assume that the circuit is a two-pole contactor, as in Figure 18–38.

1. Check for voltage between T1 and T2. It is probably zero, as neither the condenser fan motor nor the compressor are running, and it is unlikely that both have failed.

2. Assuming a zero voltage reading between T1 and T2, there are only two possible reasons:

   a. Power is available between L1 and L2 but the contactor switches are open.

   b. No power is available to L1–L2.

3. Measure the voltage between L1 and L2. If it is zero, go to the fuses or the circuit breaker box to find out if a circuit breaker has tripped. If it is 230 V, then we know that the

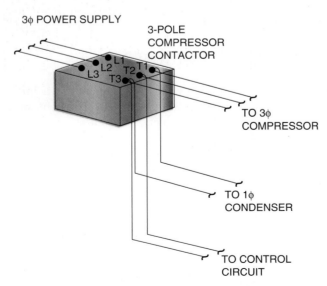

3ɸ POWER SUPPLY

3-POLE
COMPRESSOR
CONTACTOR

L1
L2
L3   T2   T1
      T3

TO 3ɸ
COMPRESSOR

TO 1ɸ
CONDENSER

TO CONTROL
CIRCUIT

FIGURE 18–40   Multiple loads on a three-phase contactor.

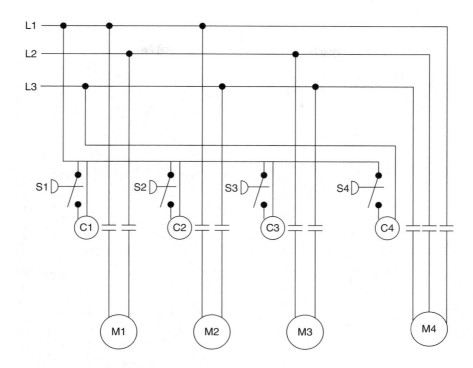

M4 = COMPRESSOR MOTOR
M1, M2, M3 = CONDENSER FAN MOTORS
C1,C2, C3, C4 = CONTACTORS
S1, S2, S3 = FAN CYCLING PRESSURE SWITCH
S4 = LOW PRESSURE CUT-OUT

FIGURE 18–42 Circuit with multiple contactors.

contactor switches are not closing. There are only two possible reasons:

a. Voltage is not being applied to the contactor coil.

b. Voltage is being applied to the contactor coil, but the contactor coil has failed.

Measure voltage across the terminals of the contactor coil. If you read 230 V, replace the contactor, if you read zero, you must determine why voltage is not getting to the coil. In this simple example, the only obvious reason is that the LPC is open. If you go to the LPC and measure 230 V across the terminals, your suspicion will be confirmed. However, you are not ready to replace the LPC quite yet. There are two possible reasons that the LPC switch might be open:

1. The LPC has failed.

2. There is insufficient refrigerant in the system. The pressure actually is below the setting of the LPC, and the LPC is merely doing its job.

The service technician can measure the pressure in the system, but this is inconvenient because the refrigerant gauges are probably not available without making an extra trip to the truck. More likely, the service technician will place a jumper wire across the LPC switch. If the system then runs and cools normally, then replacement of the switch is all that is required. If the system runs, but still doesn't appear to be cooling normally, further mechanical troubleshooting is necessary.

In order to safely jumper across the switch, the power to the unit should be turned off before the jumper is placed across the switch terminals. Whenever a jumper wire is used, the wire used must be sufficiently heavy to carry the load. In this case, the jumper will only be carrying the amps to the compressor contactor coil. A light-duty wire will be sufficient.

An alternative to placing a jumper across the switch is to mechanically close it.

Simply find the part of the moving mechanism that is getting pushed on by the pressure-sensing element, and duplicate this force with your screwdriver. Most switches can be closed in this fashion.

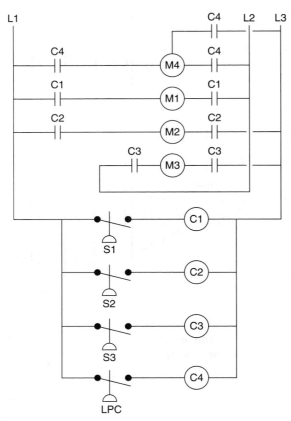

FIGURE 18–43 Multiple-contactor circuit ladder diagram.

☞ Make sure you complete this last step of troubleshooting to ensure that the fix you implement is actually going to solve the problem. It is embarrassing and bad business to quote a price for repair to the customer based on a new pressure switch, and then have to tell the customer that you were only kidding, and that you need to repair a refrigerant leak and recharge the system too!

PUMPDOWN CIRCUIT

On some refrigeration systems, especially where the evaporator is located at a higher elevation than the compressor, an automatic pumpdown circuit is used to empty all the refrigerant from the evaporator before shutting down the compressor. This prevents liquid refrigerant from migrating to the compressor during the "off" cycle, which would have the undesirable effect of diluting the compressor oil. Figure 18–44 shows how an automatic pumpdown is accomplished. During normal "freeze" operation, the sole-

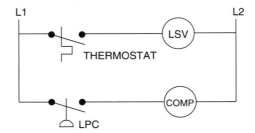

LSV = LIQUID SOLENOID VALVE
LPC = LOW PRESSURE CUT-OUT
COMP = COMPRESSOR

FIGURE 18–44 Automatic pumpdown no. 1.

noid valve in the refrigerant liquid line, **LSV** (liquid solenoid valve), and the compressor are both energized. When the thermostat becomes satisfied, it opens, de-energizing the LSV coil, causing the LSV to close. The compressor, however, continues to run. The liquid in the liquid line downstream of the LSV and the liquid in the evaporator are exposed to a decrease in pressure, and all the liquid vaporizes. When all the liquid has vaporized, the suction pressure sensed by the LPC decreases rapidly, and the LPC opens. This de-energizes the compressor.

When the thermostat once again calls for cooling, it energizes the LSV, causing it to open. The low-side pressure rises, closing the LPC, and energizing the compressor.

Figure 18–45 shows the same sequence of operation, but for a larger compressor system. Instead of the LPC operating the compressor, it operates a compressor contactor coil that, in turn, energizes a large three-phase compressor.

The liquid solenoid valve is leaking. During the "off" cycle, the refrigerant leaking through the valve

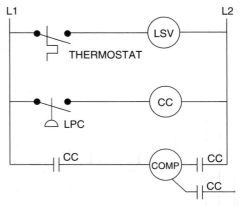

FIGURE 18–45 Automatic pumpdown no. 2.

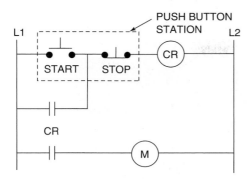

FIGURE 18–46 Seal-in circuit.

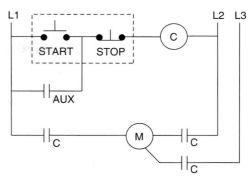

FIGURE 18–48 Push-button station with contactor and auxiliary contact.

causes the low-side pressure to rise slowly, until it finally reaches the cut-in setting on the LPC. This causes the compressor to start, but because the LSV is still closed, the low-side pressure drops to the cut-out setting within a few seconds, and the compressor turns off. When the thermostat finally closes, then the LSV opens and the compressor stays online.

SEAL-IN CIRCUIT

Figure 18–46 shows an interesting circuit that has many variations in HARV (heating, air conditioning, refrigeration, and ventilation) applications. The push-button station consists of two momentary switches. The normally open switch is labeled "Start," and the normally closed switch is labeled "Stop" (Figure 18–47). It is sometimes referred to as a "Start-Stop Push-Button Station."

When the "Start" switch is depressed, it energizes the control relay coil. When energized, the coil operates (closes) *two* switches. One switch starts a motor. The other switch is wired in parallel with the "Start" switch. When the "Start" switch is then released, it opens (remember, a normally open momentary switch only remains closed for as long

as it is being pressed). However, even though the "Start" switch opens, the control relay coil remains energized through the contacts that have *sealed-in* the circuit around the "Start" switch. In order to shut this system off, the "Stop" button needs to be pressed to de-energize the control relay coil. This will drop out the control relay contacts, and the system will not restart when the "Stop" button is released.

With this system, if there is a momentary power outage, the motor will not restart automatically when power is restored. It must go through its normal startup sequence (pushing the "Start" button).

Another way that a Start-Stop Push-Button Station may be wired is with a contactor with an auxiliary contact (Figure 18–48). To visualize the auxiliary contact, touch the three middle fingers on your left hand to the corresponding three fingers on your right hand. These are like the three poles on a contactor that carry current to the motor. Now, when you touch your middle three fingers together, also touch your pinky fingers together. Your pinky fingers are like the auxiliary switch on the contactor. It closes at the same time as the three contactor poles, but it is a **dry switch**. That is, the contactor coil makes the switch close, but the auxiliary switch is not connected electrically to any other part of the contactor.

When the "Start" button is pushed, the auxiliary switch on the contactor also closes. This switch is then used to seal in the circuit around the "Start" switch.

LOCK-OUT RELAY

A **lock-out relay (impedance relay, reset relay)** is used to provide lock-out and remote reset in refrigeration, air-conditioning, and other systems

FIGURE 18–47 Push-button station.

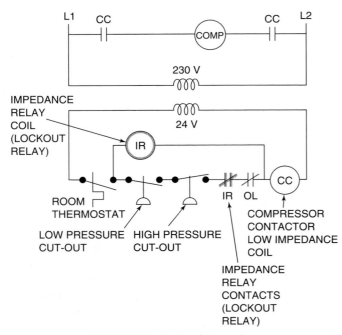

FIGURE 18–49 Lock-out relay.

FIGURE 18–50 S-9400 series liquid level switch. (Courtesy AC & R Components, Inc.)

(Figure 18–49). During normal operation, the normally closed contacts of the pressure controls and motor overloads short out the relay coil so that the contactor pulls in. If one of the pressure controls or overloads opens, the impedance relay coil is energized in series with the contactor coil. When this occurs, the high impedance of the relay coil uses most of the voltage available, leaving insufficient voltage for the contactor coil and causing the contactor to drop out. As the impedance relay pulls in, its normally closed contacts open to keep the contactor out, even though the pressure control or overload (automatic reset) remakes.

On some models, the impedance relay has an additional normally open contact that can be wired to operate a check light to signal the homeowner that the system has shut down.

LEVEL CONTROL

Figure 18–50 shows a liquid level switch. It has three wires, so we know that this is an SPDT switch. Depending on which two wires are selected, it can be used to open on a rise in level or close on a rise in level (Figure 18–51). Most level switches are a simple mechanical float type device, similar to the float that controls a toilet valve.

A level switch can be used to act as a high-level alarm, low-level alarm, or differential control using two switches. It can be installed in oil separators, oil reservoirs, liquid refrigerant receivers, suction accumulators, and compressor crankcase applications. Figure 18–52 shows an interconnection diagram for controlling the level between a certain range. When the tank is empty, both level switches are closed. This completes a circuit from L1, through the bottom switch, through the top switch, energizing the solenoid valve, back to L2. With the solenoid valve energized, it opens, allowing liquid to flow into the tank. As the liquid level rises, the bottom switch opens. However, the liquid solenoid remains energized because of the relay coil that was energized at the same time as the solenoid valve (this is a seal-in circuit). When the level reaches the top switch, it opens and the solenoid and relay coil become de-energized, closing the solenoid valve. When the liquid then drops below the bottom switch, the bottom switch closes again, opening the solenoid valve and filling the tank.

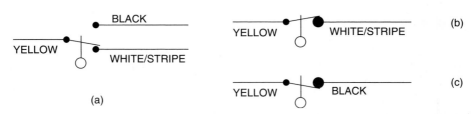

FIGURE 18–51 Level switches. (a) SPDT switch. (b) Open when level is high. (c) Open when level is low.

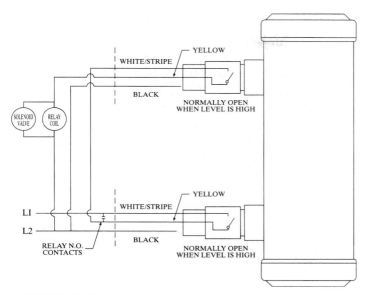

FIGURE 18–52 Interconnection diagram to control a liquid level within a range. (Courtesy AC & R Components, Inc.)

INTERLOCKS

Often, it is necessary to prevent one device from starting until after another device has started. Consider, for example, the circuit shown in Figure 18–53a. The water pump circulates chilled water through a building and a water chiller. When the room thermostat closes, it energizes both the contactor for the water chiller and the pump. This will work fine, so long as nothing goes wrong. But it is desirable to not start the water chiller until after the pump has started. In Figure 18–53b, the thermostat energizes the pump and a control relay. The chiller contactor is energized through the control relay. This is called an **electrical interlock**.

Suppose that the pump fails. The control scheme in Figure 18–53b would still allow the chiller to run, and the results would be catastrophic. The chiller would cause the still water in evaporator tubes to freeze, causing the evaporator tubes to rupture. There must be additional protection provided to prevent this damage. In Figure 18–53c, instead of using the control relay contacts to energize the water chiller, we have used a flow switch. The flow switch is located in the chilled water line and closes only if the water is actually flowing in the pipe. This is called a **mechanical interlock**. And, if the designer of the system was really serious, both the mechanical and the electrical interlock may be used Figure 18–53d.

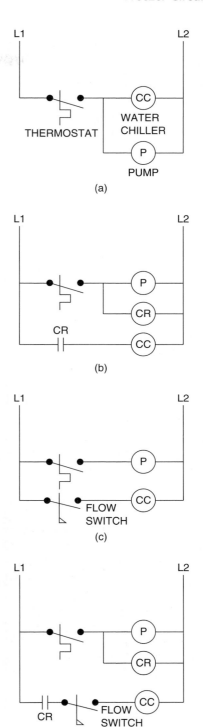

FIGURE 18–53 Electrical and mechanical interlocks.

SUMMARY

The circuits provided in this unit provide an introduction to the complete field of heating air conditioning, refrigeration, and ventilation (HARV). Make certain that you understand these simple circuits; they form a base for the circuits and systems to come later.

PRACTICAL EXPERIENCE

1. Draw the correct symbol for each of the following thermostats:

   a. To operate an attic ventilation fan during the summer months

   b. To operate an electric room heater

   c. To operate an oven

   d. To operate a radiator fan in your car

   e. A freezestat

2. Figure 18–54 shows a partial interconnection diagram for a freezer with a compressor, condenser fan, evaporator fan, overload, door switch, and cold control. Complete the wiring in accordance with the wiring diagram of Figure 18–16. For the sake of clarity, draw all wiring as vertical or horizontal lines.

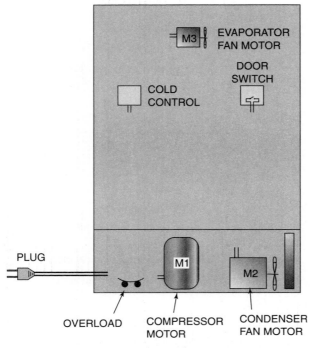

FIGURE 18–54 Exercise.

3. For each of the following applications, draw the correct symbol for the momentary switch:

   a. The dome light in your car

   b. The switch on a flashlight

   c. The door switch on a microwave oven

4. Obtain a defrost timer. Rotate the manual advance knob in a clockwise direction. You will find that it takes a lot of rotation until you hear a first "click," indicating that the switch has gone into the "defrost" position. Then, after a slight additional rotation, you will hear a second "click," indicating that the switch has gone into the "freeze" position. Use your ohmmeter to figure out the internal wiring of the defrost timer. Between which terminals is the timer motor? Which terminals art connected after the first click? After the second click?

5. Using correct symbols, draw two wiring diagrams for a 230-V compressor, one with a line-voltage thermostat, and the other with a pilot-duty thermostat.

6. Redraw the level control diagram shown in Figure 18–52 as a ladder diagram.

7. Redraw the circuit in Figure 18–31c as a ladder diagram.

REVIEW QUESTIONS

1. The compressor motor (Fig. 18–16) is running. What voltage would you measure across each of the following?

   a. Compressor

   b. Thermostat

   c. Overload

   d. Condenser fan

2. The compressor is running while the freezer door is open. What voltage reading would you obtain *across* each of the following devices?

   a. Cabinet light

   b. Momentary door switch for the cabinet light

3. You found a door switch in your truck, but it fell out of the package. It has two terminals. How can you use your ohmmeter to determine if this switch works, and whether it is a switch for a fan or for a cabinet light?

4. Referring to Figure 18–7, what voltage would you expect to measure from the right side of the thermostat to ground when the freezer temperature is below the selected temperature?

5. During the "off" cycle on an automatic pump-down circuit, the compressor starts, operates for a few seconds, and then turns off. This may happen every five minutes, until finally the compressor turns on and stays on for a reasonable period of time. What is the problem?

6. Referring to Figure 18–7, what voltage would you expect to measure from the right side of the thermostat to ground if the overload is open and the freezer box temperature is below the selected temperature?

7. Referring to Figure 18–7, what voltage would you expect to measure from the right side of the thermostat to ground if the overload were open

and the freezer box temperature was above the selected temperature?

8. Referring to Figure 18–19, you open the freezer cabinet door and note that the cabinet light is off. There is cold air blowing about the freezer. A probable cause of trouble is?

9. Compare Figure 18–25 and Figure 18–26. In which system does the evaporator fan turn on immediately after the defrost time returns the system to the freeze cycle?

10. In the circuit of Figure 18–49, What must be done to de-energize the lock-out relay after it once becomes energized?

UNIT 19

AIR-CONDITIONING AND HEATING CONTROLS AND CIRCUITS

OBJECTIVES

After completion and review of this unit, you should be familiar with the controls and circuits used in the industry, including:

- Window air conditioner.
- Window air conditioner with heat strip.
- Rooftop air conditioner.
- Packaged heating with air conditioner.
- Residential three-voltage system.
- Air switch.
- Heating coils.
- Reversing valves.
- Defrost relays.
- Bimetal thermostats.
- Mercury bulb thermostats.
- Heat and cool anticipators.
- Setback thermostats.
- Electronic thermostats.
- Two-stage thermostats.
- Temperature control switches.

Comfort cooling circuits control the same types of devices as refrigeration and freezer circuits: compressor, condenser fan, and evaporator. The differ-

ences between comfort cooling systems and refrigeration systems are

1. Comfort cooling systems (except for window air conditioners) almost always use low-voltage (24 V) controls, whereas refrigeration systems use line voltage controls.
2. Comfort cooling systems are commonly combined with furnace systems.
3. Comfort cooling systems operate with saturated suction temperature (evaporator coil temperature) above 35°F, so there is no need for defrost systems, except for heat pumps.

WINDOW AIR CONDITIONER

The window air conditioner shown in Figure 19–1 is typical of the type that plugs into a standard 115-V or 230-V outlet. The major components are the compressor and a double-shaft motor (Figure 19–2) that turns both the evaporator blower and the condenser fan. Power comes into a selector switch. Depending on the selection made, the selector switch will send the power out to one of the fan motor speed windings and the compressor. For example, if the selector switch is set for Fan only, it will complete the circuit from terminals 7–4. If it is set for High Cool, it will make 7–4 and 7–2, bringing on the compressor.

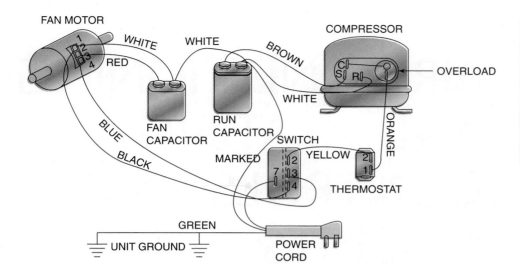

SWITCH	CONTACTS		
POSITION	2	3	4
OFF	O	O	O
NORMAL FAN	O	C	O
SUPER FAN	O	O	C
NORMAL COOL	C	C	O
SUPER COOL	C	O	C
C-CLOSED			
O-OPEN			

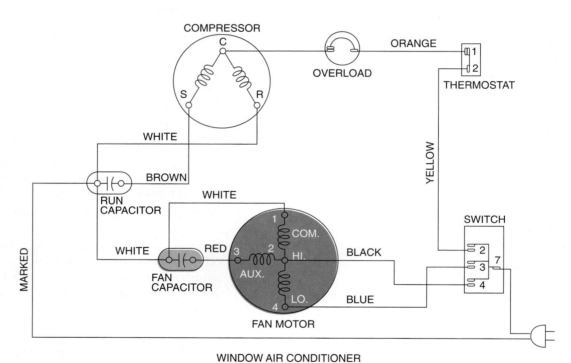

WINDOW AIR CONDITIONER

FIGURE 19–1 Window air conditioner.

FIGURE 19–2 Double-shaft motor.

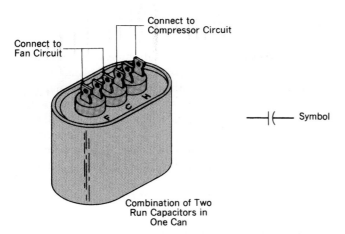

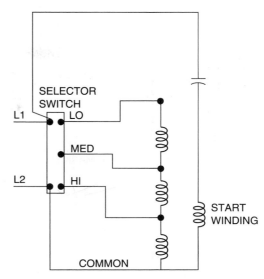

FIGURE 19–4   Three-speed motor circuit.

FIGURE 19–3   Dual capacitor for compressor and condenser fan motor.

Low Cool will cause the selector switch to make 7–3 and 7–2. When the selector switch is set for cooling, the fan will run continuously, and the compressor will cycle off and on as the room thermostat opens and closes.

The compressor motor and the fan motor each have a run capacitor. Sometimes the two capacitors will be combined into a single casing as shown in Figure 19–3. The fan is connected between terminals C (common) and F (Fan). The compressor is connected between terminals C and H (Hermetic).

In Figure 19–1, the side of the power supply that goes to the run capacitor is labeled "marked." Often, one side of the power supply lead will have a ridge on the insulation or some other marking that makes it easier for you to identify which side of the power supply you are looking at.

The fan motor that uses multiple taps on the run winding to produce different speeds will only produce different speeds when the motor is under a load. If you remove the fan from the motor shaft and operate the motor alone, it will run at full speed, even if the medium- or low-speed taps are used.

Plug and Receptacle Configurations

The standard house outlet shown in unit 1 is not always suitable for a window air conditioner. Some window air conditioners require more current than can be supplied from a standard outlet. Higher-amperage and higher-voltage outlets are available for use on larger circuits. Each different rating uses

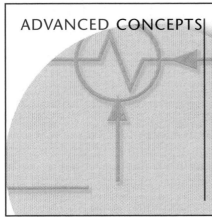

ADVANCED CONCEPTS

The fan motor shown in Figure 19–4 has two speeds. Different speeds can be achieved by tapping the run winding in different places. When the selector switch makes 7–4 (High speed), it uses only the portion of the run winding in the motor between motor terminals 2 and 1. However, when the selector switch makes 7–3 instead, another portion of run winding (between motor terminals 4–2) is inserted in series with the run winding between motor terminals 2–1. Adding this coil to the circuit reduces the current flow through the run winding and therefore reduces the strength of the magnetic field. The motor produces less torque, and the motor speed will decrease. A three-speed motor is shown in Figure 19–4. It allows yet another portion of run winding to be inserted, reducing the current flow still further and producing a still slower speed.

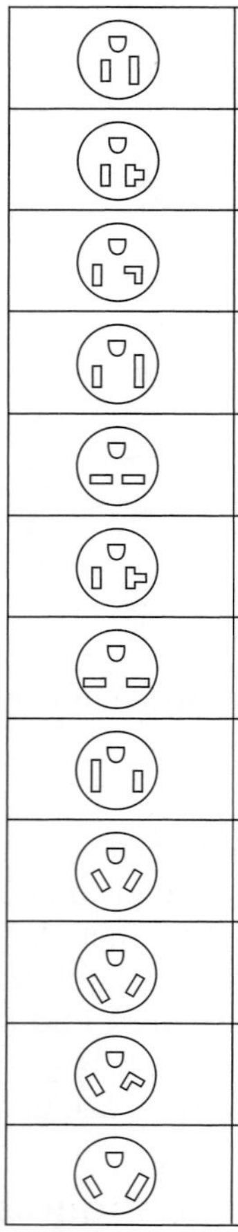

RECEPTACLE CONFIGURATION	RATING
	15 A 125 V
	20 A 125 V
	30 A 125 V
	50 A 125 V
	15 A 250 V
	20 A 250 V
	30 A 250 V
	50 A 250 V
	15 A 277 V
	20 A 277 V
	30 A 277 V
	50 A 277 V

FIGURE 19–5 Standard receptacle configurations.

a different plug and receptacle configuration to prevent a higher-voltage or amperage device from being plugged into an outlet that cannot handle the load. Figure 19–5 shows some of the standard plug and receptacle configurations.

Window Air Conditioner with Strip Heat

The window air conditioner in Figure 19–6 has, in addition to its air-conditioning components, an electric heater (commonly called a **strip heater**). The differences between this unit and a cooling-only unit are

1. There are additional switch positions in the selector switch to accommodate different heat settings.
2. The thermostat is an SPDT switch, using one side to control the heating and the other side to control the cooling.

ROOFTOP AIR CONDITIONER

One of the simplest control schemes using a control voltage that is different from the line voltage is for a packaged rooftop air conditioner. All components except for the room thermostat are located in a single unit, mounted on the roof. Electrical installation couldn't be simpler. The L1 and L2 power supply are connected to the L1 and L2 terminals in the unit and the R, Y, and G terminals on the room thermostat are connected to the R, Y, and G terminals in the air-conditioning unit, using thermostat wire.

Sequence of Operation

Figure 19–7 shows the ladder diagram of a simple packaged air conditioner. The transformer is always energized. When the room thermostat senses that the room is too warm, it closes a switch (inside the thermostat) between R-Y and another switch between R-G. The R-Y switch energizes a compressor contactor, whose contacts energize the compressor and condenser fan. The R-G contacts energize a blower relay, whose contacts energize the blower fan motor. In air-conditioning systems, the evaporator fan is commonly called a blower, and the blower relay is nothing more than an SPNO (single-pole normally open) control relay. All the cooling devices come on at the same time, and when the thermostat is satisfied, they all turn off at the same time.

The blower motor can also be operated continuously, even when the compressor turns off. This is sometimes desirable for continuous circulation, a nonchanging noise level, and equalizing temperatures between rooms that tend to have nonuniform temperatures. There is a switch on the thermostat labelled FAN-AUTO and FAN-ON. When it is set to FAN-AUTO, R-G is made whenever R-Y is made. When it is set to FAN-ON, R-G is made regardless of the position of R-Y.

PACKAGED HEATING AND AIR-CONDITIONING SYSTEM

Figure 19–8 shows the same cooling system, but the packaged rooftop also now includes a furnace. The room thermostat is also different, as it now includes the functions of both the heating thermostat and the cooling thermostat (four-wire thermostat). When the thermostat needs heating, it makes R–W only. The gas valve comes on right away, but the blower doesn't come on until a minute later, when the bonnet switch closes. When the thermostat

SWITCH POSITION	TERMINALS	READING
OFF	L1 TO 1-2-3-4-5	OPEN
COOL-LO	L1 TO 1-4	CLOSED
COOL-MED	L1 TO 1-3	CLOSED
COOL-HI	L1 TO 1-2	CLOSED
HEAT-LO	L1 TO 5-4	CLOSED
HEAT-MED	L1 TO 5-3	CLOSED
HEAT-HI	L1 TO 5-2	CLOSED
FAN	L1 TO 3	CLOSED

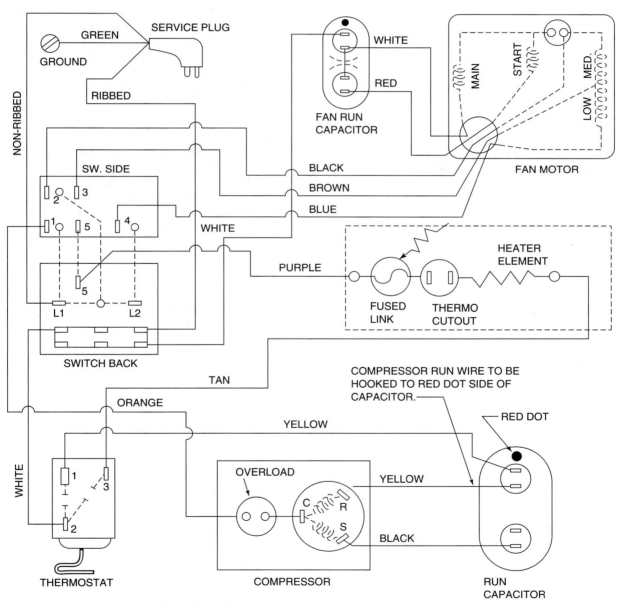

FIGURE 19–6   Window air conditioner with electric heat.

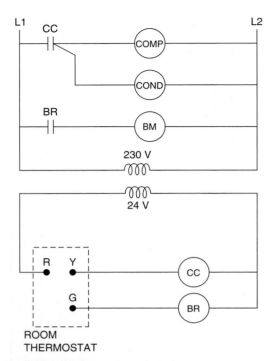

FIGURE 19–7 Packaged rooftop air conditioner.

needs cooling, it makes R–Y and R–G, and all components come on right away as described earlier.

Sometimes, the rooftop unit is furnished with a two-speed blower motor. When that is the case, the high speed is used for cooling whereas the lower speed is used for heating. This is done because on heating, air is supplied to the room at 50°F higher than the room temperature. But on air conditioning, the air is supplied to the room at only 20°F lower than the room temperature. Therefore, more air is needed for air conditioning than for heating. Figure 19–9 shows how the low-speed winding of the blower motor is energized by the bonnet switch, whereas the high-speed winding is energized by the blower relay contacts. Note that the blower relay has changed. It is now an SPDT switch, with the normally closed contacts being wired in series with the bonnet switch. If this normally closed switch were not added, there would be the following potential problem. Suppose that during heating (bonnet switch closed), an occupant switched the room thermostat from FAN-AUTO to FAN-ON. The motor would have both its High and Low speed windings energized at the same time. This would cause the motor to burn out. With the addition of the normally closed switch, when the thermostat was switched to FAN-ON, the Low speed winding would be de-energized.

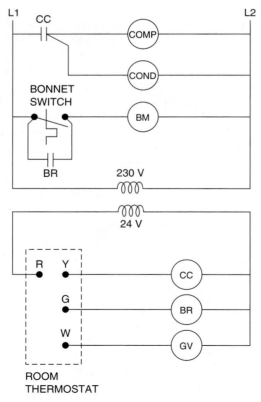

FIGURE 19–8 Packaged rooftop heating and air-conditioning unit.

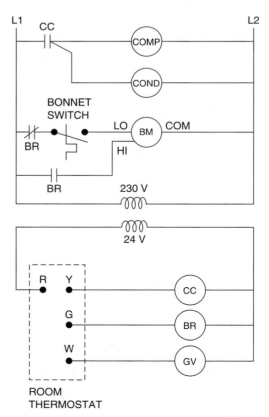

FIGURE 19–9 Packaged rooftop unit with two-speed blower.

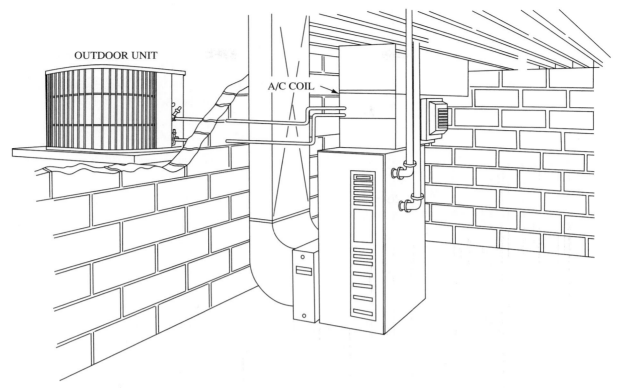

FIGURE 19–10   Three-voltage system. (Courtesy of Carrier Corp.)

RESIDENTIAL CIRCUIT NO. 1—THREE-VOLTAGE SYSTEM

Residential central heating/air-conditioning systems usually use a 115-V furnace in which the furnace fan also serves as the blower for the air conditioning. Within the furnace is a 24-V control system. Outside is the condensing unit, which is supplied from a 230-V power source. The physical location of the components for this three-voltage system is shown in Figure 19–10, and the ladder diagram is shown in Figure 19–11.

RESIDENTIAL CIRCUIT NO. 2

Figure 19–12 shows an interconnection diagram and a ladder diagram for a system that provides heating, air conditioning, humidification, and electronic air cleaning. Some of the unique features of this system are as follows.

1. Instead of the room thermostat energizing the compressor contactor when R–Y is made, the thermostat energizes a thermal delay heater. This is simply a time-delay switch similar to the one that is commonly used to provide a time delay on furnace

blower motors. It prevents the compressor from short cycling. **Short cycling** is a term used to describe a compressor that starts, runs for a few seconds, turns off, and then repeats the cycle. For example, if the refrigeration unit were low on charge, as soon as the compressor started, it would cut out on the low-pressure cut-out. After the compressor stops, the low-side pressure increases, the LPC recloses, and the compressor starts again. In this circuit, use of the time-delay heater is probably redundant to prevent short cycling because the impedance relay already provides that protection. In this circuit, the time-delay switch would prevent the compressor from restarting immediately after shutting down if somebody happened to adjust the thermostat downward. This will allow pressures to equalize within the refrigeration system before the compressor attempts to restart.

2. The impedance relay (lock-out relay) has two sets of contacts instead of one. The normally closed set of contacts provides the manual-reset feature if any of the safety controls open. The normally open set of contacts completes a circuit through a trouble-indicator

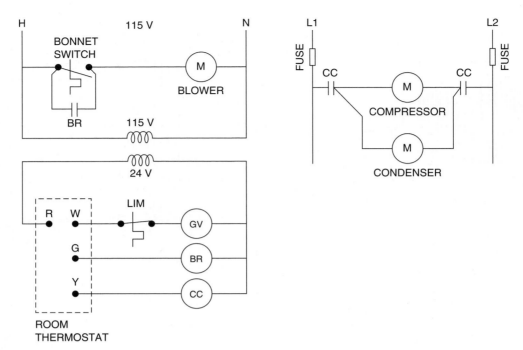

FIGURE 19–11 Ladder diagram three-voltage residential heating/air conditioning.

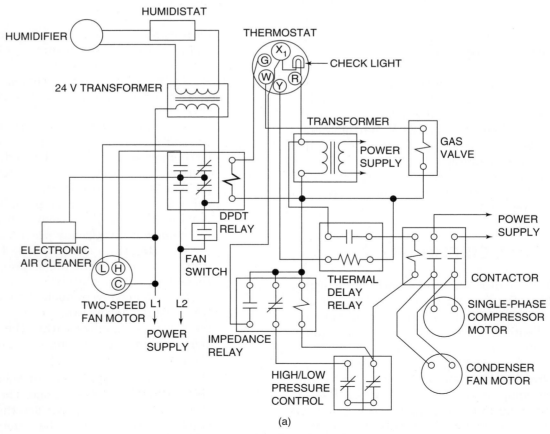

(a)

FIGURE 19–12 Residential Circuit No. 2.

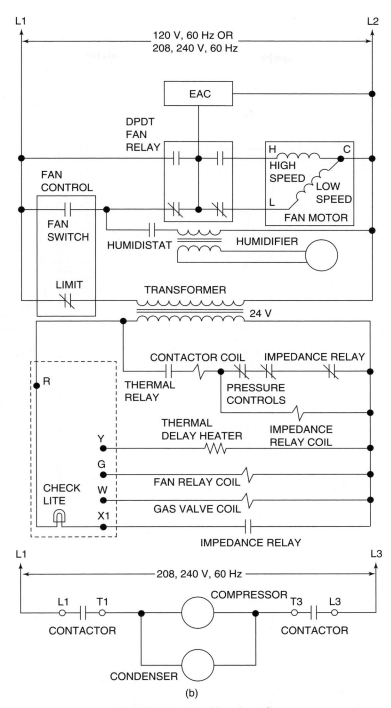

FIGURE 19–12 (Continued).

light in the room thermostat. This alerts the occupants that something is wrong with the system.

3. The humidifier operates by a low-voltage motor driving a pad through a water pan so that the room air can pass through a contin-

uously wetted pad. A separate transformer is provided to supply the correct voltage for the humidifier motor.

4. The DPDT fan relay operates the blower motor on high speed for cooling and low speed for heating.

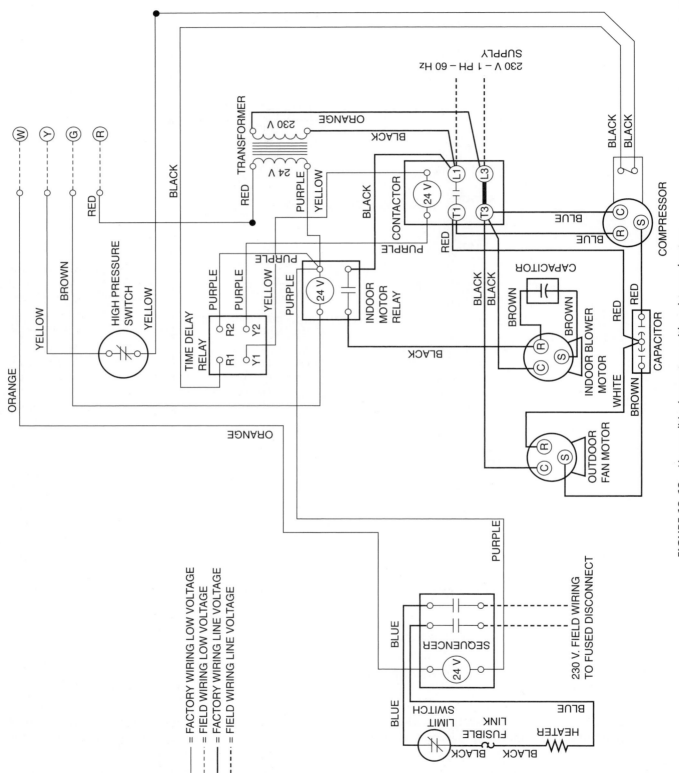

FIGURE 19–13 Air-conditioning system with resistance heaters.

= FACTORY WIRING LOW VOLTAGE
= FIELD WIRING LOW VOLTAGE
= FACTORY WIRING LINE VOLTAGE
= FIELD WIRING LINE VOLTAGE

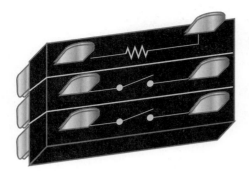

FIGURE 19–14   Heat sequencer.

RESIDENTIAL CIRCUIT NO. 3

Figure 19–13 shows a circuit for controlling a total electric packaged rooftop heating/cooling system.

Heat Sequencer

The heat is provided by an electric resistance heater located in the duct or in an electric furnace. The electric heating element is switched on and off through a **heat sequencer**. A heat sequencer (Figure 19–14) acts like a time-delay contactor. Instead of a control signal energizing a magnetic coil, which immediately closes the line-voltage switches, the control signal energizes a small heater inside the sequencer. After a time delay, a bimetal switch inside the heat sequencer operates the switch(es) that complete the circuit through the electric heating element.

Time-Delay Relay

On simpler systems, when the room thermostat makes R–Y, it completes a circuit through the compressor contactor coil. On this system, instead of energizing the contactor coil, R–Y completes a circuit through the time-delay relay (TDR), terminals R1 and R2. Some time after 24 V are applied to R1–R2 (usually about one minute), 24 V come out of the relay at terminals Y1 and Y2. The output voltage from the time-delay relay energizes the compressor contactor. This prevents short cycling of the compressor. For example, suppose an occupant adjusts the set point of the room thermostat downward. Suppose that, by chance, this causes the thermostat to make R–Y just seconds after the compressor turned off because the thermostat was satisfied. The compressor would have a hard time starting because the pressures in the high side and low side of the refrigeration system will not have had time to equalize. The result could be either an overload opening or a fuse in the supply power wiring blowing. The time delay will avoid this situation by keeping the compressor

from trying to start for at least a minute after the thermostat calls for cooling. Some of these time delays are adjustable and may be set by the service technician for any period from one to ten minutes.

Compressor Overload

The compressor overload is the switch that is shown between the two black wires emerging from the compressor. It is not known from this diagram whether the overload is internal or external. However, if it opens, it will interrupt the circuit that includes the room thermostat (R–Y), the high pressure switch, and the time-delay input.

Room Thermostat

Although the room thermostat is not shown on the wiring diagram, we can tell that an electric heat type of thermostat is required. We can figure this out, because if you look at the black wire on the red terminal of the indoor blower motor, it goes through the switch of the indoor motor relay before reaching L1. Therefore, we know that the indoor motor relay *must* be energized in both heating and cooling. The indoor relay coil gets energized from the G terminal of the room thermostat. A thermostat designed for electric heat will make R–G on both heating and cooling, whereas a thermostat not designed for electric heat will only make R–G on a call for cooling.

Heating-Cooling Thermostat

Many different types and styles of low-voltage room thermostats will be encountered. Generally, they all share the same characteristic of using a bimetal element that changes shape when its temperature rises or falls. This mechanical movement is used to operate one or more switches, which, in turn, operate various components in a heating or air-conditioning system. Figure 19–15 shows a very common

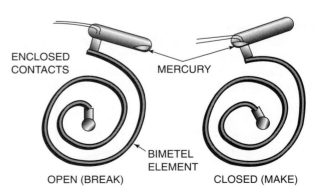

ENCLOSED CONTACTS

MERCURY

BIMETEL ELEMENT

OPEN (BREAK)   CLOSED (MAKE)

FIGURE 19–15   Thermostat using coiled bimetal element and mercury switch.

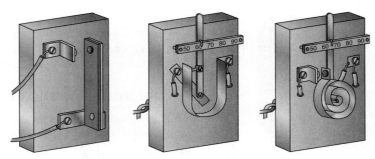

FIGURE 19–16 Thermostats with open contacts.

arrangement. The bimetal element is coiled around a fixed center. As its temperature changes (indicating a change in room temperature), the coil either winds tighter because the bimetal wants to curve more or unwinds because the bimetal wants to unwind. A mercury bulb is attached to the bimetal. A mercury bulb is a sealed bulb with mercury inside. As the mercury rolls from one end to the other, it will make contact between different pairs of wires.

Not all thermostats use a mercury bulb as the switch. Figure 19–16 shows several arrangements of switches with open contacts, as compared to the sealed contacts of the mercury bulb. The sealed contacts are used in higher-quality thermostats because their contacts are much less prone to get pitted or contaminated than open contacts. However, in applications where vibration of the thermostat is a problem (e.g., in a shipping office next to a railroad loading dock), the open-contact thermostat may actually be better because it has a more positive (spring-loaded) action.

Figure 19–17 shows a mercury bulb that acts as an SPDT switch. If the mercury rolls to the right, it makes contact between wires R and Y. If the mercury rolls to the left, it makes contact between wires R and W. As the bimetal moves due to a change in temperature, the mercury bulb tilts and makes R–W or R–Y, bringing on the heating or cooling equipment.

Figure 19–18 shows a thermostat with and without a **subbase**. The subbase (Figure 19–19) is the part that screws to the wall and contains the screws that hold the thermostat wires that emerge from inside the wall. When the thermostat is screwed to the subbase, it makes electrical connec-

tions between the thermostat and the subbase. In Figure 19–20 the connections that are made when the thermostat is screwed to the subbase are R–R1, Y–Y1, and W–W1. The subbase contains two manual switches. One is called the Fan switch. It may be set to either On or Auto. The other switch is called the System switch. It may be set to either Heat, Cool, or Off. The circles at the manual

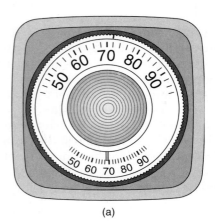

(a)

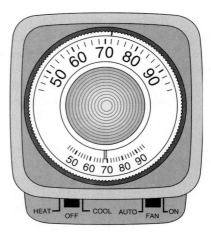

(b)

FIGURE 19–18 Thermostat with and without subbase. (a) Without subbase. (b) With subbase.

FIGURE 19–17 SPDT mercury switch.

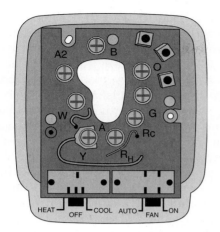

FIGURE 19–19 Thermostat subbase.

The terminal identifications used on the thermostat in Figure 19–20 are very common and should be memorized by the service technician. Red is the terminal that receives power from the transformer. It can be switched out of the thermostat through white, green, or yellow. White is most commonly connected to the wire that will energize the gas valve or other heating device. Green will be connected to the wire that energizes the blower relay. Yellow will be connected to the wire that energizes the compressor contactor. On a call for heating, the thermostat will make R–W. On a call for cooling, the thermostat will make R–Y and R–G. Note that the thermostat does not normally bring on the fan during heating. When the Fan switch is placed in the On position, it will make R–G without regard to the position of the System switch or the temperature being sensed by the thermostat.

Although the terminal identifications discussed are common, they are not universal. White-Rodgers and Honeywell use this convention. A less-used convention used by some manufacturers such as General Controls is the following:

V = Voltage (similar to R)
H = Heating (similar to W)
C = Cooling (similar to Y)
F = Fan (similar to G)

Thermostats with R, W, Y, and G terminals may also have a B terminal and an O terminal. The thermostat will make R–B whenever the System switch is set to Heat, and it will make R–O

switches in Figure 19–20 show the different connections that are made when the switches are set in their different positions. Note that when the System switch is set in the Heat position, it will make R–W on the subbase when the mercury bulb tilts to the left, calling for heat, but it will not energize any terminals if the mercury bulb tilts to the right (calling for cooling). If the System switch is set on Cool, it will make R–Y and R–G when the mercury bulb calls for cooling, but it will not energize any terminals on the subbase if the mercury bulb calls for Heat. This arrangement prevents the thermostat from calling for heating, then cooling, then heating, and so on, as the mercury bulb tilts to one side and then the other. Therefore, this thermostat can be used to control a heating and cooling system.

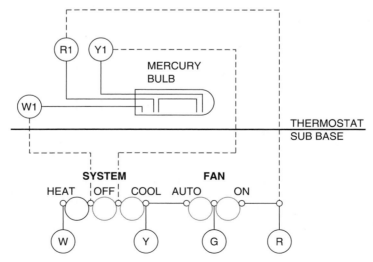

FIGURE 19–20 Dotted wiring shows connections that are made when thermostat is attached to subbase.

whenever the System switch is set to Cool. For these two terminals, it does not matter what temperature is being sensed by the thermostat. The B or the O terminal can be used to energize damper motors in either the heating or the cooling mode. But in practice, the B and O terminals are rarely used on thermostats that are not heat-pump thermostats.

Figure 19–19 shows a thermostat subbase with an R_C and an R_H terminal. Normally, the red wire coming from the transformer will be wired across both of these terminals, making them into a single terminal. The reason that these terminals are sometimes provided is for an add-on cooling system. Consider a system that has a heating-only furnace, and you are going to add a rooftop packaged cooling system. The existing furnace has a transformer and a gas valve. The existing thermostat has only R and W

terminals. You can replace the existing thermostat with one with a subbase as in Figure 19–19. The existing two wires that were attached to the old thermostat will be connected to R_H and W. The thermostat wires from the new rooftop packaged cooling system (with its own transformer) will be connected to R_C, Y, and G. The heating circuit will be electrically completely independent from the cooling circuit. The advantage of this type of thermostat is that it avoids having one thermostat for heating and a separate thermostat for the new cooling unit. Using two thermostats simultaneously on the same system has the disadvantage of potentially having the heating system and the cooling system run at the same time if the set point on the heating system thermostat is higher than the set point on the cooling system thermostat. Figure 19–21 shows a ladder diagram for an add-on cooling system.

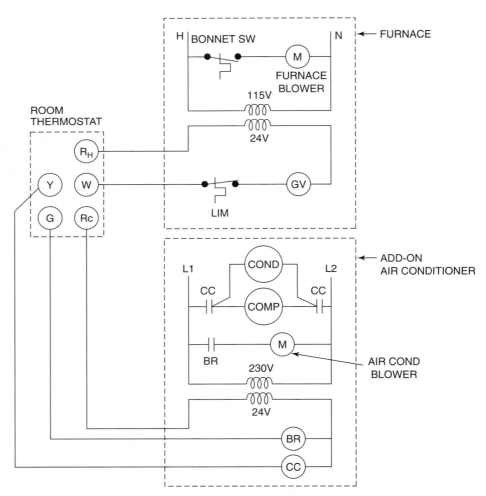

FIGURE 19–21   Add-on cooling schematic using R_C and R_H thermostat terminals.

HEATING AND COOLING ANTICIPATORS

On a heating application, once the thermostat is satisfied, the R–W contacts open. However, on most applications, this does not stop heat from being provided to the room. Typically, when R–W opens, it de-energizes the gas valve, but the fan continues to blow warm air into the room until the heat exchanger has cooled. This will cause the room temperature to "overshoot" the temperature at which the thermostat was actually satisfied. It would be nice if we could somehow anticipate when the thermostat was going to be satisfied and somehow open the R–W contacts a little sooner than they would otherwise have opened to prevent overshooting the temperature actually desired.

The **heat anticipator** does exactly that. It is a small, adjustable resistor located in the thermostat, very close to the bimetal element (Figure 19–22). It is wired in series with the R–W switch, so it is energized whenever the thermostat is calling for heat. The current passing through the anticipator generates a small amount of heat that tends to warm the bimetal element slightly quicker than the room itself would otherwise warm the bimetal. This causes the bimetal to open the R–W switch a little sooner than it would have without the anticipator.

The amount of resistance used in the heat anticipator will determine how much heat it produces. If more of the anticipator element is used, the cycles will be shorter; using less of the heat anticipator element will produce longer cycles. The correct amount of heat anticipator element to be used must be selected by the installing service technician.

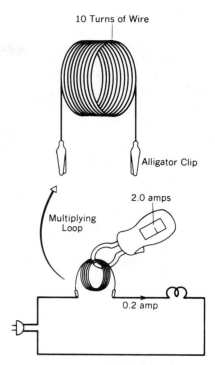

FIGURE 19–23   Multiplying loop.

To set the heat anticipator, remove the thermostat from the subbase, and place a 10-turn multiplying loop jumper wire between R–W. This jumper wire will cause the heating system to come on. Using a clamp-on ampmeter, measure the current being carried by the thermostat (Figure 19–23). Because of the 10 turns on the multiplying loop, the magnetic field produced is 10 times the field produced by one wire, and the amp reading obtained will be 10 times the actual amps being carried by the thermostat. Divide your reading by 10 to determine the actual amps. Then set the heat anticipator to a number that matches the actual amps that flow through the thermostat on the heating cycle.

On cooling, there is a similar problem. When the room gets warm, the thermostat will close R–Y and R–G, bringing on the air conditioning. However, it will take a minute or two for the refrigeration system to get down to operating temperatures and to cool the ductwork in the attic that has warmed during the Off cycle. It would be desirable to turn the air-conditioning system on prematurely, just as it was desirable to turn the heating system off prematurely.

The **cooling anticipator** (Figure 19–24) is a fixed resistor in the thermostat that is wired between

FIGURE 19–22   Heat anticipator.

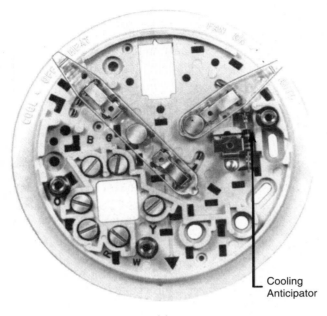

(a)

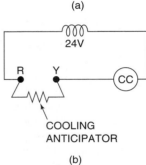

(b)

FIGURE 19–24 Cooling anticipator. (a. Courtesy of Honeywell, Inc.)

the R and Y terminals. It is energized whenever the R–Y switch is open (whenever the air-conditioning system is off). It tends to heat the bimetal and close the R–Y and R–G earlier than would otherwise happen from the rising room temperature alone. It is not adjustable, and the service technician does not have to do anything about the cooling anticipator.

AUTO-CHANGEOVER THERMOSTAT

The most common heating-cooling thermostats have a System switch that allows three different selections: Heat, Cool, and Off. The system is always in either the heating mode or the cooling mode, but cannot switch to the other unless the System switch is manually operated. This presents some problems in office environments. Often, setting of the thermostat can be a source of disagreement among office

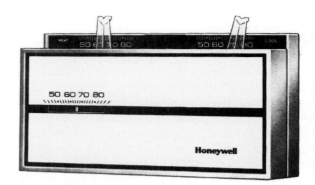

FIGURE 19–25 Auto-changeover thermostat.

occupants. Owners will sometimes place a locked box around the thermostat so that it cannot be adjusted by the office personnel. However, if it is set for Heating, and there is an exceptionally warm day on which cooling is needed to maintain the set temperature, there will be a problem.

Figure 19–25 shows an autochangeover thermostat. The System switch has four selections: Heat, Cool, Auto, and Off. When the System switch is set in the Auto position, the thermostat can provide heating when the room temperature is below a set point and cooling when the room temperature is above a set point. Note that there are two different set point levers on the auto-changeover thermostat. The lower set point is for heating, and the higher set point is for cooling. Typically, the heating set point will be color coded red, and the cooling set point will be color coded blue. They may be set independently, but a mechanical separation between them prevents them from being set so there is less than 2°F between them. They may be set as far apart as desired, for energy savings. For example, if the red lever is set to 70°F and the blue lever is set for 75°F, the thermostat will make R–W, bringing on the heating whenever the room temperature is below 70°F, and the thermostat will make R–Y and R–G whenever the room temperature is above 75°F. If the room temperature is between 70°F and 75°F, neither heating nor cooling will be used to change the room temperature.

SETBACK THERMOSTAT

Figure 19–26 shows a thermostat that looks similar to the auto-changeover thermostat because it also has two set point levers. But the function of these two set points is different from the auto-changeover thermostat. This is a setback thermostat. It has a clock and colored pins (red and blue) that may be placed on a 24-hour wheel at a place that corre-

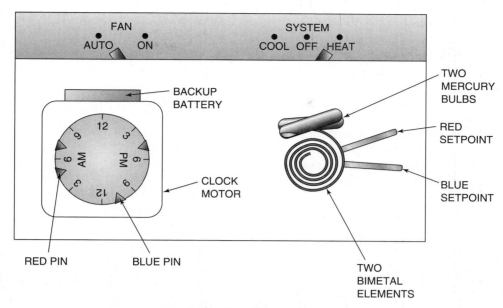

FIGURE 19–26 Setback thermostat.

sponds to a particular time of day. Whenever it is a time corresponding to a blue pin, the thermostat set point will correspond to the set point indicated by the blue lever. That will be the thermostat set point until the time corresponding to the next red pin occurs. Then, the set point of the thermostat will automatically change to the temperature indicated by the red lever. It is common to use four pins, set up to provide a schedule that uses the higher set point starting at 6 a.m., or shortly before everyone gets up in the morning. Another is set at 7:30 a.m. when everyone has left for work or school, the lower set point. Then, at 4 p.m. when people start returning home, the higher set point is used, until 11 p.m., when everybody has gone to sleep. Of course, in an office or commercial application, you would only need two programs, not four. Only two pins would be used, and the extra pins would be stored in a holder on the thermostat.

The setback thermostat must have a clock, and the clock is a load that must receive power. Several methods are employed to provide this power to the clock.

1. Older setback thermostats used an 18-V clock motor. There were two clock terminals on the subbase, and two extra thermostat wires went back to the furnace where they

were connected to a 120-V to 18-V transformer that was provided for the sole purpose of providing power to the clock. This was separate from the 120-V to 24-V transformer that was used as control voltage.

2. Newer thermostats use a clock that operates on 24 V. The power source for the clock is the same transformer that provides 24-V control voltage. Some of these thermostats have a terminal labelled C on the subbase (for Common). This terminal is connected to the other side of the transformer (the one opposite the side connected to the red terminal on the thermostat).

It is important that you never attach a jumper wire to this Common terminal. It can cause a short circuit that can ruin the thermostat in a "flash."

Some "smarter" thermostats with a 24-V clock circuit do not need a Common terminal. They recognize that there is always at least one open switch in the thermostat, and there will be 24 V available across that switch. The thermostat can use that 24 V potential to power the clock.

ADVANCED CONCEPTS

Using all of the preceding options, you can see that there are potentially 56 different set points that might be programmed into the microelectronic thermostat (4 programs per day times 7 days per week times 2 modes of operation = 56).

MICROELECTRONIC THERMOSTAT

The mechanical setback thermostat just described has a number of shortcomings.

1. Although there can be four different programs, there can be only two different set points. For example, if you wanted a minimum of 66°F when you were sleeping, 62°F when nobody was home during the day, you could not do it with the mechanical setback thermostat.

2. The mechanical setback thermostat must have the pins reset for different times when you change from heating to cooling. In heating, you want the warmer set point to be used during the occupied periods of the day, but for cooling, you want the warmer set point to be used only when there is nobody home.

3. There is no provision to have a different program for the weekends when the occupancy pattern is different from weekdays.

The microelectronic thermostat in Figure 19–27 addresses all these problems and more. It has the same terminal identifications on the subbase, but the temperature sensing and switching is all based on microprocessors similar to those found in a computer. Many microelectronic thermostats have the following features.

1. A different set point can be selected for each of the four daily programs.

2. A different program (times and/or temperatures) can be used for each day of the week.

3. Separate programs are used for heating mode and cooling mode.

The microelectronic thermostats have a battery that is used to store the program in case there is a power outage. The method of programming each thermostat varies, depending on the manufacturer.

Before attempting to install one of these thermostats for a customer, you should review the instructions and practice programming until you can do it quickly and easily. Then, when the customer tells you what his or her daily routine is like, you can program it quickly and easily, as well as teach the customer how the programming is done.

TWO-STAGE THERMOSTATS

Figure 19–28 shows a thermostat that has two mercury switch elements attached to a single bimetal element. They are mounted slightly out of parallel, so that as the bimetal tilts, it will close one mercury switch, and then, if the bimetal continues to tilt more in the same direction, it will close the other mercury switch. In this way, we can provide two stages of heating, or two stages of cooling, or both. In a system with two stages of heat, instead of there being a W terminal on the subbase, there will be a W1 and a W2. If there are two stages of cooling, instead of a Y terminal, there will be a Y1 and Y2.

Here's how a two-stage cooling thermostat might work. A rooftop air conditioner has two completely separate refrigeration circuits, but with a single evaporator fan and a single condenser fan (Figure 19–29). The thermostat set point is 73°F. When the room temperature rises to 74°F, the thermostat makes R–Y1, bringing on the first compressor, and R–G, bringing on the blower motor. Most times, this will be sufficient to cool the room back down to 72°F, and the thermostat will become satisfied and open R–Y1, turning off the first stage of cooling. However, on a warm day, the first stage of cooling may not be sufficient to stem the rise of room temperature, and even after the first-stage compressor turns on, the room temperature continues to rise. When the room temperature rises to 75°F, the thermostat makes R–Y2 (in addition to R–Y1, which has already been made). Then, the second stage of cooling operates along with the first stage. The second stage of cooling is the second to come on, but when the room temperature drops, it is the first to go off.

HEAT PUMP THERMOSTATS

With the rising popularity of heat pumps, thermostats have been designed to facilitate the wiring of heat pumps. Heat pump thermostats are different from the standard heating–cooling thermostats in two respects:

1. On heating, the heat pump thermostat makes R–Y and R–G, compared to a standard thermostat that makes R–W and uses a separate furnace switch to bring on the indoor fan.

2. The B or the O terminal is always used (to operate the reversing valve), whereas on the standard heating–cooling thermostat, it is almost never used, and often is not even supplied.

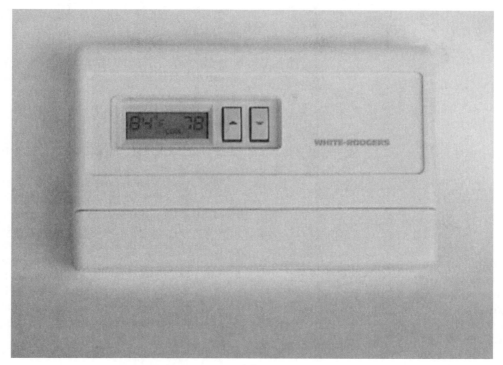

(a)

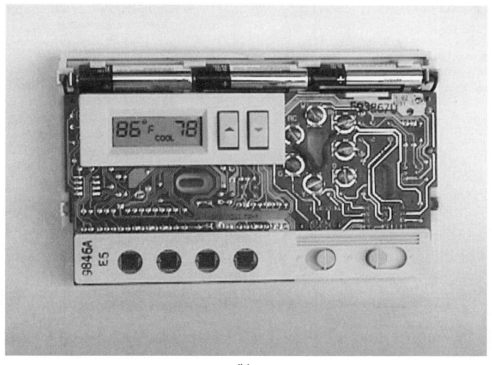

(b)

FIGURE 19–27   (a) Electronic thermostat, (b) with cover off. (Courtesy of BET Inc.)

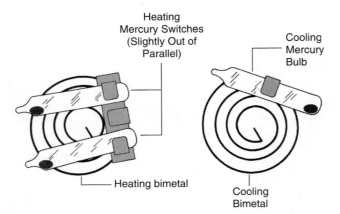

FIGURE 19–28   Two-stage thermostat element.

3. The heat pump thermostat will normally have a second stage of heating available, because auxiliary heat will be required in all but the mildest climates. The second stage heat terminal may be labelled W2. There is no W1. The first stage of heating is accomplished by bringing on the compressor.

ELECTRIC HEAT THERMOSTATS

An electric furnace consists of a blower that circulates room air over one or more electric resistance heating elements. Electric heat is different from gas heat in that we cannot tolerate the time delay for

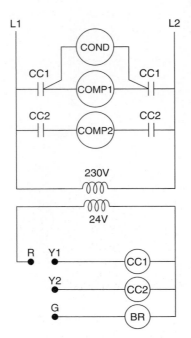

FIGURE 19–29   Packaged rooftop air conditioner with two stages of cooling.

the operation of the furnace fan. The electric heat thermostat makes R–G at the same time that it makes R–W. The R–G switch will bring on the furnace fan at the same time that R–W is bringing on the electric resistance heating elements. On cooling, R–G is made at the same time as R–Y, just as with a standard thermostat.

MILLIVOLT THERMOSTATS

Millivolt systems use a pilot generator that provides an operating voltage of between 185 and 600 millivolts (1000 mV = 1 V). Because of these very low voltages, the millivolt thermostat is designed to provide less resistance than a standard 24-V thermostat. Even the slightest resistance on contact points or resistance of a heat anticipator can render a millivolt system inoperable. When replacing a millivolt thermostat, make sure that the new thermostat is rated for millivolt service.

CALIBRATING THERMOSTATS

There can be two calibrations that you might need to do on a room thermostat:

1. When the thermostat lever is set for 72°F, the thermostat may actually be controlling the room at a different temperature. The customer complaint will be something like, "we need to set the thermostat at 78°F in order to maintain the room at 72°F."

2. The thermometer on the thermostat (not technically part of the thermostat) reads an incorrect temperature.

The first problem may be caused if you have a mercury bulb thermostat that is not mounted in a level position. There is usually one slotted screw hole for the screw holding the subbase to the wall. This allows you to move the thermostat and hold it in a level position (using a carpenter's level) while you tighten the screw (Figure 19–30). If the thermostat is level, and it is still controlling around a different temperature than the set point lever, the bimetal element must be rotated. Sometimes a special thermostat tool is required.

☞ If you don't have the tool necessary to rotate the bimetal element on its mounting, you may be able to compensate for an out-of-calibration thermostat by actually mounting it out-of-level to compensate. If the

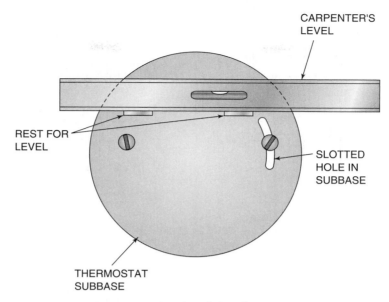

CARPENTER'S
LEVEL

REST FOR
LEVEL

SLOTTED
HOLE IN
SUBBASE

THERMOSTAT
SUBBASE

FIGURE 19–30 Leveling a thermostat.

calibration is not too out-of-line, this may be acceptable, especially if the thermostat is round.

Heat-Limit Thermostats

Safety devices called **limit controls** are used with most heating systems. Limit controls (switches) are used in different areas of the heating system. The limit switch is designed to open if the temperature at the sensing point is above a predetermined limit.

In the furnace, a limit switch is located in the plenum chamber (Figure 19–31). If the heat in the plenum rises to a dangerous level, the limit thermostat opens, removing power from the electrically operated controls. The limit switch may be connected in either the 24-volt control circuit or the 120-volt line circuit.

Limit switches, also called high-limit controls, are connected in oil-burner-heating and strip-heating systems in much the same manner. Excessive temperature in the plenum chamber causes the limit switch to open, removing electrical power from the oil burner control valves or the heat strip contactors.

Bimetal Snap Limits (Trade Name KLIXON)

Electric heat-strip units used with modern air-conditioning systems usually have a bimetal limit switch built into the heat-strip assembly. If for any reason the temperature at the heat-strip unit rises above a safe limit, the bimetal switch snaps open, removing power from the heat strip. The bimetal limit switch is connected in series with the heating element (Figure 19–31).

Heat Limit

FIGURE 19–31 Heat strip. (Courtesy of BET Inc.)

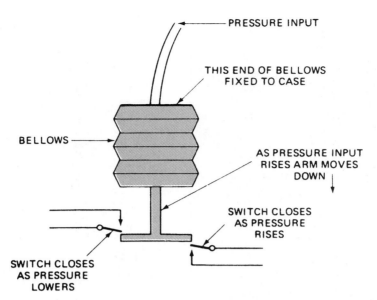

FIGURE 19–32 Pressure-operated switch.

As an added protection, a special fuse is included as an integral part of the heat strip. Should the bimetal limit switch fail to open at the high-temperature limit, the fuse link will melt, removing power from the heat strip.

Pressure-Control Switches

Pressure-control switches are used as safety and/or temperature-sensing devices. The heat-limit switch shown in Figure 19–32 is a pressure switch that is sensitive to temperature change.

Pressure switches, regardless of type, have a common element, the bellows. A bellows is an expanding-contracting element that operates the switch or switches. The heat-limit switch shown in Figure 19–32 is an example of a bellows-operated switch.

The switches in Figure 19–32 are both normally open. However, they could have both been normally closed, or one closed and one open. Switch functions NO or NC or any combination are found as required by the application of the device.

Temperature-Sensitive Pressure Switch When the pressure input (capillary) tube to a pressure switch is closed at the end and is filled with an inert gas or liquid, it becomes sensitive to temperature (Figure 19–33). Often a remote bulb is soldered to the input tube to hold a larger supply of inert material in the miniature closed system. As the temperature rises, the inert material expands, causing the bellows to move. As the temperature falls, the inert pressure decreases, and the bellows move in the opposite direction. As required by the design, switching is controlled by the moving end of the bellows.

Limit Pressure Controls Limit pressure controls are used to connect or disconnect electrical power from condenser system components when the pressure being monitored is above or below the preset limit value. The pressure switch used may be a high-pressure, low-pressure, dual-pressure, or pressure-differential switch. Figure 19–34 shows examples of pressure switches.

High- and low-pressure switches are similar in appearance. The pressure range at which the switch bellows mechanism operates determines whether it is a high- or low-pressure control. Low-pressure controls

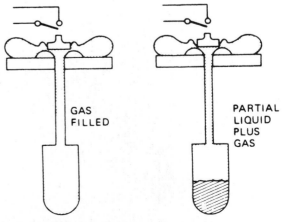

FIGURE 19–33 Temperature-sensitive switches.

FIGURE 19–34 Pressure switches. (Courtesy of BET Inc.)

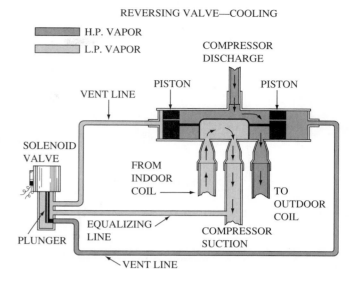

FIGURE 19–36 Reversing valve cutaway. (Reprinted from Refrigeration and Air Conditioning, 3rd Ed, by Air Conditioning and Refrigeration Institute.)

in the air-conditioning and refrigeration industry usually have a gauge (psig) range of from 12 inches of mercury (Hg) to 50 pounds per square inch. High-pressure controls usually range from 10 to 425 psig.

Often, the pressure at which the switches actuate is adjustable. There is a difference in the pressure range at which a system should be operational, depending on the altitude at which the system is located or on the desired temperature range.

Heat Pump Reversing Valves

The operation of the heat pump is the same as an air conditioner, except for the addition of a reversing valve and a defrost cycle.

The reversing valve is shown in Figure 19–35. A cutaway view is shown in Figure 19–36. Hot gas from

FIGURE 19–35 Reversing valve. (Courtesy of Ranco North America.)

the compressor enters the top port and is directed out to either the indoor coil or the outdoor coil, depending on the position of the piston inside the barrel. Whichever coil is receiving the hot gas from the compressor acts as the condenser. The other coil acts as the evaporator. Figure 19–37 shows the direction of refrigerant flow for heating and for cooling.

When the position of the reversing valve changes, the operation of the unit changes from cooling to heating. In the cooling mode, the indoor coil is cold and absorbs heat from the indoor air. The outdoor coil is warm, and the heat is rejected to the outside air. In the heating mode, the outdoor coil is

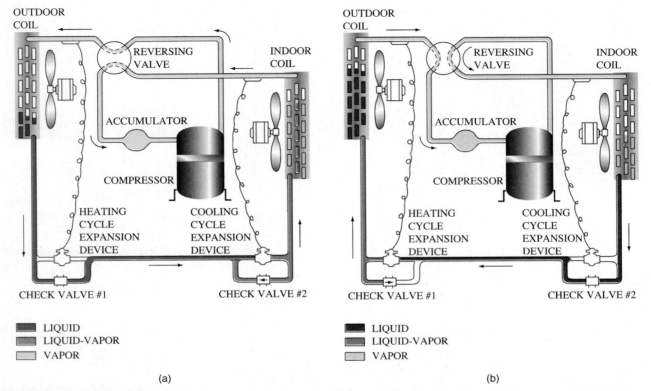

FIGURE 19–37 Heat pump refrigerant flow. (a) Cooling cycle. (b) Heating cycle. (Reprinted from Refrigeration and Air Conditioning, 3rd Ed, by Air Conditioning and Refrigeration Institute.)

cold, and it absorbs heat from the outside air. The indoor coil is warm, and the heat is rejected into the room. Some manufacturers energize the solenoid coil on the reversing valve to put it into the heating position. Other manufacturers energize the solenoid coil in the cooling mode. There is no standard. Some manufacturers believe it is more important to provide heat in the event of a solenoid coil failure, whereas others believe that it is more important to provide cooling in the event of a solenoid coil failure. It will be up to the service technician to determine which method is being used.

There are two general schemes for controlling the reversing valve. They are shown, in general terms, in Figure 19–38. SW1 is a switch that is operated by the operation of the room thermostat. In Figure 19–38a, it may be a normally open switch that is closed when the thermostat calls for cooling (usually through a control relay called the reversing valve relay), or it may be a normally closed switch that is opened when the room thermostat calls for heating. SW2 is a switch that is normally open and closes when the unit needs to go into defrost cycle (usually through a control relay called a defrost relay). The load, L, may be the reversing valve solenoid

coil, or a control relay coil whose contacts will operate the reversing valve. In Figure 19–38a, the reversing valve will be energized when the room thermostat calls for cooling or when the defrost control says that it's time for a defrost. When the reversing valve

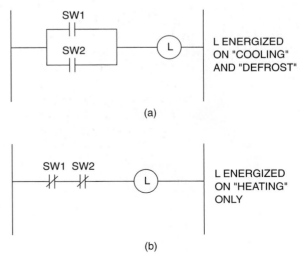

FIGURE 19–38 Two schemes for controlling the reversing valve.

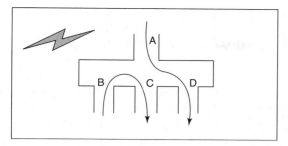

FIGURE 19–39   Reversing valve label.

is de-energized, the heat pump will deliver heat. In Figure 19–38b, if the room thermostat needs cooling, it opens SW1. If the defrost control wants a defrost, it opens SW2. Either of these two switches opening will cause L to become de-energized. Therefore, L will be energized only in the heating mode.

> In most cases, if the reversing valve relay switch and the defrost relay switch are wired in parallel, the reversing valve is de-energized on heating (Figure 19–39). If the reversing valve relay switch and the defrost relay switch are wired in series with the reversing valve, the reversing valve is energized on heating.

HEAT PUMP THERMOSTAT

The room thermostat used to control a heat pump might be of the same type that is used to control a standard gas-heating electric-cooling unit. But more likely, it is a special thermostat designed specifically for heat pump service. Some of the differences between a heat pump thermostat and a standard gas-heating electric-cooling thermostat are as follows.

1. The heat pump system uses the B or the O terminal on the thermostat subbase. Although a standard thermostat may have these terminals, they are usually not used. The thermostat makes R to B when the system switch on the thermostat subbase is set to Heat. It makes R to O when the system switch is set to Cool (to help you remember. . . R to O on Cool).

2. The standard thermostat makes R to Y only on a call for cooling to energize the compressor contactor. The heat pump thermostat makes R to Y on a call for heating as well as on a call for cooling. This is necessary because the heat pump system needs to run the compressor for heat.

3. The standard thermostat makes R to G (energizing the blower relay) only on a call for cooling, or when the fan switch is moved from Auto to On. The heat pump thermostat makes R to G on a call for heating, as well as on a call for cooling. This is necessary because on heating, the indoor fan acts as the condenser fan. It must start at the same time as the compressor, or the head pressure will go too high.

HEAT PUMP DEFROST CYCLE

Unlike a conventional heating/cooling system, the heat pump requires a defrost cycle. A standard air conditioner does not require a defrost because under normal conditions, the evaporator should never operate at a temperature below freezing. However, when the heat pump operates in the heating mode, it might be cooling outdoor air that is 40°F or less, and the outdoor coil that is cooling the outdoor air could be well below freezing. Moisture from the outside air condenses on the cold outdoor coil and freezes. Without a means to defrost the outdoor coil, the airflow would be blocked within a few hours of operation on cool days.

During the defrost cycle, the following happens.

1. The reversing valve is energized or de-energized as required to return the system to the cooling mode of operation. The hot gas from the compressor will be directed to the outdoor coil, melting the ice from its fins.

2. The outdoor fan is switched off. This will allow the coil to get warmer, without losing heat to the cool outside air.

3. Supplemental electric heat will be energized in order to offset the room cooling that is being done by the cold indoor coil. (In exceptionally mild climates, where defrost seldom occurs, heat pumps are often supplied without supplemental heat.)

HEAT PUMP CIRCUITS

Various manufacturers use different methods to accomplish control of the heat pump. However, they all accomplish essentially the same things. Some of the differences are as follows.

1. Some heat pumps energize the reversing valve on cooling and defrost; others energize the reversing valve on heating.

2. Some heat pumps use a heat pump thermostat; others use relays to accomplish the same thing with conventional thermostats.

3. Some heat pumps use the B terminal on a heat pump thermostat; others use the O terminal.

4. Some reversing valves are in the 24-V circuit; others are in the 230-V circuit.

5. Many different methods are used to sense when the heat pump requires a defrost and to put the system into the defrost mode of operation.

HEAT PUMP CIRCUIT NO. 1 (STANDARD THERMOSTAT, AIR SWITCH DEFROST)

Figure 19–40 shows an early design of heat pump circuit that was used before heat pump thermostats became popular. In order to understand the sequence for any heat pump system, your first analysis should be to determine whether the reversing valve solenoid is energized on heating or on cooling. In this case, the RVR switch and the DFR switch are in series with the RVS, so you suspect that the reversing valve will

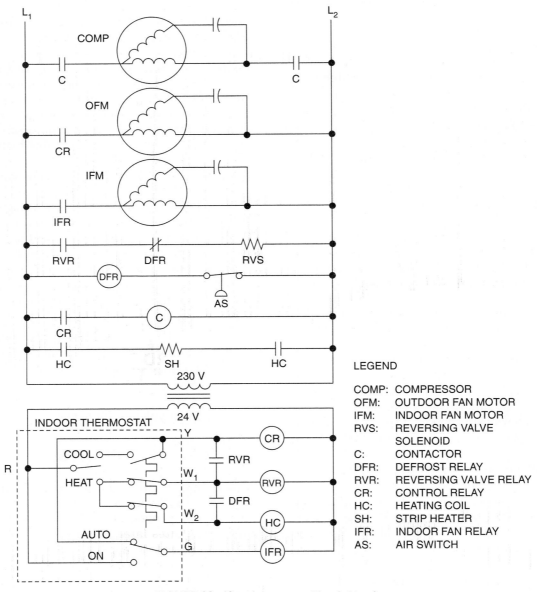

FIGURE 19–40 Heat pump Circuit No. 1.

LEGEND

COMP: COMPRESSOR
OFM: OUTDOOR FAN MOTOR
IFM: INDOOR FAN MOTOR
RVS: REVERSING VALVE SOLENOID
C: CONTACTOR
DFR: DEFROST RELAY
RVR: REVERSING VALVE RELAY
CR: CONTROL RELAY
HC: HEATING COIL
SH: STRIP HEATER
IFR: INDOOR FAN RELAY
AS: AIR SWITCH

be energized on heating (see earlier practice pointer). To confirm your suspicion, you note that when the room thermostat makes R–W1 (heating), it energizes the RVR coil, closing the RVR switch, and energizing RVS. Your suspicion is confirmed.

Sequence of Operation

On a call for cooling, the room thermostat makes R–Y and R–G. The Y terminal energizes CR coil, operating two sets of contacts. One set of CR contacts energizes the compressor contactor, whereas another set of CR contacts brings on the outdoor fan motor. The G terminal energizes the indoor fan relay (IFR) coil, bringing on the indoor fan motor in the 230-V circuit. The unit runs the same as if it were a conventional air conditioner.

On the heating cycle, the room thermostat only closes the R–W1 contact, which energizes the reversing valve relay (RVR). The RVR is simply a control relay with multiple sets of contacts. One set of RVR contacts is wired in parallel with the R–Y switch to energize the CR and operate the outdoor unit (compressor and outdoor fan). It also energizes the IFR. A second set of RVR contacts in the 230-V circuit complete a circuit to energize the reversing valve solenoid (RVS). The unit operates the same as in the cooling mode, except that the reversing valve has operated to direct the hot gas to the indoor coil instead of the outdoor coil. If the heating capacity of the heat pump is insufficient to maintain room temperature, the room thermostat will make the second stage of heating, R–W2. This will energize the heating contactor (HC), which will, in turn, bring on the supplemental electric strip heater (SH).

The defrost cycle is initiated by an air pressure switch (AS) that senses when the airflow across the outdoor coil is becoming blocked due to ice. When the air pressure at the suction side of the outdoor fan decreases, it closes AS, energizing DFR, and de-energizing RVS. This places the reversing valve in the position that will direct hot gas to the outdoor coil. At the same time that the DFR energizes, a second set of DFR contacts in the 24-V circuit closes, energizing the heating contactor, HC1. This will provide supplemental heat during the period of defrost. When the temperature-sensing portion of the AS switch determines that the outdoor coil has been defrosted, it opens, de-energizing the defrost relay.

Outside Air Override

Figure 19–41 shows an almost identical circuit to the previous circuit, but with one important change.

A thermostat sensing outside air temperature (OT) has been added in series with the contactor for the strip heater (HC1). The purpose of the outdoor air thermostat is to keep the strip heater off whenever the temperature outside is warm enough that the heat pump operation alone should be sufficient to maintain comfort conditions. Without the outdoor air thermostat, the following undesirable result can occur.

Suppose a home owner turns the heater off at the thermostat upon leaving in the morning. It is 55°F outside, so the temperature in the house drifts down to 67°F over a period of several hours. When the home owner returns, she sets the thermostat switch to Heat. The set point is 72°F. The thermostat is set up so that it will make R–W1 if the room temperature is below 71°F, and it will also make R–W2 if the room temperature is below 70°F. Therefore, with the room at 67°F, the thermostat will bring on both the heat pump and the electric strip heater. It is far more expensive to bring the house up to 72°F using the strip heater, than to keep the strip heater off and simply let the heat pump bring the house temperature up to set point.

The outdoor air thermostat is typically set for 25°F to 45°F. So long as the temperature outside is above this set point, the electric heat is locked out. The heat pump, operating alone, should be able to bring the house temperature up to set point. It will take longer, but it will be much less expensive. When the outside air drops below the outdoor air thermostat set point, the heat pump might not be able to satisfy the room demand by itself, and the strip heater is allowed to operate.

HEAT PUMP CIRCUIT NO. 2 (STANDARD THERMOSTAT, OLD STYLE TIMER)

Figure 19–42 shows another heat pump circuit, similar to the previous circuit, except for the operation of the defrost cycle.

This defrost cycle is operated by a time clock, DTM (defrost timer motor). It has two sets of contacts (DT) that are operated off a cam that is driven by the DTM. Every few hours the DTM will cause the normally open contact to close momentarily, and 30 minutes later, the DTM will cause the normally closed contact to open momentarily. The defrost thermostat (DFT) senses whether or not a defrost is needed by sensing the temperature of the outdoor coil. If the DFT is open, indicating that a defrost is

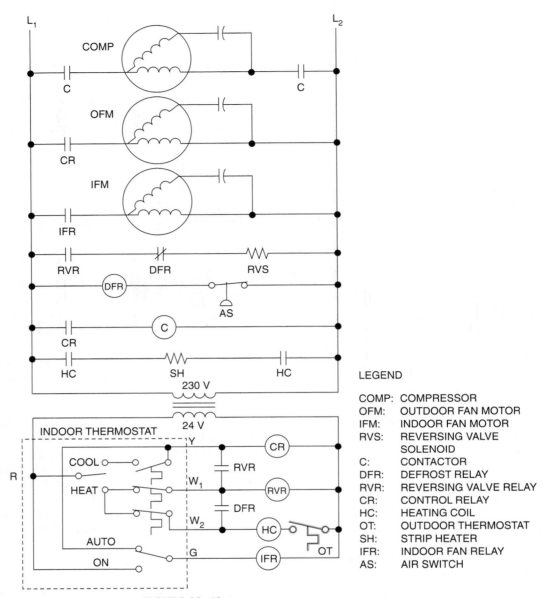

FIGURE 19–41   Circuit with outside air override.

not required, then the operation of the DTM and its switches has no effect. However, when the DFT is closed, the defrost relay (DFR) becomes energized as soon as the normally open DT switch closes. The DFR is simply another control relay with two sets of contacts. The DFR remains energized, even after the normally open DT contact reopens, because it gets "sealed-in" by the normally open DFR contacts.

With the DFR energized, the normally closed DFR contacts open, de-energizing both the RVS and the OFM. Hot gas is directed to the outdoor coil,

accomplishing a defrost. At the same time, the indoor coil is receiving cold refrigerant, so the room temperature drops. The second stage heat on the room thermostat makes R–W2, energizing a heating contactor (HC), which energizes the supplemental heater. When the outdoor coil has been defrosted, it is sensed by the DFT which opens, de-energizing the DFR, and the system returns to normal operation. If, for some reason, the DFT does not open, the defrost will terminate anyway when the DTM causes the momentary opening of the normally closed DT contact.

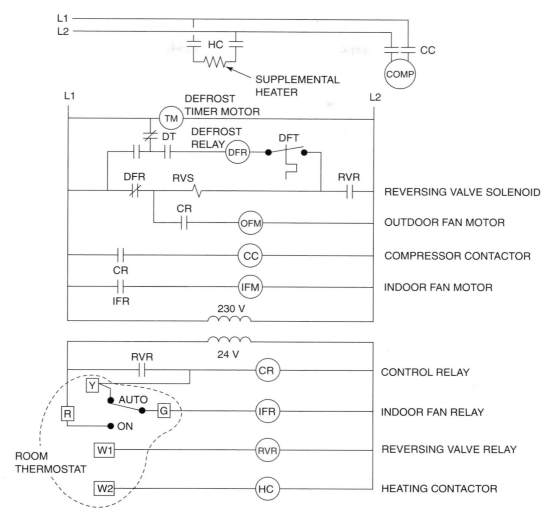

FIGURE 19–42 Heat pump with standard thermostat and old-style timer.

TYPES OF MODERN DEFROST CONTROLS

Generally, the modern controls that place the heat pump into the defrost mode fall into two categories. The mechanical type (Figure 19–43) consists of a clock motor that operates an SPDT switch that toggles between outdoor fan operation and placing the reversing valve in the cooling mode so that it will send hot gas to the outdoor coil for defrost. An alternative mechanical type of defrost control (Figure 19–44) can use the air pressure difference across the outdoor coil to initiate the defrost, instead of a clock.

The electronic type of defrost control (Figure 19–45) consists of solid-state controls on a printed circuit board (a "black box" control). The printed circuit board logic may be one of two types.

1. Time-initiated, temperature-terminated: The board receives 24 V power. There is a fixed time interval between defrost cycles (there are usually several different time intervals available, one of which is selected). There is a defrost temperature sensor on the outdoor coil. When sufficient time has elapsed for the circuit board to place the heat pump into defrost, it energizes a defrost relay. When the defrost temperature sensor determines that the outdoor coil is defrosted, it interrupts the 24-V circuit to the printed circuit board, and the defrost relay switches return to their normal position (heating).

2. Demand defrost: The printed circuit board for the demand-defrost system has two permanently attached thermistors that sense

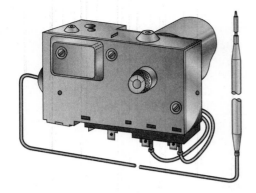

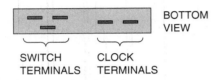

BOTTOM VIEW

SWITCH TERMINALS   CLOCK TERMINALS

FIGURE 19–43 Mechanical defrost control.

temperature on the outdoor coil and the outdoor ambient temperature. The board does not put the system into defrost on any regular schedule. It energizes a defrost relay (mounted on the board) whenever it is required.

HEAT PUMP CIRCUIT NO. 3 (SPDT MECHANICAL DEFROST CONTROL)

The reversing valve for this system (Figure 19–46) is de-energized in the heating mode and energized in the cooling mode and the defrost mode.

The mechanical defrost control switch is normally in the 2–1 position, allowing the outdoor fan to operate whenever the compressor contactor is

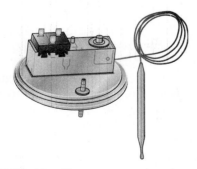

FIGURE 19–44 Air temperature/pressure switch.

FIGURE 19–45 Electronic defrost control.

energized. Every few hours of time-clock operation, the defrost control will analyze the temperature signal being received from the thermal bulb (mounted on the outdoor coil) to determine if a defrost is required. If the sensing bulb is above freezing, nothing happens, at least for another few hours. If the sensing bulb indicates that a defrost is needed, the switch will move from 2–3 to 2–1. This energizes the

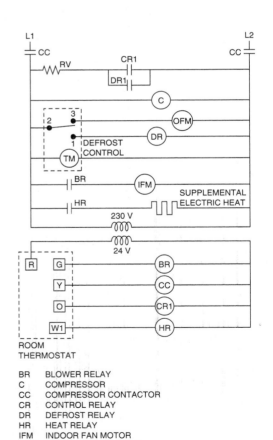

BR	BLOWER RELAY
C	COMPRESSOR
CC	COMPRESSOR CONTACTOR
CR	CONTROL RELAY
DR	DEFROST RELAY
HR	HEAT RELAY
IFM	INDOOR FAN MOTOR
OFM	OUTDOOR FAN MOTOR
TM	TIMER MOTOR

FIGURE 19–46 Heat pump Circuit No. 3.

defrost relay (DR), energizing the reversing valve to send hot gas to the outdoor coil. It also shuts off the outdoor fan. When defrost is complete, the switch in the defrost control returns to the 2–3 position, and normal heating operation resumes.

HEAT PUMP CIRCUIT NO. 4 (ELECTRONIC DEFROST CONTROL)

Figure 19–47 shows an interconnection diagram for a packaged heat pump with a heat-pump thermostat and electronic defrost control. It provides one stage of cooling and two stages of heating (as an optional accessory). This diagram can be intimidating to the new service technician, so it has been redrawn in a somewhat simplified ladder diagram format in Figure 19–48. In many cases, you can troubleshoot the circuit and determine the problem without attempting to understand the entire sequence of operation. For example, if the compressor and outdoor fan won't run, you can always go to the compressor contactor to determine if the coil is energized. If it is not, it is simple to determine from the wiring diagram that the contactor coil is in series with the high-pressure switch and the room thermostat. Those two switches can be checked for voltage drop to determine which is interrupting the power to the contactor coil.

But sometimes you must determine the whole sequence of operation in order to diagnose the problem. The cooling portion of the circuit is similar to any air conditioner. When the thermostat makes R–Y and R–G, the compressor contactor and blower relay coil are energized, starting all of the cooling components. One question that always arises on a heat pump circuit is whether the reversing valve is energized to place the system in cooling mode or in heating mode. There are two independent ways of arriving at that answer.

1. The reversing valve relay (RVR) is energized when the room thermostat makes R to B, which happens when the room thermostat system switch is set to Heat. When the RVR is energized, it closes contacts to energize the reversing valve (RV). Therefore, the reversing valve is energized during the heating mode.

2. During defrost, the reversing valve goes to the position that will produce cooling (so

that hot gas is directed to the outdoor coil, defrosting it). When the defrost relay (DFR) coil is energized, calling for defrost, it opens the normally closed contacts between terminals 1 and 2, de-energizing the reversing valve. Therefore, during defrost or cooling, the reversing valve is de-energized, so it must be energized during the heating mode.

Defrost Control

The defrost control is an electronic "black box." If you are not familiar with the particular model being used (of which there are many), you have to figure out its operation based on what you know it should do. The box has 24 V applied to it continuously between the lower 24-V terminal and the COMM terminal. The defrost sensor switch senses when a defrost is needed (it might be a thermostat that senses when the outdoor coil is excessively cold, or it might be a pressure switch sensing when there is excessive pressure drop across the outdoor coil). When that happens, the box should provide 24 V to the defrost relay coil. This should de-energize the reversing valve, causing it to switch to cooling mode. If you come upon this unit, and the outdoor coil is covered with ice, you would check the defrost control as follows.

1. Check to see that you have 24 V to the defrost control between the lower 24-V terminal and the COMM terminal. If you do . . .

2. Check to see that the defrost sensor is closed (zero voltage across it). If it is . . .

3. Check to see if the defrost control is putting out 24 V to the defrost relay (between the upper 24-V terminal and the DEF RELAY terminal). If it is, then the defrost control is doing its job, and you will need to check the defrost relay coil, contacts, and reversing valve. If there is not 24 V to the DFR coil from the defrost control, then the defrost control has failed because all of the required inputs to the defrost control are present, and you are not getting the proper output.

Supplemental Heat

When the unit goes into defrost, it will deliver cold air off the indoor coil. To offset the cooling effect in the room, the 230-V heat relay coil becomes energized

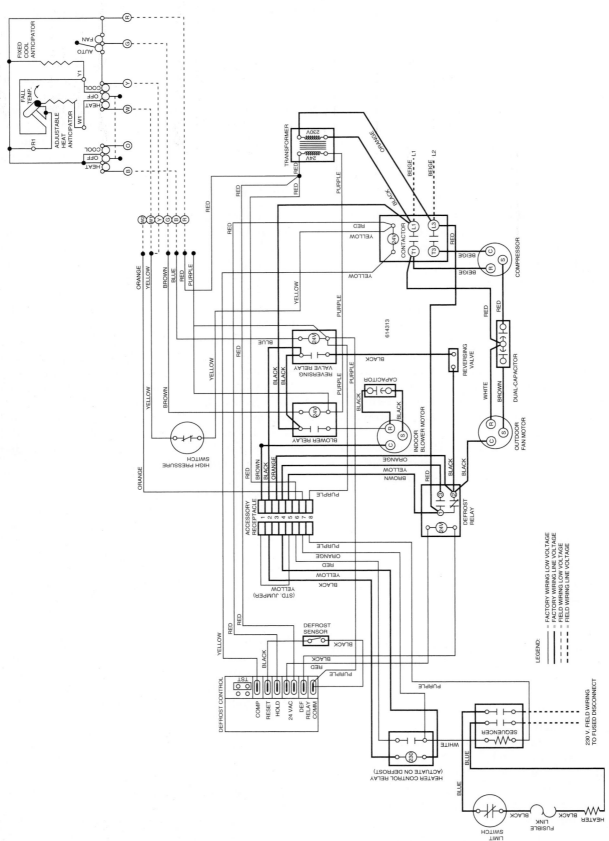

FIGURE 19–47 Packaged heat pump.

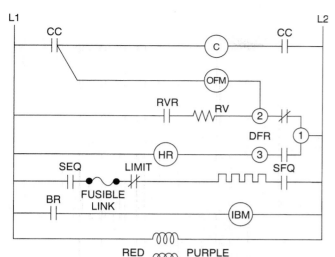

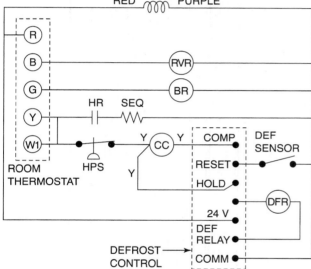

FIGURE 19–48 Heat pump Circuit No. 4.

through the normally open contacts of the defrost relay. This closes the normally open heat relay switch, energizing the sequencer heater element. In this circuit, the supplemental heat is not used, except during defrost.

☞ If you have difficulty redrawing an interconnection diagram as a ladder diagram, try this. With one color, mark all the wires on the interconnection diagram that are connected to what would be the left side of the ladder diagram (e.g., L1). Using a different color, mark all the wires connected to what would be the right side of the ladder diagram (e.g., L2). If you are interested in both the power wiring and the control wiring, make a copy of the wiring diagram. Use one to color the

power wiring, and the other to color the control wiring. From this, it should be much easier to construct the ladder diagram. Also, when constructing the ladder diagram, don't try to show all the detail. For example, if the interconnection diagram shows compressor windings, start relay, start and run capacitors, and an overload, you can simply use a circle labeled COMP on the ladder diagram to represent the compressor and all of its associated components.

HEAT PUMP CIRCUIT NO. 5

Figure 19–49 shows the same circuit as Figure 19–47 with the following additions:

1. The room thermostat has two bulbs to provide two-stage heating. If the room temperature continues to drop after the room thermostat has made R–W1, the thermostat will then make R–W2, energizing the small heater element inside the heat sequencer. The switches inside the sequencer will then close, bringing on the heater (through the safety devices).

2. A PTC has been added around the compressor run capacitor for added starting torque (refer to unit 15 for a discussion on the PTC device).

3. There is a five-wire control power transformer. The extra tap is provided in the event that the line voltage is 208 V instead of 230 V.

4. A crankcase heater is provided. It is energized whenever there is power available to the unit.

HEAT PUMP CIRCUIT NO. 6—SPLIT SYSTEM HEAT PUMP

Figure 19–50 is the wiring diagram that applies for a condensing unit only. It is similar to an air-conditioning condensing unit, except that it also contains a black box defrost controller, a defrost relay, and a reversing valve.

Sequence of Operation

When the room thermostat makes R–Y (on a call for heating or cooling), it energizes the R1–R2 terminals of the time-delay relay (assuming that the HPC and LPCs are satisfied), energizing the compressor

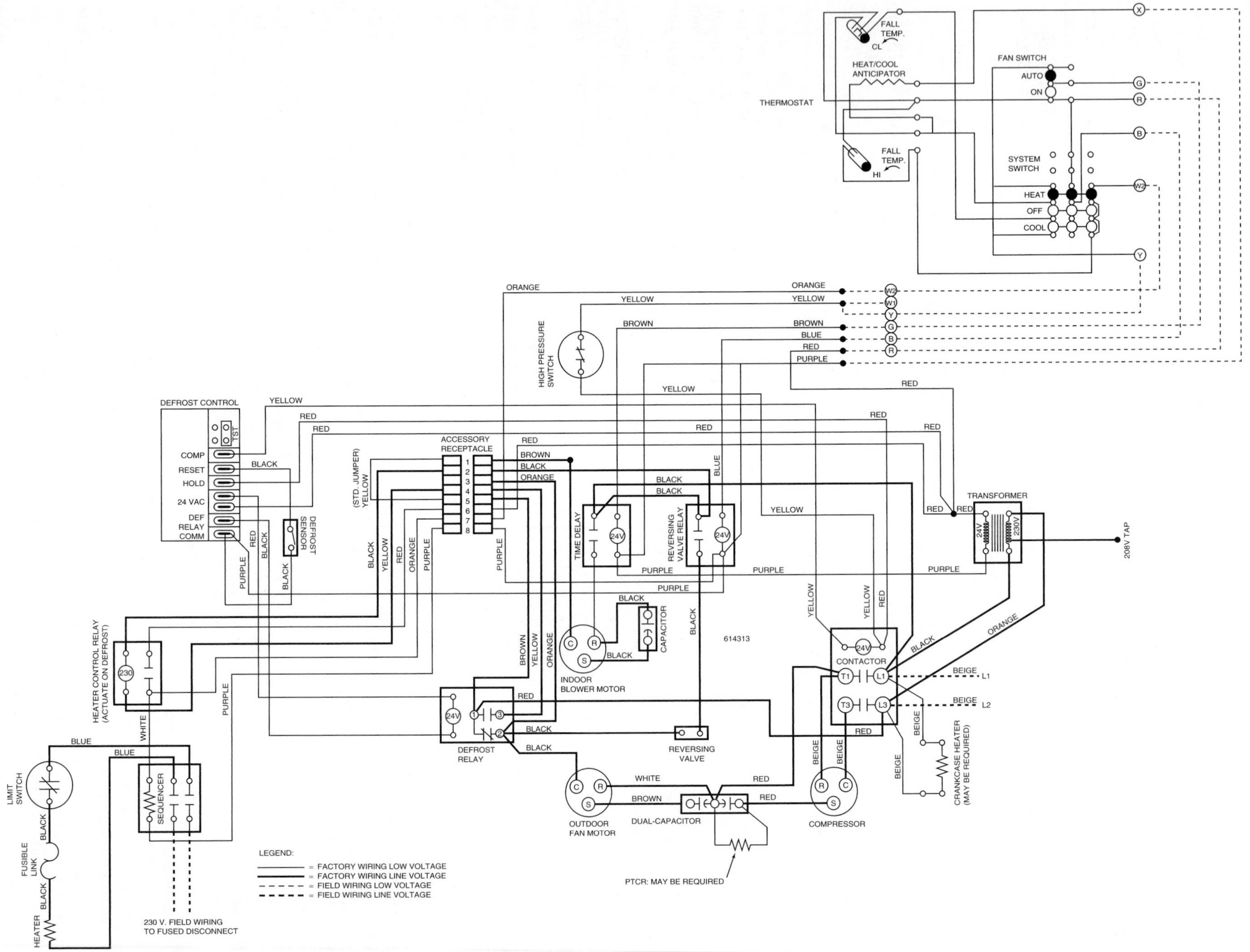

FIGURE 19–49   Heat pump Circuit No. 5.

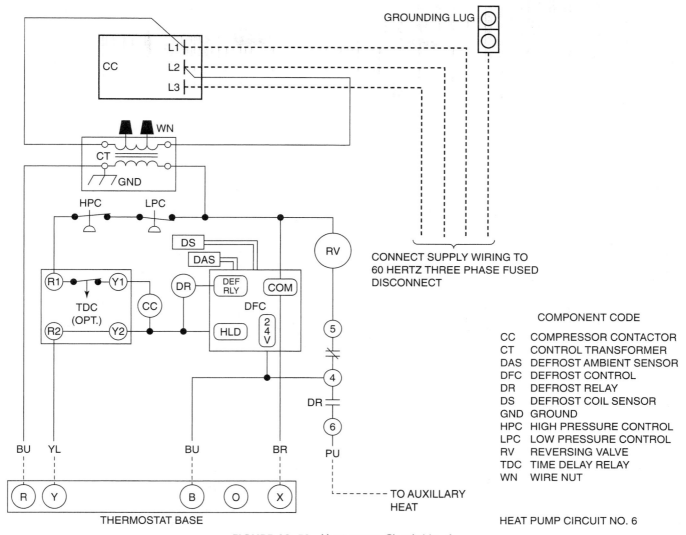

FIGURE 19–50 Heat pump Circuit No. 6.

contactor. If the thermostat is set for Heat, R–B is also made, energizing the reversing valve coil through the normally closed contacts of the defrost relay.

The defrost control (DFC) gets input signals from a defrost coil sensor (DS) and a defrost ambient sensor (DAS). These are thermistors that sense the temperatures of the outdoor coil and the outdoor air. When the DFC decides to initiate a defrost, it energizes the defrost relay (DR), de-energizing the reversing valve from its terminal 5 and energizing an auxiliary heater from its terminal 6.

WATER-SOURCE HEAT PUMP

Figure 19–51 is a ladder diagram for a packaged water-source heat pump. Instead of using outside air as the medium to reject heat during the summer

and draw heat from during the winter, this system uses well water. When the room thermostat calls for heating or cooling, it makes R–Y1, energizing the pump relay and energizing the pump motor. R–Y1 on the thermostat also brings 24 V to the W1 terminal of the heat pump, energizing the compressor contactor through the time delay, the high-pressure cut-out, the low-pressure cut-out, and the normally closed contacts of the lock-out relay. The time delay in this circuit is really remarkable, as it is only a two-wire device, as compared to the previously discussed four-wire time delays. You can visualize the operation of the two-wire time delay as follows: when 24 V are applied to the one terminal, some time later, it comes out from the second terminal.

The time delay (TD) is another "black box" electronic device. When power is first applied to

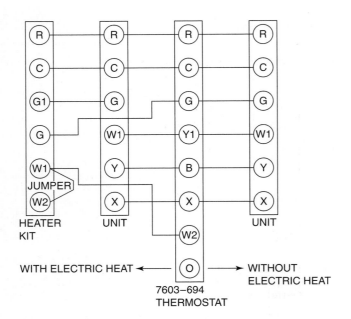

HEATER
KIT

UNIT

WITH ELECTRIC HEAT ← → WITHOUT
ELECTRIC HEAT

7603–694
THERMOSTAT

UNIT

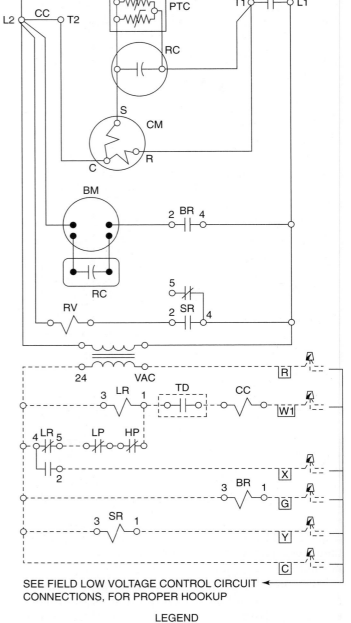

WELL WATER PUMP INSTALLATION

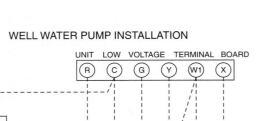

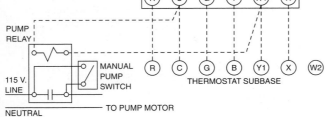

NOTE: WHEN USING 230V. PUMP, USE 2 POLE RELAY
AND BREAK BOTH SIDES OF LINE

SEE FIELD LOW VOLTAGE CONTROL CIRCUIT ←
CONNECTIONS, FOR PROPER HOOKUP

LEGEND

BM	BLOWER MOTOR	RC	RUN CAPACITOR
BR	BLOWER RELAY	RV	REVERSING VALVE
CC	COMPR. CONTACTOR	SC	START CAPACITOR
CM	COMPR. MOTOR	SR	SLAVE RELAY
HP	HI PRESSURE SW.	STR	START RELAY
LP	LOW PRESSURE SW.	TR	TRANSFORMER
LR	LOCKOUT RELAY	TB	TERMINAL BOARD
LT	LOW TEMP. T-STAT	TSB	T-STAT SUBBASE

FIGURE 19–51 Water-source heat pump.

the TD, it presents high resistance in series with the compressor contactor coil (CC) circuit. The current path is, left to right, LR, LP, HP, TD, and CC. Initially almost all the voltage appears across the TD. After a fixed time delay a triac (see Figure 17–20) inside the TD is actuated. The voltage across the TD drops to about 0.5 volts. The remaining voltage appears across the CC coil, energizing it.

The lock-out relay will become energized if either the HPC or LPC opens (see unit 18 to review the operation of the lock-out relay). In addition to locking out the compressor by opening the switch between terminals 4–5, the lock-out relay also closes a switch between terminals 2–4. This completes a circuit through the X terminal of the room thermostat, which will cause a trouble light to illuminate.

SUMMARY

The different types of electrical and mechanical controls available for service in air conditioning and refrigeration are very extensive. It is impractical to attempt to cover them all in a basic textbook. It is therefore recommended that the technician study catalogs and specification sheets of control manufacturers in detail for further information; manufacturers' catalogs and "spec" sheets are an invaluable source of technical information.

PRACTICAL EXPERIENCE

Using the procedure given in the paragraph on Heating and Cooling Anticipators, measure the current draw of relays and/or contactors suitable in the control of heat systems.

Obtain manufacturers' information on different reversing valves. Study the material noting the differences.

1. Redraw the rooftop air conditioner components shown in Figure 19–52. Add the required wiring.
2. Redraw the component location diagram in Figure 19–53. Using the ladder diagram of Figure 19–11 as a guide, add the interconnection wiring to the physical components, and answer the following questions:
   a. If the furnace is unplugged, will the compressor contactor still work?
   b. If the 230-V circuit breaker has tripped, will the thermostat be able to operate the compressor contactor?
3. Redraw the interconnection diagram in Figure 19–6 as a ladder diagram. It is not necessary to show the internal parts of the fan motor or compressor. Show all wire colors.
4. Redraw the wiring diagram in Figure 19–13 as a ladder diagram. Include all wire colors and terminal identification.

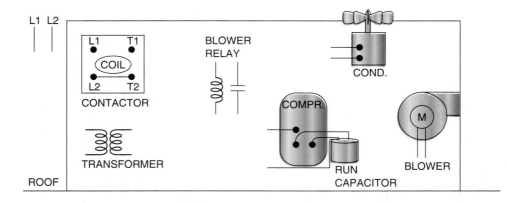

FIGURE 19–52 Exercise.

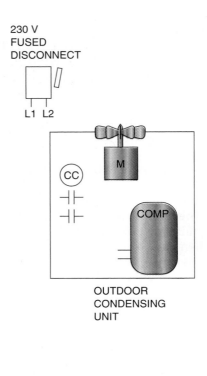

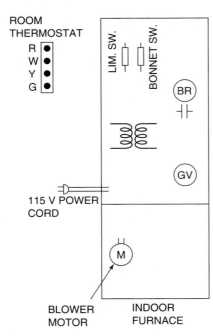

FIGURE 19–53   Exercise.

Conclusion   To understand air-conditioning and refrigeration systems operation, it is necessary to understand how the controls operate.

REVIEW QUESTIONS

1. Open thermostats are less prone to trouble than mercury switch types.

   T _____   F _____

2. In a temperature-sensitive bellows system, as temperature decreases, pressure increases.

   T _____   F _____

3. Anticipators are used in thermostats to decrease the frequency of compressor operation.

   T _____   F _____

4. A room thermostat with a properly operating heat anticipator will provide limits on the variation in room temperature. Would this temperature variation be larger or smaller than without a heat anticipator? (Explain)

5. Why must the heating anticipator be adjustable, whereas the cooling anticipator is nonadjustable?

6. What will be the customer complaint if the heat anticipator is burned out (open)? What will be the customer complaint if the cooling anticipator is burned out?

7. In Figure 19–40, if the defrost relay contacts between W1 and W2 should fail in the open position, the system would malfunction. What complaint would you expect to hear from the customer?

8. In wiring diagram Figure 19–47, the voltage measured at the Defrost Control between 24 vac and DEF is 24 vac. What voltage would you expect to measure across the heater control relay coil? Explain your answer.

9. In the circuit of Figure 19–47, will the outdoor fan operate during the defrost cycle? Will the indoor blower operate? Explain your answers.

10. In Figure 19–51, is the compressor contactor energized before or after the lock-out relay? Explain your answer.

UNIT 20

COMMERCIAL SYSTEMS

OBJECTIVES

After completion and review of this unit, you should be familiar with the components and circuits in commercial systems, such as:

- Defrost timers.
- Hot-gas defrost.
- Special defrost circuits.
- Overloads.
- Motor starters.
- Condensing units.

Commercial systems are distinguished from residential systems because they are generally larger and more expensive, and they sometimes must deal with more rigorous requirements and provide more flexibility than residential systems. Commercial systems dealt with in this chapter include walk-in coolers and freezers, commercial packaged refrigeration systems, and air-conditioning systems used on commercial type occupancies.

COMMERCIAL DEFROST TIMER CIRCUIT NO. 1

Figure 20–1 shows a defrost timer that is used in commercial freezer applications, rather than the type shown in a previous unit, which is limited to smaller units. The commercial defrost time clock looks far more complicated, especially when you look at the wiring diagram that is pasted on the timer enclosure door (Figure 20–2). However, a little investigation will reveal to you that the circuit for the commercial defrost timer shown in Figure 20–3 is exactly the same as for the smaller defrost timers (Figure 20–21). The commercial defrost timer has an outer ring that rotates once each 24 hours and an inner ring that rotates once each two hours. There are multiple trippers located on the outside ring and a single tripper located on the inner ring. The trippers on the outer ring determine the start times for "defrost." The tripper

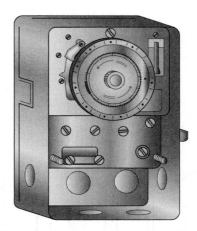

FIGURE 20–1   Commercial defrost timer.

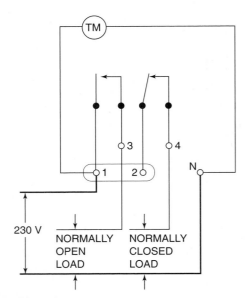

FIGURE 20–2   Factory-supplied wiring diagram.

on the inner ring determines the duration of the "defrost."

Although the circuit of this commercial defrost time clock is similar to the smaller defrost timers, there are several differences:

1. By the placement of tripper pins on the outer ring (they unscrew), the technician may choose how many times the "defrost" will be initiated in a day and at what times they will occur. The intervals do not need to

be the same. If the time of "defrost" initiation is important, then the timer must be set to the correct time of day. Find the present time on the outer scale. Rotate the inner wheel clockwise, until the outer wheel has moved to a position where the present time is aligned with the time pointer.

2. The duration of the "defrost" may be adjusted by moving the pointer on the inner wheel to the duration desired (push in and rotate the pointer).

3. The switches are accessible. The whole "works" of the timer may be pulled out by removing one screw at the top, and two posts at the bottom (they unscrew).

☞ If you find that a switch is not making contact, and you want to get the freezer running while you run for a replacement timer, you can clean the contacts. With the power off, hold a nonglossy business card or matchbook cover between the contacts. Press the contacts together, and move the business card back and forth. It will provide a gentle cleaning for the contacts. The contacts should now work until you are able to return with a replacement time clock. [NOTE: Contacts that have been cleaned have not been repaired. They may last for days or even weeks, but the protective silver coating on the original contacts has been worn away. The cleaned contacts will deteriorate rapidly.]

COMMERCIAL DEFROST TIMER CIRCUIT NO. 2

All the defrost timer circuits that have been presented to this point have been time-initiated, time-terminated. That is, it was only the passage of time that caused the timer to place the SPDT switch into the "defrost" mode, and it was only the passage of time that caused the timer to return the SPDT switch to the "freeze" mode. The circuit shown in Figure 20–4 is typical of the manufacturer's diagram for a time-initiated, *temperature*-terminated system. When the evaporator is defrosted, the system will go back to the "freeze" mode, without waiting for the rest of the "defrost" mode to time out. There are two differences from the prior circuit.

1. The defrost timer includes a solenoid coil. When it is energized, it pulls on a trip-lever that returns the SPDT switch to the "freeze" mode.

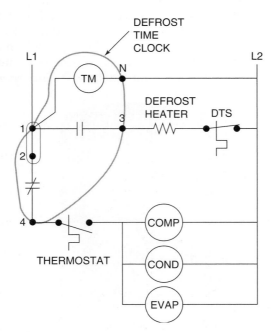

FIGURE 20–3   Commercial defrost timer Circuit No. 1.

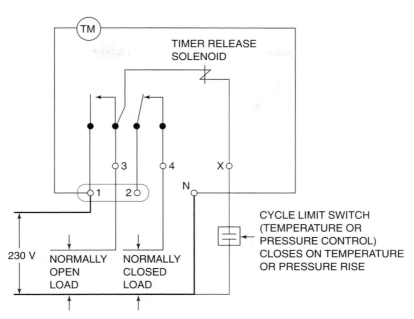

FIGURE 20–4 Defrost time clock with timer release solenoid.

2. The defrost termination thermostat is a three-wire, SPDT switch, usually with a remote sensing bulb (see Figure 20–5).

The manufacturer's diagram may be confusing to you, but except for the solenoid, it is the same as the previously discussed defrost timers. In Figure 20–6, it has been redrawn as a ladder diagram that should be easier to follow. During the "freeze" cycle, the normally closed switch is closed between terminals 2–4, and the normally open switch is open between 1–3. When the "freeze" cycle has timed out, the gears in the defrost timer operate the switches into "defrost" mode, opening 2–4 and closing 1–3.

FIGURE 20–5 Defrost termination switch with remote sensing bulb.

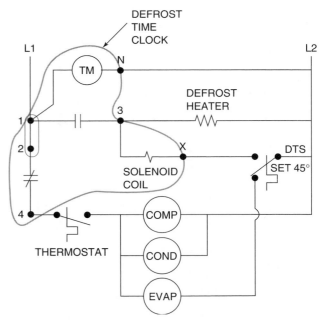

FIGURE 20–6 Defrost timer Circuit No. 2.

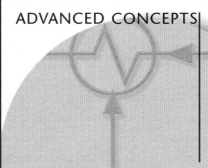

ADVANCED CONCEPTS

Note that the defrost timer can return to the "freeze" mode on *either* time or temperature. If the duration time setting times out before the evaporator coil reaches 45°F, the "freeze" cycle will resume. Eventually, the coil will become covered in ice. But the time-termination is necessary as a safety feature in case the solenoid coil or the defrost termination switch fails. Without this time-termination feature, a failed solenoid coil would prevent the compressor from operating. A failed defrost termination switch would be devastating. If its contacts were welded in the "cold" position, the heaters would stay on, and all the product in the box would become warm and spoil quickly.

The freezing components are de-energized, and the electric heater is energized. After some time, when the evaporator temperature rises to 45°F, the defrost termination switch operates, completing a circuit through the solenoid coil of the defrost timer through terminal X. However, the solenoid coil only stays energized for a second. As soon as the solenoid is energized, it returns the normally open and normally closed switches to their normal positions ("freeze" mode), once again de-energizing the solenoid coil. The compressor and condenser will run, but the evaporator fan will remain de-energized until the defrost termination switch senses that the evaporator coil temperature has returned to normal operating conditions.

Note that the power supply in Figure 20–6 is labeled L1 and L2 instead of H and N. This just indicates that it is a 230-V system rather than a

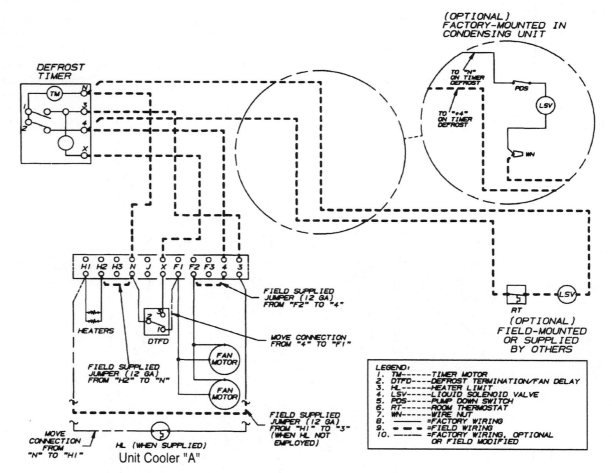

FIGURE 20–7   Factory supplied wiring diagram of defrost timer circuit. (Courtesy of Heatcraft.)

115-V system. Commercial defrost timers are also available for use on 115-V systems. When replacing a commercial defrost timer, make sure that you purchase the correct replacement. The clock motors are different to correspond to the available voltage. Figure 20–7 is an example of a factory supplied wiring diagram of a defrost circuit.

EXAMPLE 1

Assume you have a walk-in freezer with the compressor and condenser fan running, evaporator fans not running. There is a three-wire defrost timer mounted on the side of the casing. As there are multiple evaporator fans not running, you suspect that the defrost termination switch has failed. However, there are three wires emerging from the switch. You would like to take a voltage reading across the "cold" switch of the defrost termination switch to see if it has failed, but unfortunately, the actual wires on the DTS are not identified. You don't know which wire is which.

It's 20°F in the freezer, and you'd like to get out of there sooner rather than later. With the power turned off, take the three wires of the DTS, and put them all together in the same wire nut (or jumper them together). When you restore the power, this should energize the evaporator fans. If it does, replace the DTS.

Solution

Place the defrost timer in the "freeze" mode. Label the wires that were connected to the failed DTS as #1, #2, and #3. Connect them together, two at a time, until you find the pair that makes the evaporator fans run. The one wire not involved in making the evaporator fans run must be the one that comes from terminal X in the defrost timer. For ease of discussion, let's say that you found that to be wire #2. Next, go to the defrost timer, and turn the knob until it goes into the "defrost" mode. Return to the evaporator and measure voltage between wire #2 and each of the other two wires. One of the pairs will give you a voltage reading equal to the line voltage. Let's say that a line-voltage reading was obtained between wires #2 and #3. That means that wire #1 must be the one that comes from the evaporator fan, and wire #3 must be the one that connects to L2. Connect wire #1 to terminal B on the replacement DTS, connect wire #2 to terminal A, and connect wire #3 to terminal C. ∎

ELECTRONIC REFRIGERATION CONTROL

The newest trend in controlling walk-in boxes and other commercial refrigeration systems is to use a "black box" electronic control similar to the Beacon control system shown in Figure 20–8. This black box is factory mounted on the blower coil unit inside the box. It has temperature inputs that tell the controller what is happening. Temperature is sensed by thermistors, whose resistance changes with the temperature. The controller then has outputs to the electrical components (including an electronic

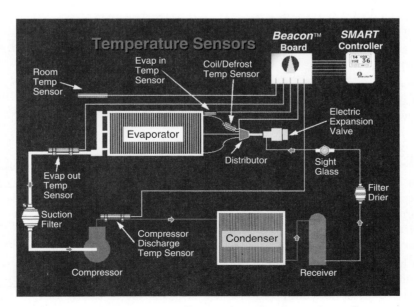

FIGURE 20–8 Beacon system diagram featuring Smart Controller. (Courtesy of Heatcraft.)

expansion valve) that accomplish all the conventional functions of refrigeration, and more. Following are some of the functions of this controller.

1. The controller can cause the electronic expansion valve to close, without regard to superheat, allowing the expansion valve to operate like a liquid solenoid valve. There is no need for a separate LSV to accomplish a pumpdown.

2. During pull-down from a warm start, the electronic expansion valve will be modulated so that the suction pressure (as sensed by saturated suction temperature) is limited to a preset maximum. This eliminates the need for a crankcase pressure regulator, which might otherwise be required to prevent overloading the compressor on start-up.

3. If the discharge line reaches 225°F for four minutes, the superheat setting will be temporarily lowered to keep the discharge temperature from going any higher. If the discharge line reaches 275°F for four minutes, the compressor will be shut down and locked out. These are safety functions that were rarely (if ever) done prior to the advent of electronic controls.

4. There is a minimum run time of four minutes for the compressor to eliminate the possibility of short cycling.

5. Pump down can be accomplished by closing the electronic expansion valve, and then shutting down the refrigeration system one minute later. This eliminates the need for a low-pressure switch (although one can be used if desired).

6. The controller can be used to cycle the condenser fan off if needed to get sufficient flow through the evaporator. This is more accurate than conventional head pressure control, which is only concerned with maintaining a minimum high-side pressure.

7. Status indicator lights on the controller board provide information as to what the system is doing. It also performs an error check to determine if any of the temperature sensors are open or shorted.

8. The controller has a set of **dry contacts** that are normally closed. The term *dry contacts* refers to a switch that is not connected electrically to any other device within the controller. It is simply a set of contacts that can be used in a different circuit. In this case, during normal operation, the controller will open the dry contacts. However, if an alarm condition exists, the dry contacts will return to their normally closed position to activate a light, buzzer, or bell to indicate trouble.

Optional Smart Controller

The Smart Controller connected to the controller in Figure 20–8 provides a means for remote programming of the controller. The following functions can be set by this controller:

1. Box temperature
2. Defrost frequency, start times, fail-safe times, and termination temperature
3. Evaporator superheat setting
4. Alarm set points

Additionally, with the push of a button, the Smart Controller allows the operator to monitor box temperature, actual superheat, the expansion valve position, the evaporator coil temperature, compressor discharge temperature, outdoor temperature, compressor run time, compressor cycling rate, and accumulated defrost time. These are powerful diagnostic tools that can indicate when a problem is developing long before an emergency situation arises. For example, if maintenance personnel note that the total defrost time each day is gradually increasing, it might indicate that the door seals or the door closer needs to be checked.

USING AN LPC INSTEAD OF A THERMOSTAT

In walk-in boxes, it is very common to use a low pressure cut-out (Figure 20–9) instead of a thermostat to control the operation of the compressor. Note the

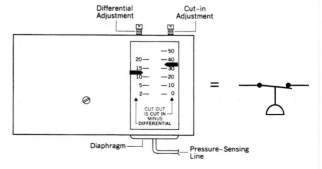

FIGURE 20–9 Low-pressure cut-out.

symbol used to represent an LPC. The switch closes on a rise in sensed pressure. The LPC senses the pressure on the low side of the refrigeration system. When the compressor is off, the evaporator is at the same temperature as the box. The refrigerant inside the evaporator is at the saturation pressure that corresponds to the box temperature. Therefore, an LPC can be used to sense pressure instead of using a thermostat to sense temperature. When the box temperature rises to its maximum allowable temperature, the evaporator pressure rises to the cut-in pressure setting of the LPC, starting the compressor. This method of control is popular for two reasons.

1. The LPC is located at the condensing unit, sensing low-side pressure at the compressor suction. It is not subject to the mechanical damage that might knock a thermostat off the wall as the refrigerated box is being loaded or unloaded.

2. The wiring cost is lower with the LPC because it is unnecessary to run wire between the condensing unit and the box. See Figure 20–10 for a comparison of the two methods of wiring.

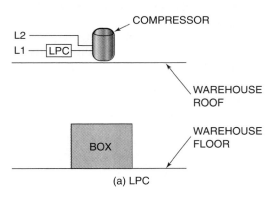

(a) LPC

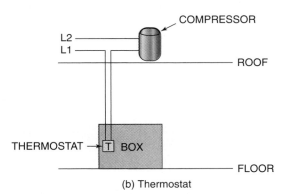

(b) Thermostat

FIGURE 20–10 Comparison of LPC and thermostat control. (a) LPC. (b) Thermostat.

When an LPC is used to sense box temperature, the evaporator fan is usually on a separate circuit, and the evaporator fan runs continuously. This ensures good heat transfer between the box and the refrigerant inside the evaporator coil. The pressure of that refrigerant will then accurately reflect the actual box temperature.

FREEZER WITH HOT-GAS DEFROST

Figure 20–11 shows a wiring diagram for a walk-in freezer that uses hot gas from the compressor to accomplish the defrost. This is different from the more popular electric defrost as follows.

1. During the defrost, the compressor continues to run.

2. Instead of energizing an electric heating element to provide heat for defrost, we energize a hot-gas defrost valve (Figure 20–12) to allow hot gas from the compressor to bypass the condenser and metering device and reenter the system between the metering device and the evaporator.

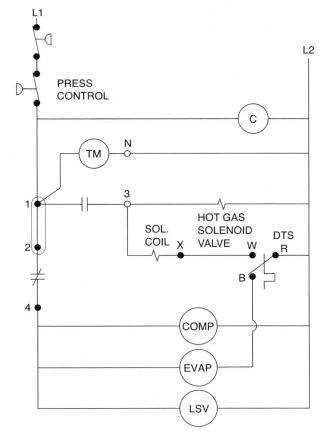

FIGURE 20–11 Hot-gas defrost.

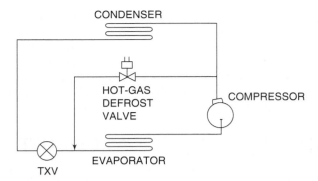

FIGURE 20–12 Location of hot-gas defrost valve.

During the freeze portion of the timer operation, the timer switch 2–4 is closed, bringing L1 pressure to terminal 4 in the defrost timer. This powers terminal 4 on the terminal block of the evaporator coil unit and the temperature control. Terminal 4 energizes the fan motors, and the temperature control energizes the liquid solenoid valve, which will cause the compressor contactor C to energize through the LPC pressure control (automatic pumpdown system). Note that the defrost termination switch makes R–B when it senses a cold evaporator and R–W when the evaporator is warm. Also, note the symbol for the pressure control. It shows two sensing elements operating a single switch. It senses pressure on both the high side and the low side of the refrigeration circuit. If the high-side pressure gets too high or if the low-side pressure gets too low, the switch will open, shutting down the compressor.

When the timer motor advances the switches to the "defrost" mode, switch 2–4 opens, and switch 1–3 closes.

When the hot gas from the compressor completely defrosts the evaporator, the defrost termination switch opens R–B and closes R–W. The R–W switch completes a circuit through the terminating solenoid in the defrost time clock (through terminal X). The solenoid mechanically opens switch 1–3, and closes switch 2–4. This restarts the compressor, but not the evaporator fan. The evaporator fan will not restart until the defrost termination switch makes R–B.

You have been called to repair a walk-in freezer. It has been down for many hours, and the box is at 75°F. When you start the system after you have completed your repair, you note that the evaporator fans do not run. Before you start troubleshooting the system again to find out why, give the compressor time to bring down the evaporator temperature. If you give the system a few minutes to run, the defrost termination switch will close, and the evaporator fans should start.

OTHER DEFROST SCHEMES

In Figure 20–13, the compressor is operated through a 115-V liquid solenoid valve and a box thermostat, whereas the electric heaters are operated on a separate 230-V circuit. The time clock and the fan motor both operate on 115-V.

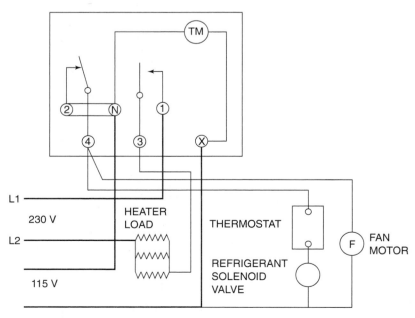

FIGURE 20–13 230-V system with 115-V liquid solenoid valve.

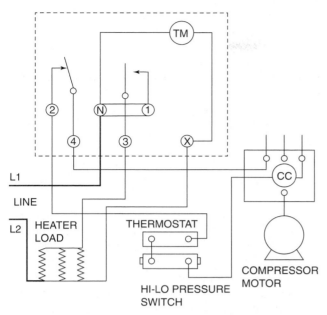

FIGURE 20–14   Time clock with jumper moved.

Figure 20–14 shows the same defrost time clock, but with the electrical bridge moved from terminals 2–N to terminals N–1. With this scheme, the time clock motor, the compressor contactor, and the electric heaters all operate on the same voltage.

Figure 20–15 shows a slightly different defrost timer, in which one of the internal switches is SPDT whereas the other is SPST. This allows the defrost timer to open both legs of the electric heater circuit when the defrost system is not being used.

OVERLOADS

Three-Phase Line Duty

Many walk-in refrigeration systems use a semihermetic three-phase compressor with a line-duty overload (Figure 20–16). This arrangement can be confusing to those technicians unfamiliar with it. The fusite connector looks exactly like the C, S, R or L1, L2, L3 power connections on a welded hermetic compressor. However, if you remove the wires from the fusite connector and use your ohmmeter on those pins to check the compressor windings, you will read infinite ohms, and you are likely to condemn a perfectly good compressor.

The compressor windings are connected in a Y-shape, but the connection point where each winding joins together is brought external to the compressor through the fusite connector. The windings are then joined together in the overload, but not before first going through an overload switch in each leg. The three-phase overload operates just like the single-phase overload, operating on current or temperature. The only difference is that when the three-phase overload senses an overload condition, it opens three switches instead of one switch.

Three-Phase Pilot Duty

A pilot-duty overload can also be used on three-phase systems (Figure 20–17). Three separate overloads are used. Each overload senses the current on

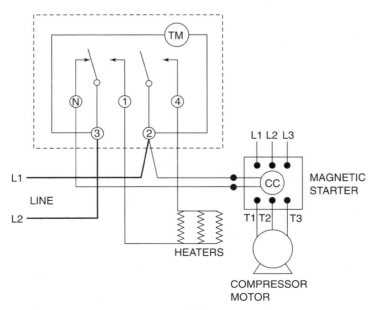

FIGURE 20–15   Time clock breaks both L1 and L2 in heater circuit.

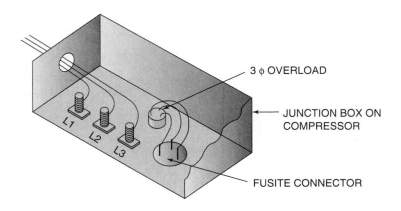

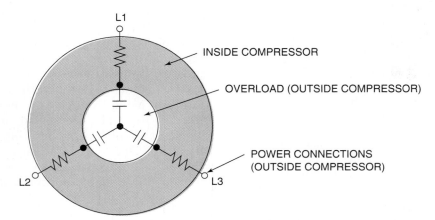

FIGURE 20–16 Three-phase compressor with line-duty overload.

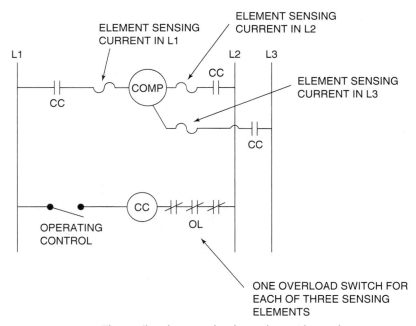

FIGURE 20–17 Three pilot-duty overloads used on a three-phase compressor.

a different leg of the power supply to the compressor. But the three different overload switches are wired in series with the compressor contactor. If any of the three legs draws higher than the design current, it will be sensed by the overload on that leg and will open the corresponding switch. If any one of the overload switches opens, the compressor contactor will be de-energized.

STARTERS

Some larger applications will use a motor starter. Sometimes called a magnetic starter, it is really nothing more than a combination of a contactor and a pilot-duty overload combined into a single unit (Figure 20–18). A schematic for a motor starter is given in Figure 20–19. This type of starter has overload switches that are spring loaded to open. When a reset button is pressed, the overload switches are closed, and they are held closed against the spring pressure by a ratchet mechanism. The ratchet mechanism is held secure by an enclosed solder pot.

When the operating control closes, it energizes the starter coil, closing the starter contacts and energizing the motor. The current in each power leg passes through a heater, which is wrapped around the solder pot in the starter. As long as the current draw of the motor is within limits, the heater will

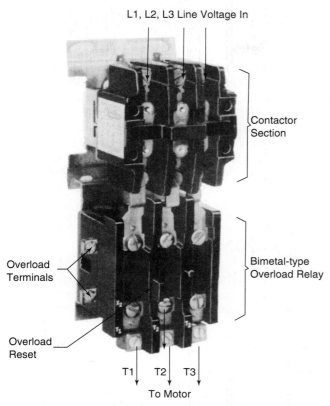

FIGURE 20–18 Magnetic starter. (Courtesy of Siemens & Furnace Controls Business Unit, a division of Siemens Energy & Automation.)

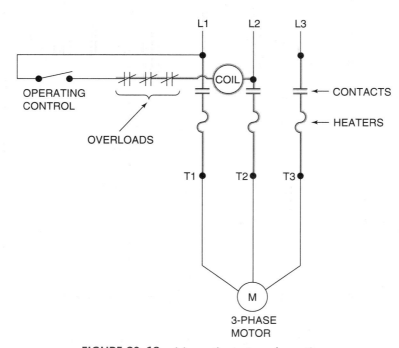

FIGURE 20–19 Magnetic starter schematic.

not generate sufficient heat to have any effect. However, if the motor current in any leg exceeds the motor rating, that current will be sufficient to cause the heater to generate enough heat to make the solder in the solder pot melt, thus causing the spring-loaded overload switch to open. The starter coil will be de-energized, and the motor will stop until the overload switch is manually reset.

One size starter is suitable for use with a range of motor sizes. The starter is matched to a particular motor by the selection of the proper heater. Sometimes, on new installations, the wrong heater size has been selected by the electrical contractor. This can cause the starter overload to trip, even though the motor is operating within its acceptable range of amp draw. In this case, you need to advise the electrician on the job to check the heater for proper sizing.

SMALL CONDENSING UNIT

Figure 20–20 shows the ladder diagram for a condensing unit used on air-conditioning systems in the 7.5-ton range. When the room thermostat makes R–Y1, the compressor contactor coil CC is energized. This starts the three-phase compressor motor, as well as the condenser fan (OFM) when it is required. The fan cycling control (FCC) senses high-side pressure. During periods when it is cold outside, the FCC will turn off the condenser fan if the head pressure drops too low. When the head pressure rises, the FCC will turn the fan back on.

LARGE CONDENSING UNIT

Figure 20–21 shows the ladder diagram for a condensing unit used on air-conditioning systems in the 15-ton range. The numbers on the wires

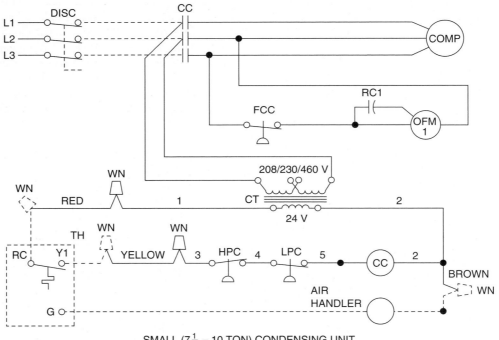

SMALL (7½ – 10 TON) CONDENSING UNIT

COMPONENT CODE		WIRING INFORMATION
CC	COMPRESSOR CONTACTOR	1. LINE VOLTAGE
COMP	COMPRESSOR	FACTORY STANDARD
CT	CONTROL TRANSFER	FACTORY OPTION
FCC	FAN CYCLE CONTROL	FIELD INSTALLED
HPC	HIGH PRESSURE CONTROL	2. LOW VOLTAGE
LPC	LOW PRESSURE CONTROL	FACTORY STANDARD
OFM	OUTDOOR FAN MOTOR	FACTORY OPTION
RC	RUN CAPACITOR	FIELD INSTALLED
TH	THERMOSTAT	
WN	WIRE NUT	

FIGURE 20–20 Small (7 1/2–10 ton) condensing unit.

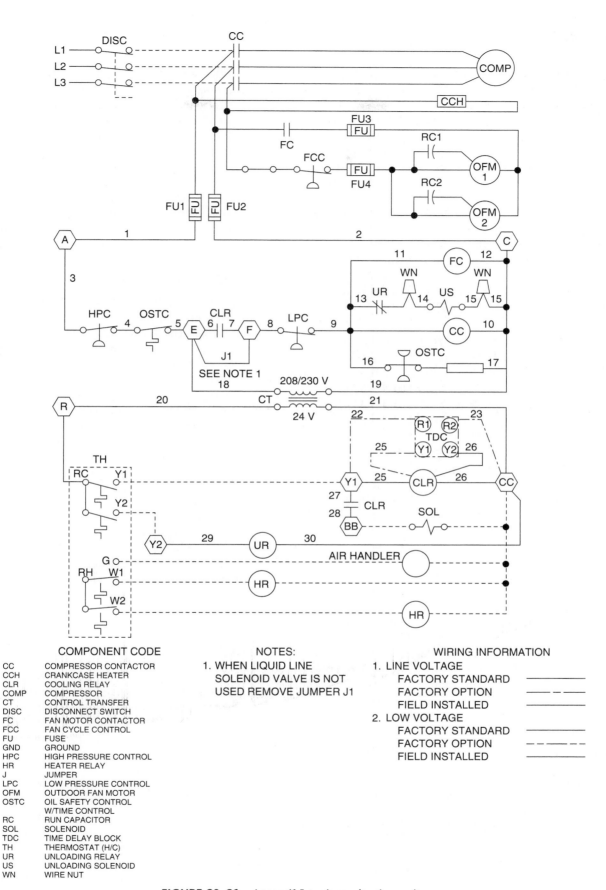

FIGURE 20–21 Large (15-ton) condensing unit.

COMPONENT CODE

CC	COMPRESSOR CONTACTOR
CCH	CRANKCASE HEATER
CLR	COOLING RELAY
COMP	COMPRESSOR
CT	CONTROL TRANSFER
DISC	DISCONNECT SWITCH
FC	FAN MOTOR CONTACTOR
FCC	FAN CYCLE CONTROL
FU	FUSE
GND	GROUND
HPC	HIGH PRESSURE CONTROL
HR	HEATER RELAY
J	JUMPER
LPC	LOW PRESSURE CONTROL
OFM	OUTDOOR FAN MOTOR
OSTC	OIL SAFETY CONTROL W/TIME CONTROL
RC	RUN CAPACITOR
SOL	SOLENOID
TDC	TIME DELAY BLOCK
TH	THERMOSTAT (H/C)
UR	UNLOADING RELAY
US	UNLOADING SOLENOID
WN	WIRE NUT

NOTES:
1. WHEN LIQUID LINE SOLENOID VALVE IS NOT USED REMOVE JUMPER J1

WIRING INFORMATION
1. LINE VOLTAGE
   FACTORY STANDARD ———————
   FACTORY OPTION ——— ——— ———
   FIELD INSTALLED ———————
2. LOW VOLTAGE
   FACTORY STANDARD ———————
   FACTORY OPTION ——— — — — ———
   FIELD INSTALLED ———————

263

correspond to labels that are actually attached to the wires in the unit. This makes it especially easy for the service technician to identify wires on the unit that correspond to the wiring diagram.

There is a three-phase compressor motor, a crankcase heater that is energized all the time wired between L1 and L3, two single-phase condenser fan motors wired between L2 and L3, and a control power wired between L1 and L2. The single-phase loads are all wired to different pairs of legs of the supply power in an attempt to equalize the loads between legs as much as possible. The following information about some of the components will assist you in understanding the sequence of operation.

1. The **unloader solenoid** (US) is located on the compressor. When it is energized, it mechanically holds open the suction valve on one or more of the compressor cylinders. This renders those cylinders ineffective, as no compression can take place if the suction valve doesn't close.

2. The oil safety time control (OSTC) is shown in two places. Between wires no. 16 and 17, there is a pressure switch sensing two pressures. The crankcase pressure acts on the top of the switch, and the oil pump discharge pressure acts on the bottom of the switch. The unlabeled rectangle in series with the pressure switch is a small resistance heater located in the OSTC. If this heater remains energized for more than about one minute (45 to 90 seconds, depending on the model), it will open the bimetal switch of the OSTC located between wires no. 4 and 5, de-energizing the compressor contactor. However, if sufficient oil pressure is established soon enough, it will open the pressure switch, de-energizing the heater, and the compressor contactor will remain energized.

3. The solenoid between terminals BB and CC (terminals are identified by hexagons) is a liquid line solenoid valve. When it is de-energized, it shuts off the flow of refrigerant in the liquid line. This is used to provide automatic pumpdown each time the room thermostat is energized.

4. The time-delay control (TDC) is a factory option. You can tell by looking at the key for wiring information on the diagram.

When it is used, instead of the thermostat energizing the cooling relay (CLR) directly, the thermostat energizes the TDC. The TDC may be an on-delay or an off-delay. We can't tell from simply looking at the wiring diagram. The output from the TDC, in turn, will energize the CLR. When the TDC is used, the wire between the Y1 terminal and the CLR coil and the wire between the CC terminal and the CLR coil do not exist.

Sequence of Operation

On a call for cooling, the room thermostat (TH) makes contact from RC to Y1 (first-stage cooling). This energizes the CLR coil, closing CLR contacts between wires 27 and 28, thus energizing the liquid line solenoid valve. When the liquid solenoid valve opens, refrigerant enters the low side of the system from the receiver, and the low-side pressure rises. This causes the low-pressure control (LPC) between wires 8 and 9 to close, energizing the compressor, the unloader solenoid, and the fan motor contactor. The compressor starts (in an unloaded condition). Some time later, after the head pressure reaches the cut-in setting of the FCC, the condenser fans will start.

If the first stage of cooling is insufficient to keep the room temperature from rising, the room thermostat will make RC to Y2 (second-stage cooling). This will energize the unloader relay, de-energizing the unloader solenoid, thus allowing the compressor to operate at full capacity.

Field Wiring

The dashed lines indicate field wiring, that is, wiring that is installed by you, the installing technician. The manufacturer of the condensing unit anticipates that you will be installing a room thermostat that has an RC and RH terminal, so that a separate control power transformer can be used to power the heating controls.

The condensing unit manufacturer has also made it possible to operate this system without the automatic pumpdown. If automatic pumpdown is not used, there would be no liquid solenoid valve, and the jumper wire (J1) between terminals E and F would be removed. This allows the CC coil to be operated directly from the CLR contacts between terminals E and F.

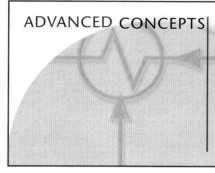

The oil pressure safety control is more commonly called an oil pressure control or oil pressure cut-out (OPC). A schematic of the OPC is shown in Figure 20–22. A diagram typical of that usually found on the OPC itself is shown in Figure 20–23. There is a "dropping resistor" wired in series with the heater. If the controller is to be used on a 230-V system, the entire length of the dropping resistor is used. However, if used on a 110- to 120-V control system, only half of the dropping resistor is used in the circuit. In this way, the same amount of current is allowed to flow through the heater, and the same time delay is produced for either control voltage.

CONDENSING UNIT WITH MECHANICAL TIMER AND PUMP-OUT RELAY

Figure 20–24 shows an older model condensing unit. It has the following unique features:

1. Mechanical **anti-recycle** circuit
2. Automatic pumpdown

An antirecycle circuit is one that prevents the compressor from restarting for a period of time after it has shut down. This is important on air-conditioning systems that use a PSC compressor. The PSC compressor does not have a lot of starting torque, and it may not be able to start against the pressure of the high side. The system requires some "off time" to allow the refrigerant pressures to equalize between the high side and the low side. Without the antirecycle timer, if an occupant moves the thermostat to a lower temperature just after the compressor has shut off, it could cause the compressor to trip out on overload or, worse, blow a fuse.

The antirecycle timer is a unique device, designed specifically for this type of circuit. The timer starts with the switches made between A–A2 and B–B2. The operation of the timer is that if the timer motor is continuously energized, the switches remain in that position for 15 seconds, then switch down to A–A1 and B–B1 for 10 minutes. After that, the switches return to the original position. But in the circuit, the timer motor does not stay energized all the time.

Sequence of Operation

When the room thermostat closes, it energizes the CR coil. The liquid line solenoid valve gets energized through the CR normally open contacts, the internal thermostat on the compressor, the two overload switches, the high-pressure switch, a second set of CR normally open contacts, and then through the solenoid valve coil. When the solenoid valve opens, refrigerant is allowed to pass into the low side, closing the low-pressure switch. This energizes the timer relay (TR) coil through A2–A. The timer motor TM is also energized, through switch B2–B. After the timer motor runs for 15 seconds, the switches change position to make A1–A and B1–B. This energizes the compressor contactor through the normally open TR contacts. The pump-out relay (POR) is energized in parallel with the contactor coil, and its contacts seal in around the first set of CR contacts. The timer motor becomes de-energized, because the normally closed TR contacts are open. The compressor continues to run (and the timer motor continues to not run) as long as the thermostat continues to call for cooling.

When the thermostat opens, the compressor continues to run (through the normally open POR contacts), but the solenoid valve becomes immediately de-energized as the second set of CR contacts opens. With the compressor running and the liquid solenoid valve closed, the low-side refrigerant gets pumped into the receiver, and the low-side pressure drops. When the low-pressure switch opens, the compressor stops, the TR coil is de-energized, and the timer motor

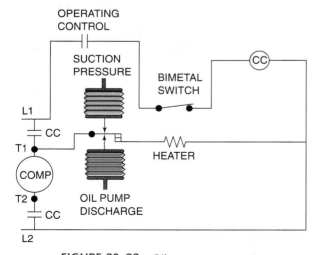

FIGURE 20–22 Oil pressure cut-out.

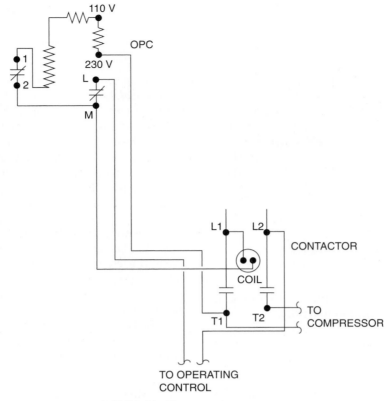

FIGURE 20–23 Oil pressure cut-out.

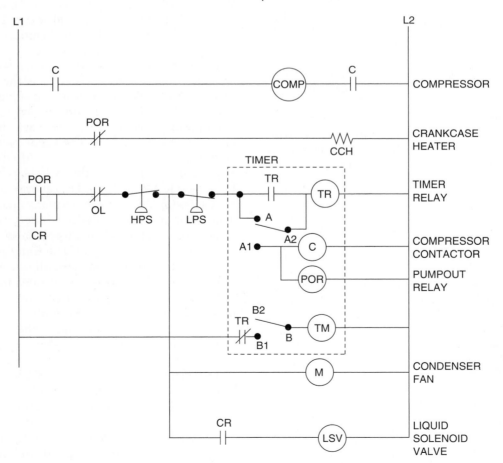

FIGURE 20–24 Condensing unit with mechanical antirecycle timer.

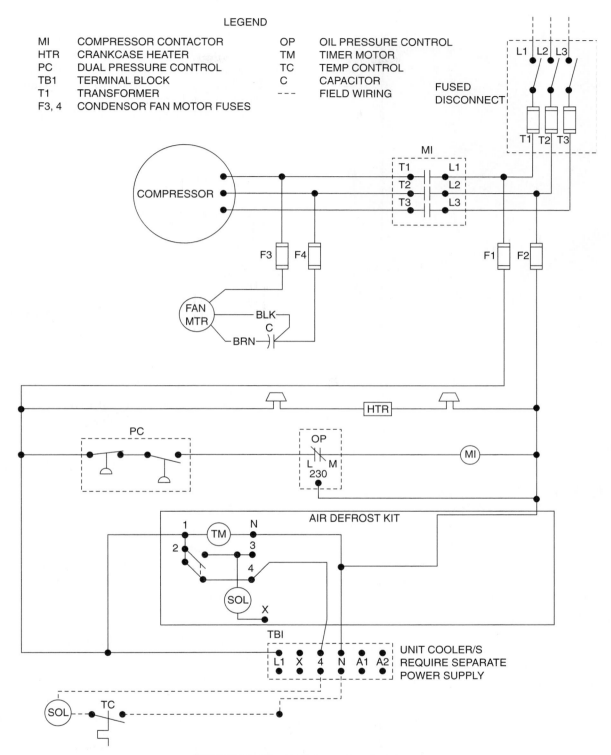

FIGURE 20–25   Walk-in cooler-air defrost.

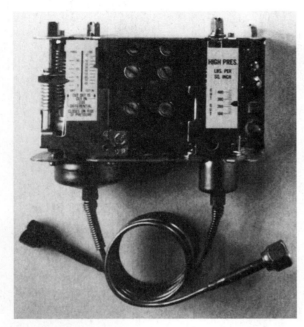

FIGURE 20–26 High-low pressure cut-out. (Courtesy of Johnson Controls/PENN.®)

starts. After the timer motor runs for 10 minutes, the switches will return to the A2–A and B2–B positions, ready to start again. The compressor cannot restart during that 10-minute period, even if the thermostat recloses.

WALK-IN COOLER—AIR DEFROST

Figure 20–25 shows a split system that might be used in a walk-in cooler (a cooler is a box that operates above freezing temperatures).

Sequence of Operation

When the box temperature rises above the set point of the temperature control (TC), TC closes, energizing the liquid line solenoid valve (SOL). This allows refrigerant to pass from the receiver into the low side, allowing the low-pressure switch in the combination high-low pressure control (PC) to close (see Figure 20–26). This will energize the compressor contactor M1, which in turn, energizes the three-phase compressor motor and the single-phase condenser fan motor.

Air Defrost

The air-defrost kit is an optional accessory. Most coolers depend on having sufficient off-cycle time

for the evaporator coil to defrost. The defrost kit ensures that a defrost period will occur at the frequency dictated by the defrost time clock. The timer motor TM is always energized between T1 and T2. When the cams and switches inside the defrost time clock determine that it is time for a defrost, the switch between terminals 2 and 4 open, de-energizing the liquid line solenoid valve and throwing the condensing unit into pumpdown. Note that terminals 3 and X in the defrost time clock are unused with an air defrost.

WALK-IN FREEZER WITH ELECTRIC DEFROST

Figure 20–27 shows a wiring diagram similar to the previous system, but it is used for a split system on a freezer. Therefore, it requires an active defrost system. Following are the differences between this diagram and the previous diagram.

1. When the defrost time clock initiates the defrost, making the switch between terminals 2–3, it applies power to terminal A1 on terminal board TB1. There is an auxiliary contact on the compressor contactor M1 that takes the voltage on terminal A1 and applies it to terminal A2, where it energizes contactor M2. When the M2 contacts close, the electric defrost heaters will be energized.

2. When the defrost has continued long enough for the indoor coil to reach 45°F (approximately), the defrost termination switch (TC1) opens the switch between terminals B and N and closes the switch between terminals X and N. This energizes the solenoid valve inside the defrost time clock, mechanically returning the switches to the "freeze" position (2–3 open, 2–4 closed).

SUMMARY

Although many components in commercial systems have the same name and function as those in residential systems, they are usually more substantial in construction. Commercial systems may remain in operation for extended periods. Failure of a commercial system could result in losses in many thousands of dollars. The components used in commercial systems are usually larger and more expensive than their residential equivalents.

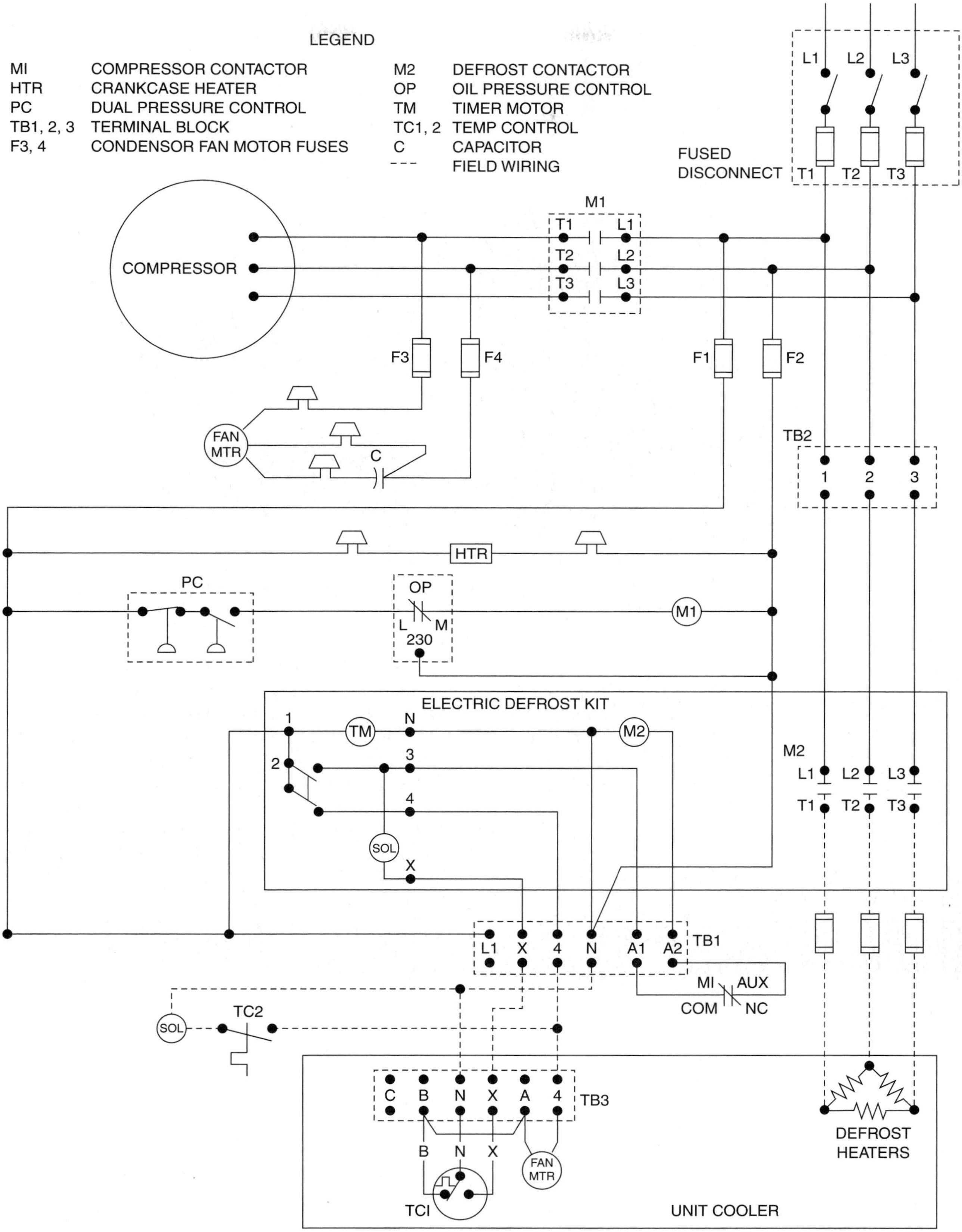
LEGEND
MI COMPRESSOR CONTACTOR
HTR CRANKCASE HEATER
PC DUAL PRESSURE CONTROL
TB1, 2, 3 TERMINAL BLOCK
F3, 4 CONDENSOR FAN MOTOR FUSES
M2 DEFROST CONTACTOR
OP OIL PRESSURE CONTROL
TM TIMER MOTOR
TC1, 2 TEMP CONTROL
C CAPACITOR
--- FIELD WIRING
FUSED DISCONNECT
L1 L2 L3
T1 T2 T3
M1
COMPRESSOR
F3
F4
F1
F2
FAN MTR
C
TB2
1 2 3
HTR
PC
OP
L M
230
M1
ELECTRIC DEFROST KIT
TM
M2
1 2 N 3 4 X
SOL
M2
L1 L2 L3
T1 T2 T3
L1 X 4 N A1 A2
TB1
MI AUX
COM NC
SOL
TC2
C B N X A 4
TB3
B N X
FAN MTR
TCI
UNIT COOLER
DEFROST HEATERS

FIGURE 20–27 Walk-in cooler with electric defrost.

PRACTICAL EXPERIENCE

1. Redraw the wiring diagram in Figure 20–7 as a ladder diagram. Show all the terminals on the terminal strip. Answer the following questions:
   a. Where does the power come into this system? (It is not shown on the diagram.)
   b. What would you guess is the purpose of the heat limiter (HL)?

2. Draw a line encircling all the portions of Figure 20–11 that would be found inside the defrost time clock.

3. For each of the manufacturer's time clock wiring diagrams in Figures 20–13, 20–14, and 20–15, draw an equivalent ladder diagram.

4. You have been called to diagnose a three-phase compressor with a line-duty overload. Describe how you would check to determine if the overload has failed.

5. Redraw the OPC diagram in Figure 20–23 similar to the format used in Figure 20–22. It will be slightly different. Show all terminal identification. [HINT: The switch between terminals 1–2 in the OPC is the switch that opens when sufficient oil pressure is established.]

6. Redraw the walk-in cooler–air defrost diagram Figure 20–25 as a ladder diagram.

7. Redraw the refrigeration condensing unit Figure 20–29 as a ladder diagram.

REVIEW QUESTIONS

1. You have removed the nonfunctioning defrost termination of Figure 20–6. Your new DTS comes with the terminals A, B, and C identified as follows:
   a. A–C makes on a rise in temperature.
   b. B–C makes on a fall in temperature.

   There are three unidentified wires hanging out of the evaporator unit, which were connected to the failed DTS. Which wires will you connect to which terminals on the replacement DTS?

2. When the condensing unit is running (Figure 20–21) at maximum capacity, what voltages would you measure:
   a. Between terminals Y1 and CC
   b. Between terminals A and F
   c. Across the unloader relay contacts
   d. Between terminals R and Y2

3. a. When the condensing unit Figure 20–20 is running, what voltages would you measure
      1. Across the HPC?
      2. Across CC coil?
   b. What type of motor is the OFM?

4. When the condensing unit Figure 20–21 is running at maximum capacity, what voltages would you measure
   a. Between terminals Y2 and CC?
   b. Between terminals A and E?

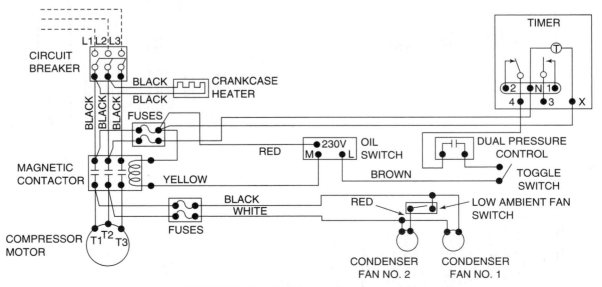

FIGURE 20–28   5–7.5-ton condensing unit.

c. Across the unloader solenoid?

d. Between terminals R and CC?

5. a. If the POR coil Figure 20–24 failed, how would the operation of this system change?

b. If the solenoid coil failed, what would you see operating when you arrived on the job?

6. In Figure 20–25, walk-in cooler–air defrost,

a. What type of motor is used to drive the condenser fan?

b. What is the purpose of the wire between the OP and L2?

c. What is the SOL located inside the air-defrost kit?

7. a. When does the fan motor in the unit cooler Figure 20–27 reclose following the defrost?

b. What temperature is sensed by TC1?

c. You have found the compressor and fans running, but the box is too warm because the evaporator coil is covered with ice. You move the defrost time clock into the defrost position. The compressor and fans stop,

but the ice on the coil does not begin to melt. You measure the following voltages at TB1:

L1–X = 0 V, L1–3 = 0 V, L1–4 = 230 V, L1–N = 230 V, L1–A1 = 0 V, L1–A2 = 230 V. What is the problem?

8. See the wiring diagram in Figure 20–28 for a refrigeration condensing unit in the 5- to 7.5-hp range.

a. Is this condensing unit for a walk-in cooler or a walk-in freezer? How do you know?

b. What is the purpose of the low-ambient fan switch?

c. What is the purpose of the toggle switch?

d. How many switches (other than the disconnect) must be closed for the magnetic contactor to become energized?

9. See the wiring diagram in Figure 20–29 for a refrigeration condensing unit.

a. The fan motor in the unit cooler (inside the walk-in) does not run. If you measured 230 V

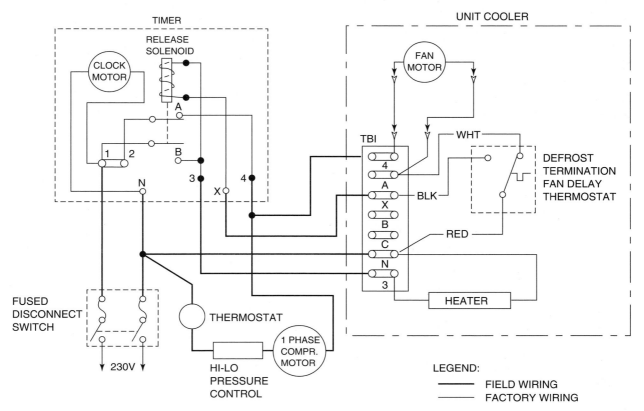

FIGURE 20–29 Refrigeration condensing unit.

between terminals A and 4 on TB1, what voltage would you measure between terminals 4 and N? Explain your answer.

b. Is the problem with the fan motor or with the defrost timer? How do you know?

c. What switch must close on the defrost termination fan delay thermostat in order to energize the release solenoid (identify the wire colors that become connected)?

10. In Figure 20–29, during the defrost cycle, is the fan motor operating when the release solenoid actuates? Explain your answer.

UNIT 21

STANDING PILOT FURNACES

OBJECTIVES

After completion and review of this unit, you should be familiar with:

- Standing pilot furnaces.
- Single-function gas valves.
- Combination gas valves.
- Thermocouples.
- Limit switches.
- Fan switches.
- Time-delay fan switches.
- Inducer fan operation.
- Pilot generators (powerpile).

A standing pilot furnace is one in which there is a small pilot flame burning continuously in a pilot burner. It is located physically close to the main burner. When main gas is allowed to enter the main burner, it is ignited by the pilot flame. This heats a heat exchanger. When the room-air side of the heat exchanger (sometimes referred to as the bonnet) is sufficiently warmed (usually around 130°F), a fan turns on to circulate room air over the heat exchanger. The fan is not turned on until the heat exchanger is warmed so that there will not be an initial blast of cool air that could be uncomfortable for the occupants.

The variations in standing pilot furnace control schemes are as follows.

1. Different methods are used to prove that the pilot flame is actually burning before main gas is allowed to be introduced into the main burner.
2. Different methods are used to determine when to turn on the fan that will circulate the room air over the heat exchanger.

FURNACE CIRCUIT NO. 1

Figure 21–1 shows the wiring for a central standing pilot gas-fired furnace for a residence. The modern trend is away from the use of standing pilot systems, in favor of electronic ignition furnaces (see unit 22). But it will be many years until all the standing pilot furnaces have been replaced.

The devices in this diagram that have not been previously described are as follows.

1. A **transformer** is used to supply 24 V.
2. A gas valve that has a 24-V coil. When the coil is energized, the gas valve opens.
3. The thermostat is designed to operate in the 24-V circuit. It is called a **low-voltage thermostat**. 24 V is a standard low voltage in heating and air-conditioning systems.

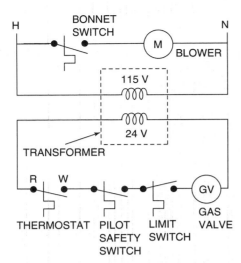

FIGURE 21–1 Furnace Circuit No. 1.

4. A **pilot safety switch** is located at the pilot flame. If it does not sense heat (indicating that the pilot flame is present), the contacts will open.

5. A **limit switch**, located in the **bonnet** of the furnace. The bonnet is the side of the heat exchanger through which the room air passes. If the temperature of the air reaches approximately 180°F, something is wrong and the limit switch will open.

6. A **bonnet switch** located in the bonnet of the furnace.

Transformer Review

The transformer (Figure 21–2) works on the principle of electric induction. You will recall that when current passes through an electric wire, it creates a magnetic field. When the wire is wound into a coil, the magnetic field increases with each additional turn. You will also recall that when a moving magnetic field passes through a coil of wire, a voltage is induced in that coil of wire. In the transformer, there are two coils of wire. The first coil is called the **primary winding**. It is in the 115-V circuit. Because the 115-V circuit is alternating current, the magnetic field around the primary winding builds and collapses 60 times each second. A second coil (the **secondary winding**) is placed physically close to the primary coil. As the magnetic field from the primary coil builds and collapses, it induces a voltage in the secondary coil. Depending on the number of turns in the primary and the secondary, the

voltage in the secondary may be higher or lower than the voltage in the primary.

When the secondary voltage is higher than the primary, it is called a **step-up transformer**. When the secondary voltage is lower than the primary, it is a **step-down transformer**.

As more loads are added in parallel on the secondary side, the current increases in both the secondary winding and the primary winding. If there is a short circuit in the circuit being powered by the secondary winding, the most likely result is that the primary winding will burn out. This is because the primary winding is designed for lower current than the secondary winding and therefore is constructed from thinner wire than the secondary winding.

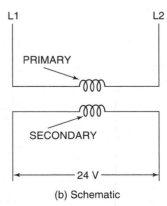

(a) Transformer

(b) Schematic

FIGURE 21–2 Control transformer. (a) Transformer. (b) Schematic. (a. Courtesy of Honeywell, Inc.)

ADVANCED CONCEPTS

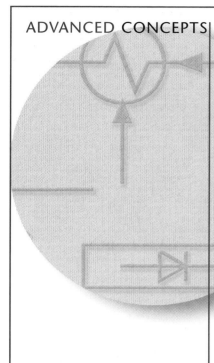

The ratio of the primary voltage to the secondary voltage is equal to the ratio of the turns. For example, if the primary has 100 turns, and the secondary has 20 turns, the voltage ratio will be 100/20 = 5. If the voltage on the primary is 120 V, then the output voltage from the secondary will be 120 V/5 = 24 V. The formula for this relationship is

$$\frac{Turns_{pri}}{Turns_{sec}} = \frac{Volts_{pri}}{Volts_{sec}}$$

where

$Turns_{pri}$ = number of turns in the primary coil
$Turns_{sec}$ = number of turns in the secondary coil
$Volts_{pri}$ = voltage applied to the primary coil
$Volts_{sec}$ = voltage output from the secondary coil

The current in the primary and the secondary follow the same ratio, but inversely. In the above example, the 24-V secondary coil would produce five times more current than the primary coil. The power output from the secondary will theoretically equal the power input to the primary, as follows.

$$Volts_{sec} \times Amps_{sec} = Volts_{pri} \times Amps_{pri}$$

In actual practice, the power output from the secondary will be slightly lower because some power is wasted because of the heat being produced by the wiring.

Gas Valve

The gas valve in this circuit is called a single-function gas valve (Figure 21–3). Other gas valves are **combination gas valves** (Figure 21–4). When the coil of a single-function gas valve is energized, the valve will open, allowing main gas to flow, without checking to see if a pilot flame is present. The single-function gas valve must be used with some other device (such as a pilot safety switch) to check whether or not a pilot flame is present before the main gas valve is allowed to open.

The combination gas valve actually consists of two valves in series (Figure 21–5). It uses a **thermocouple** (Figure 21–6) to sense heat from the pilot flame. The thermocouple converts this high temperature into a small electrical voltage (12–18 mV, where 1000 mV = 1 V). The thermocouple output is used to energize a small coil that holds the pilot valve open. In a system using a combination gas valve, the pilot safety switch is not used. The pilot valve is pushed open manually to allow pilot gas to flow. While the pilot valve is being manually held open, the pilot flame must be lit manually. After approximately one minute, the thermocouple generates suf-

ficient voltage to hold the pilot valve open. The button that was manually pressed (or pulled, or pushed or twisted) may then be released, and the pilot valve will stay open for as long as the thermocouple stays hot from the pilot flame. Neither the thermocouple nor the coil on the pilot safety valve shows up on the wiring diagram.

If the main valve in a combination gas valve gets energized while there is no pilot flame, the main valve will open but no gas will flow. Main gas cannot flow unless *both* the pilot valve and main valve are open.

Room Thermostat

The low-voltage room thermostat (Figure 21–7) has two terminals, R (for red) and W (for white). The room thermostat is located in the room, remote from the furnace. It is connected to the furnace by a pair of thermostat wires inside a common sheath (Figure 21–8). Thermostat wire is commonly referred to as two-wire, three-wire, four-wire, and so forth, depending on how many wires are inside the sheath. There may be as many as ten conductors, each with

FIGURE 21–3   Single-function gas valve. (Courtesy of Johnson Controls, Inc.)

a different color insulation. For a heating-only system, two-wire would be used.

Pilot Safety Switch

The pilot safety switch is usually combined with the pilot burner (Figure 21–9). The heat from the pilot flame impinges on a bimetal element that then bends, closing a switch (the pilot safety switch). It is popular on Carrier, Payne, Day Nite, and Bryant furnaces. Most other standing pilot furnaces use the combination gas valve system with a thermocouple.

Another type of pilot safety switch that is sometimes used is the **thermopilot relay** (Figure 21–10). It uses a thermocouple, like the combination gas valve that uses a thermocouple to hold open a valve. But with the thermopilot relay, the thermocouple holds a switch closed. When you manually press the button, you close the pilot safety switch and, at the same time, allow pilot gas to flow. After the pilot flame has been lit and allowed time to heat the thermocouple, the button is released. If the pilot flame then goes out, the switch will be pushed open by an internal spring.

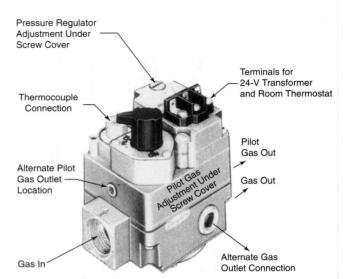

FIGURE 21–4   Combination gas valve. (Courtesy of White-Roagers Div., Emerson Electric Co.)

Limit Switch

The **limit switch** is a safety control that should remain closed during the entire operating life of the furnace, unless there is some malfunction that causes it to open. Several types of limit switches are shown in Figure 21–11.

The limit switch is a thermostat that senses the temperature on the room-air side of the heat exchanger. Normally, this temperature should not exceed 130°F to 150°F. However, if the flow of room air is restricted (or stopped, as in the case of a failed fan motor) or if the furnace fires at higher than its normal Btu rating, the room–air temperature side of the heat exchanger will rise. If the limit switch reaches its set temperature (usually 180°F to 200°F), it will open, de-energizing the gas valve and extin-

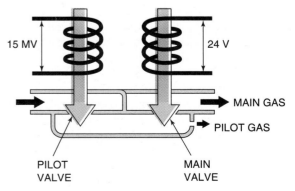

FIGURE 21–5   Cutaway-combination gas valve.

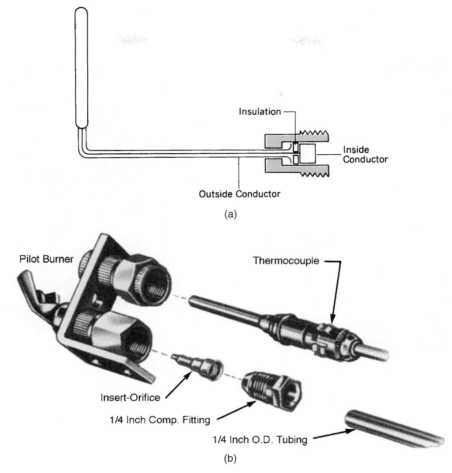

Insulation

Inside Conductor

Outside Conductor

(a)

Pilot Burner

Thermocouple

Insert-Orifice

1/4 Inch Comp. Fitting

1/4 Inch O.D. Tubing

(b)

FIGURE 21–6   Thermocouple. (b. Courtesy of Honeywell, Inc.)

guishing the main flame. Most limit switches are **automatic reset**. That is, when the sensing portion of the limit switch cools, the contacts will reclose automatically. Some limit switches are

manual reset. Once tripped, they will not reclose until someone pushes on the reset button.

Bimetal Fan Switch

The bimetal fan switch senses the same bonnet temperature as the limit switch. Its appearance can also be the same as a limit switch. However, its

FIGURE 21–7   Low-voltage room thermostat. (Courtesy of Honeywell, Inc.)

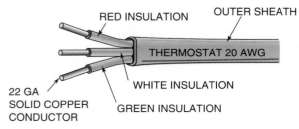

RED INSULATION

OUTER SHEATH

THERMOSTAT 20 AWG

22 GA SOLID COPPER CONDUCTOR

WHITE INSULATION

GREEN INSULATION

FIGURE 21–8   Thermostat wire.

FIGURE 21–9   Pilot safety switch.

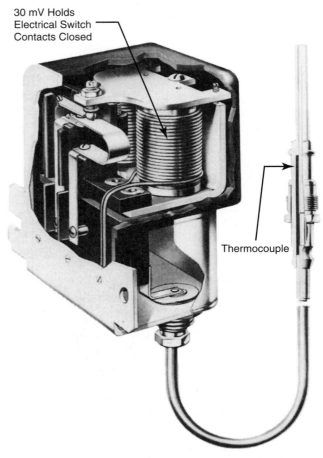

30 mV Holds
Electrical Switch
Contacts Closed

Thermocouple

FIGURE 21–10   Thermopilot relay. (Courtesy of Johnson Controls, Inc.)

function is the opposite of a limit switch. A bimetal fan switch *closes* when the bonnet temperature rises. It is usually set for an "On" temperature of approximately 130°F and an "Off" temperature of approximately 100°F.

Sometimes, as a manufacturing cost savings, the limit switch and the fan switch are combined together into a single unit, sharing a single sensing element. This is sometimes referred to as a **fan-limit switch** (Figure 21–12). Even though the two switches share the same sensing element

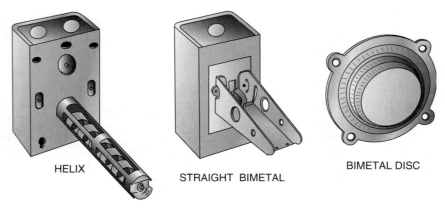

HELIX

STRAIGHT BIMETAL

BIMETAL DISC

FIGURE 21–11   Several types of limit switches.

FIGURE 21–12 Fan-limit switch. (Courtesy of Honeywell, Inc.)

(Figure 21–13), the switches are completely separate electrically, and each has its own set point.

> 👉 Note that the bimetal element movement is what causes the round face plate to rotate. Do not attempt to rotate the face plate by hand. You may ruin the calibration of the device, so that it will not actually operate at temperatures that match the set temperatures.

Sequence of Operation

When the room gets cold, the room thermostat closes, completing a circuit through the gas valve. If the pilot flame is lit, the pilot safety switch will be closed, and the solenoid coil of the gas valve will be energized. The gas valve opens, admitting main gas into the combustion chamber, where it ignites from the standing pilot flame. After approximately one minute, the bimetal fan switch senses that the heat exchanger temperature has risen to approximately 130°F and closes, energizing the blower motor to circulate warm air through the occupied space.

When the room is sufficiently warmed, the room thermostat opens, de-energizing the main gas valve.

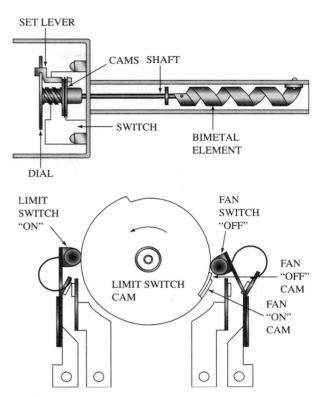

FIGURE 21–13 Two separate switches share one sensing element. (Reprinted from Refrigeration and Air Conditioning, 3rd Ed., Air Conditioning and Refrigeration Institute, copyright, Prentice Hall, Inc. Reprinted by permission.)

However, the blower motor continues to run for approximately another minute. When the bonnet temperature falls to 100°F, the bonnet switch opens, and the blower stops.

FURNACE CIRCUIT NO. 2

The fan-limit switch shown in Figure 21–14 is supplied from the factory with a jumper between one side of the fan switch and one side of the limit switch. It says, "Remove jumper for 24-V limit." Figure 21–15 shows how this limit switch would be used when the jumper is not removed. If the limit switch opens, it will de-energize the transformer. The 24 V on the secondary of the transformer will be lost, and the gas valve will be de-energized. In this way, the limit switch in the 115-V circuit shuts down the gas valve just as surely as it would if the limit switch were wired in series with the gas valve.

If the jumper on the fan-limit switch is removed, then the fan switch and the limit switch may be wired independently into the 115 V and 24 V circuits, as in Figure 21–16.

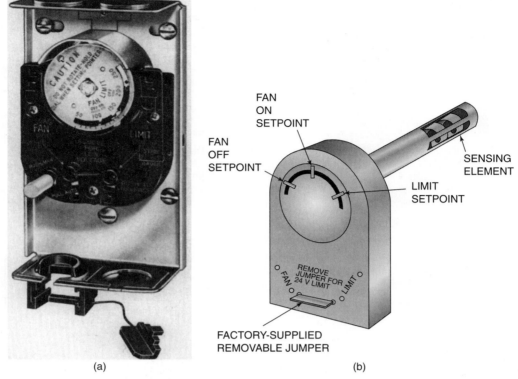

FIGURE 21–14 Fan-limit switch with jumper. (a. Courtesy of Honeywell, Inc.)

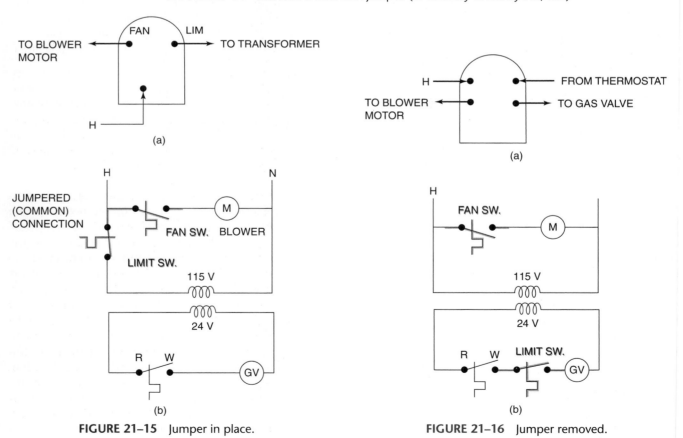

FIGURE 21–15 Jumper in place.

FIGURE 21–16 Jumper removed.

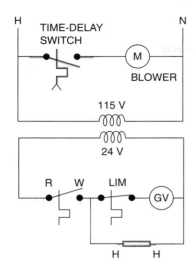

FIGURE 21–17 Time-delay fan switch circuit.

FURNACE CIRCUIT NO. 3

The circuit shown in Figure 21–17 shows another way that the operation of the blower motor is controlled. It uses a **time-delay fan switch** (Figure 21–18). This device contains a bimetal switch and a 24-V heater. The bimetal is wrapped with **Nichrome** wire. Nichrome is a material that has a significant resistance, and when a current is passed through it, it heats up because of the I × R drop. There is no electrical connection between the heater wire and the switch. Note that the ladder diagram in Figure 21–17 shows the heater and the switch in completely

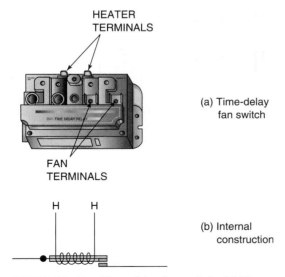

FIGURE 21–18 Time-delay fan switch. (a) Time-delay fan switch. (b) Internal construction.

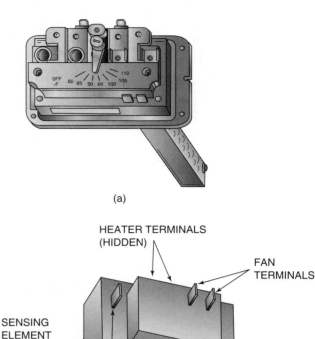

FIGURE 21–19 (a) Time-delay fan switch with temperature sensing element. (b) Time-delay fan/limit switch.

different parts of the diagram, even though they are contained within the same device (Figure 21–17). Remember, the ladder diagram does not attempt to show you any of the physical relationships of the locations of electrical devices.

The heater is wired in parallel with the gas valve. Whenever the gas valve becomes energized, the heater will also become energized. Approximately one minute after the heater wire is energized, it will have generated enough heat to cause the bimetal switch to close, turning on the blower.

Other variations of the time-delay fan switch are shown in Figure 21–19. In Figure 21–19a, a bonnet-sensing temperature element has been added. This switch will close after a fixed time delay *or* whenever the bonnet temperature reaches the set point of the switch, whichever occurs first. In Figure 21–19b, a limit switch has also been added, sharing the same bimetal sensing element. There is no electrical connection between the limit switch, the fan switch, or the heater element.

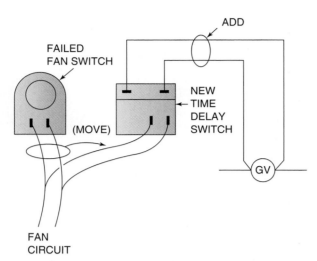

FIGURE 21–20   Replacing a bonnet temperature sensing fan switch with a time-delay switch.

The time-delay fan switch is a valuable part to carry on your truck. There are many different types and sizes of bonnet switches. When a bonnet switch fails, it is convenient to replace it with a time-delay switch. Simply leave the old bonnet switch in place. Attach the time delay switch to any convenient place on the furnace front panel. Move the wires that are on the failed fan switch to the switch terminals on the time-delay switch (Figure 21–20). Then wire the heater element in parallel with the gas valve. In this way, you can carry a single part that can be used to replace many different types of fan switches. In fact, this method can even be used where the fan switch portion of a fan-limit switch fails. The wires on the failed fan switch can be moved without upsetting the function of the existing limit switch.

TIME DELAY WITH THREE-WIRE LIMIT SWITCH

Figure 21–21 shows a wiring scheme used on unit heaters that are mounted near the ceiling. In a limit condition caused by a gas valve stuck in the open position, normally the room thermostat would open and the fan would turn off. The flame would continue to burn, but the occupants would be unaware of the problem because the heat would stratify near the ceiling. However, with the three-wire limit switch shown in Figure 21–21, the fan will remain on during the limit condition, even if the room thermostat turns off. This will alert the occupants (who will become too warm) that there is a problem with the heater.

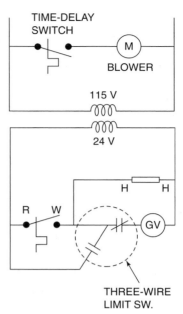

FIGURE 21–21   Three-wire limit switch.

INDUCER FAN

In some furnace applications (Figure 21–22), the manufacturers have tried to gain better efficiency by increasing the number of passes that the flue gas makes through the heat exchanger. This allows the room air to absorb more heat from the products of combustion, but it also increases the pressure drop through the heat exchanger. At the same time, the cooler flue gas has less buoyancy, and the result is a furnace that needs an **inducer fan** to help pull the products of combustion through the furnace. When an inducer fan is used, generally the following sequence of operation is used.

1. On a call for heat, the room thermostat energizes an inducer fan relay, bringing on the inducer fan.
2. The operation of the inducer fan is proved by a pressure switch, a sail switch, centrifugal switch, or other means, which, in turn, supplies power to the ignition sequence. A simplified schematic is given in Figure 21–23.

MILLIVOLT FURNACES

A millivolt furnace is one that uses a **pilot generator** (also referred to as a **powerpile**) to sense the pilot flame and provide a source of power to the millivolt gas valve (Figure 21–24). The pilot generator

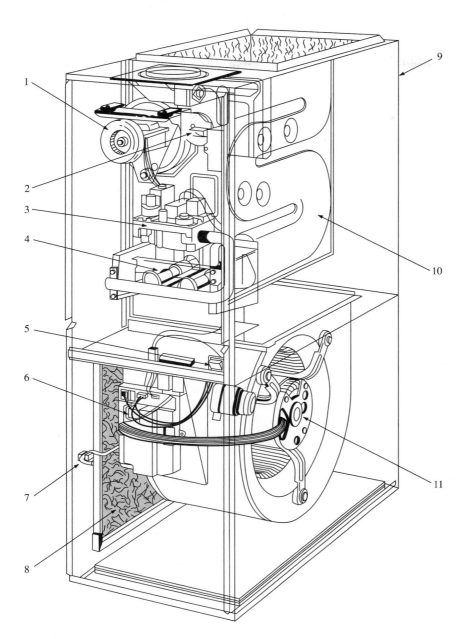

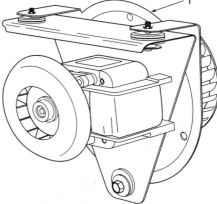

①	INDUCER ASSEMBLY	⑦	AIR FILTER RETAINER
②	PRESSURE SWITCH	⑧	AIR FILTER
③	GAS CONTROL VALVE	⑨	WRAP-AROUND CASING
④	BURNER ASSEMBLY	⑩	HEAT EXCHANGER
⑤	BLOWER DOOR SAFETY SWITCH	⑪	BLOWER AND BLOWER MOTOR
⑥	CONTROL BOX		

FIGURE 21–22   Furnace with inducer fan. (Courtesy of Carrier Corp.)

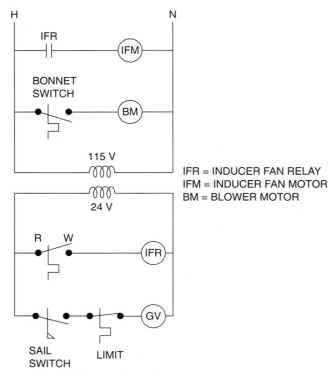

IFR = INDUCER FAN RELAY
IFM = INDUCER FAN MOTOR
BM = BLOWER MOTOR

FIGURE 21–23 Inducer fan circuit.

works on the principal that two dissimilar metals, when joined together, will generate a small voltage that is dependent on the temperature of the junction. The hotter the junction, the higher the voltage output. In a pilot generator, a number of these junctions are wired together in series to produce an

output as high as 600 mV (less than one volt). The millivolt furnace does not require any external source of power and is therefore well suited for use on wall furnaces and floor furnaces.

The inexperienced technician may mistake the pilot generator for a thermocouple because they are each positioned next to the pilot flame. The end of the pilot generator is physically larger than the end of the thermocouple. The voltage is transmitted over two wires, protected by metal shielding. The end of the pilot generator usually has normal wire terminations rather than the screw-in arrangement familiar to the thermocouple.

The gas valve used on a millivolt system is also very different from the 24-V gas valve. Although similar in appearance, the millivolt valve is a diaphragm-type valve, shown schematically in Figure 21–25. When the valve is closed, there is inlet gas pressure available on both the top and the bottom of the diaphragm. In fact, the area of the top of the diaphragm that "sees" the gas pressure is larger than the area on the bottom. Therefore, the gas pressure actually tends to help the small spring hold the valve closed. When the room thermostat closes, the millivoltage from the pilot generator is supplied to the coil, creating a magnetic field. The top of the small pilot valve is pulled toward the coil. The passage from the gas inlet

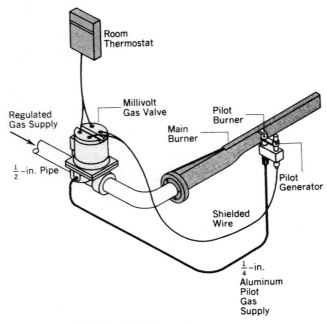

FIGURE 21–24 Millivolt system.

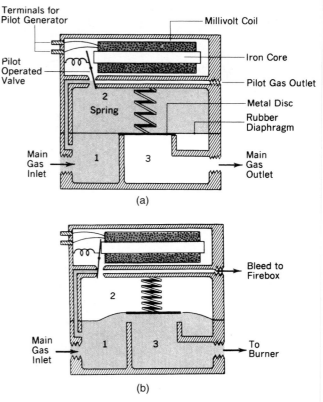

FIGURE 21–25 Diaphragm-type millivolt gas valve.

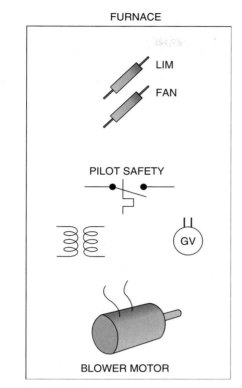

FIGURE 21–26 Exercise.

is shut off, while the trapped gas pressure above the diaphragm is allowed to bleed off through the vent tube. The only remaining pressure on the diaphragm is from chamber #1, lifting the diaphragm, and opening the main gas valve. In this way, the small voltage available from the pilot generator can be used to operate a valve simply by directing the forces already available from the gas pressure.

SUMMARY

- Many types of control devices are used in gas furnaces.
- The manufacturers' service data provide the best sources of maintenance information for gas-furnace controls.

PRACTICAL EXPERIENCE

Required equipment Gas furnaces, manufacturers' maintenance data

Procedure

1. In the maintenance data, determine the system that controls the furnace.
2. Follow the circuit diagram while at the same time locating the operating and control elements of the furnace within the furnace.

Procedure Redraw the furnace components shown in Figure 21–26. Add wiring to complete the diagram. For this and all future exercises using two voltages, use one color for the line voltage (115 V in this exercise) wiring, and a different color for 24-V wiring.

Procedure Figure 21–27 shows a furnace manufacturer's diagram with some unusual symbols. Redraw as a ladder diagram, using conventional symbols. What device is being used as a pilot safety?

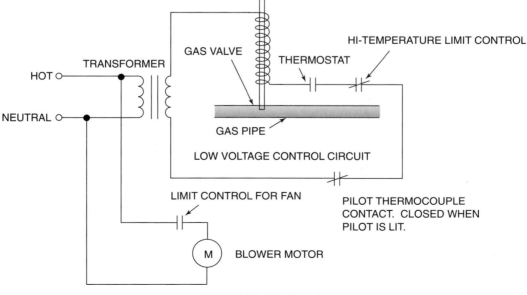

FIGURE 21–27 Exercise.

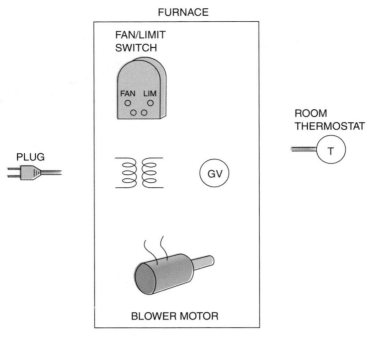

FIGURE 21–28 Exercise.

Procedure Redraw the furnace components shown in Figure 21–28. The fan/limit switch is the type shown previously in Figure 21–14. Add the necessary wiring, with the limit switch operating in the line voltage circuit. Repeat, with the jumper on the fan/limit switch removed, and the limit switch operating in the 24-V circuit.

Procedure Redraw the furnace components shown in Figure 21–29. Add the necessary wiring

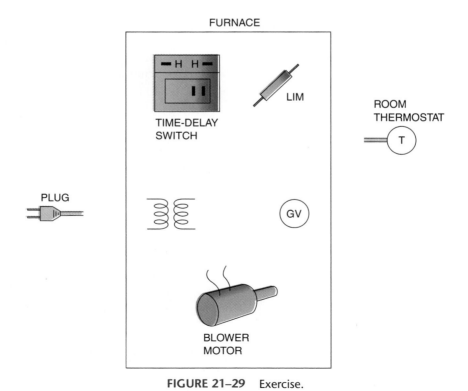

FIGURE 21–29 Exercise.

for a fully operational furnace with a time-delay fan switch.

REVIEW QUESTIONS

1. Referring to Figure 21–5, which coil of the gas valve is energized first, the pilot or main valve?

2. Referring to Figure 21–1, describe what must happen before the blower motor operates.

3. What is the purpose of the limit switch in Figure 21–1?

4. Describe the operation of thermocouple controlled gas valve.

5. In the circuit of Figure 21–1, if the bonnet switch fails and does not close, what might you expect to happen next?

6. What customer complaint might you expect if the heater in Figure 21–17 were to open?

7. What is the purpose of the time delay switch in Figure 21–17?

8. Referring to Figure 21–24, will the gas valve open if the inducer fan motor fails? Explain your answer.

9. In normal operation, will the blower motor run if the IFM inducer fan motor fails? Explain your answer.

10. Will a standard replacement 24-volt gas valve function properly with a millivolt furnace?

UNIT 22

ELECTRONIC IGNITION GAS-FIRED FURNACES

OBJECTIVES

After completion and review of this unit, you should be familiar with electronic ignition systems, including:

- Intermittent pilot systems.
- Direct spark systems.
- Sensing methods.
- Troubleshooting methods.

In the 1980s, energy shortages around the country moved manufacturers to start designing furnaces that did not use a standing pilot flame. The theory was that keeping a pilot flame lit when heat was not required was wasting a lot of energy. Several methods were developed to accomplish ignition using a spark instead of a pilot flame. Generally, these methods can be classified as either intermittent pilot or direct spark systems. A description of the operation of each follows.

INTERMITTENT PILOT SYSTEMS

Intermittent pilot systems use two valves in one body (Figure 22–1). The first valve, when opened, will allow pilot gas to flow through the pilot tubing to a pilot burner. It is called the pilot valve and is usually abbreviated in wiring diagrams as PV. The second valve,

when opened at the same time that the pilot valve is already open, will allow gas to flow into the main burner where it can be ignited if there is a flame at the pilot burner. This second valve is the main valve and is usually abbreviated MV in wiring diagrams.

The general sequence of operations for these systems follows.

1. The room thermostat calls for heat, closing its contacts.

2. The pilot valve is opened, supplying gas to the pilot burner. At the same time, an electronic device provides a spark between a wire and the pilot burner hood. This will cause the pilot gas to ignite.

3. The pilot flame will be sensed by one of several different methods. If the pilot flame has, in fact, lit, then the main valve will be energized, allowing the main flame to come on.

4. If the pilot flame were not lit, or if it were not sensed, some systems will allow the pilot valve to remain energized and the spark to keep sparking. These are called non-100% shut-off systems. Other systems are 100% shut-off. They will only allow the trial for ignition to continue for a fixed time (maybe 30 seconds). After that, they will go into **lockout**. That means that the pilot valve will be de-energized, the spark will stop, and the system will not try to relight again until it is

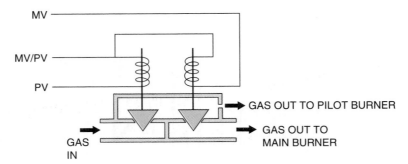

FIGURE 22–1   Gas valve for an intermittent pilot furnace.

reset. Most systems can be reset from a lock-out condition by simply turning the room thermostat down so that its contacts open. When the thermostat is then returned to a higher setting, the light-off sequence will be initiated again.

DIRECT-SPARK SYSTEMS

Direct-spark systems differ from intermittent pilot systems in the following ways.

1. There is no pilot burner.
2. The spark is made at the main burner. The spark lights the main flame directly, without using the intermediate pilot flame.
3. Instead of sensing pilot flame, the system senses main flame.
4. All direct-spark systems must be 100% shut-off. While a steady flow of unburned pilot gas in an intermittent pilot system would simply vent itself through the flue stack, a much larger steady flow of unburned main gas is too unsafe to be tolerated. A direct spark system will go into lock-out if the main flame is not established within a very short time (maybe 10 seconds).

INTERMITTENT PILOT FURNACE WIRING

Figure 22–2 shows a typical furnace manufacturer's wiring diagram for an intermittent pilot system. Some of the features of this wiring scheme are

1. There is an ignition control (IC) that performs the functions described earlier.

2. A vent fan assists with the removal of flue gas. In modern furnaces, the heat exchangers are more efficient, but have higher pressure drop. Sometimes gravity venting of the products of combustion is not sufficient by itself. The vent fan is operated through a relay (RVF) that is energized on a call for heat. When the centrifugal switch on the vent fan motor closes, then the ignition sequence is allowed to proceed.

3. There is a **roll-out switch** (identified in the diagram as S_{FR}). This is common on furnaces that use a vent fan for the products of combustion. If the vent fan does not properly remove the products of combustion, the flame may roll out of the heat exchanger and endanger the wiring on the front panel of the furnace. The roll-out switch senses if the flame reaches outside the heat exchanger where it doesn't belong. It the roll-out switch opens, the ignition sequence is interrupted.

DIRECT-SPARK FURNACE WIRING

Figure 22–3 shows the wiring for two different Honeywell direct-spark modules. The difference between them is that the S87 module lights the main flame using an ignitor-sensor that creates a spark to the pilot burner hood (Figure 22–4). The S89 module lights the main flame using a hot surface ignitor (Figure 22–5). The Alarm terminal is only used on some models. It provides power to an alarm to notify occupants if the module has gone into lock-out. The vent-damper plug is also available only on some models. It provides an output signal to a 24-V motor that closes a damper in the flue stack to conserve heat energy during the off-cycle.

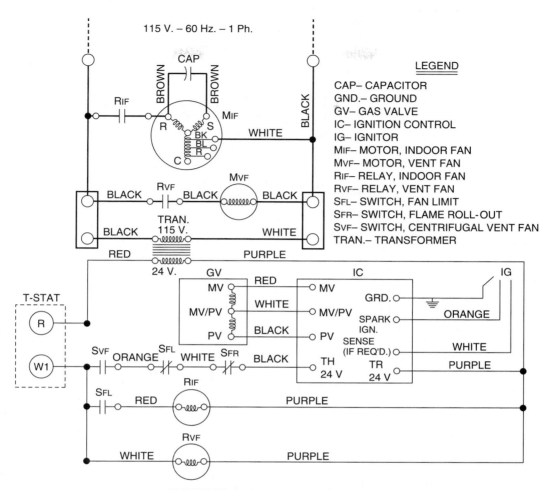

FIGURE 22–2   Intermittent pilot system.

SENSING METHODS

Standing pilot systems usually use a thermocouple to prove the existence of a pilot flame before main gas is allowed to be supplied to the burners. Intermittent pilot systems, depending on the manufacturer of the ignition system, use either a **bimetal pilot safety switch** (Figure 22–6), a mercury-filled bulb that operates a switch (Figure 22–7), or an electronic means of sensing such as flame conduction or **rectification** (Figure 22–8). The bimetal and mercury bulb systems are easy to understand. When the pilot flame is established, expansion of the bimetal or expansion of the mercury in a sensing bulb causes an SPDT switch to operate. For the bimetal switch, the common terminal is yellow, the normally closed terminal (when the bimetal is cool) is green, and the normally open terminal is white.

However, to understand the principles of flame conduction and rectification, you must understand what happens in a gas flame. The flame is actually a series of small explosions that cause the immediate atmosphere to become ionized. This ionization causes the atmosphere to become conductive. The flame can be thought of as a switch. This "switch" is located between the pilot burner tip and the flame sensor. When there is no flame between the pilot tip and the flame sensor, the switch is open. When a flame is in contact with the pilot tip and the flame sensor, the switch is closed.

JOHNSON CONTROLS ELECTRONIC IGNITION

The G-60 (Figure 22–9) is the first generation of electronic ignition controls manufactured by Johnson Controls. It is available in 24-V and 115-V models, 100% or

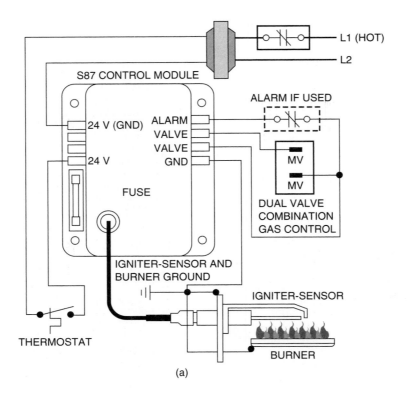

(a)

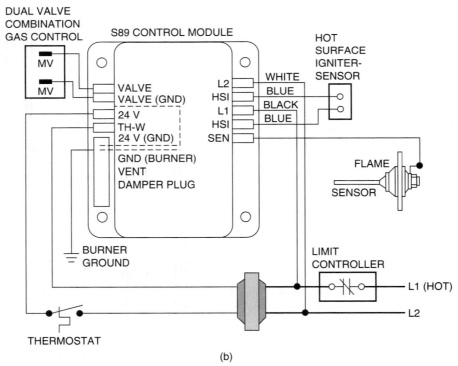

(b)

FIGURE 22–3 Honeywell direct spark systems.

FIGURE 22–4   S87 components. (Courtesy of Honeywell, Inc.)

non-100% shutoff, and with an optional vent damper plug. The G65 and G66 are newer models. The sequence of operation is similar for all of these controls.

1. When the temperature control calls for heat, the ignition control's spark transformer and pilot valve are energized.

2. The spark lights the pilot.

3. The flame sensor proves the presence of the flame. The ignition control then shuts off the spark. At the same time, the main burner valve is opened. Some models permit the spark to continue for a short period of time after the main burner lights. A 100% lock-out model provides for shutdown of the

entire system when the pilot does not light within some fixed period of time (usually 30 seconds).

4. The main burner ignites and the system continues normal operation.

5. When the temperature control is satisfied, the main burner and pilot valves are de-energized, shutting off all gas flow.

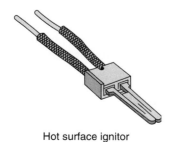

Hot surface ignitor

FIGURE 22–5   Hot surface ignitor.

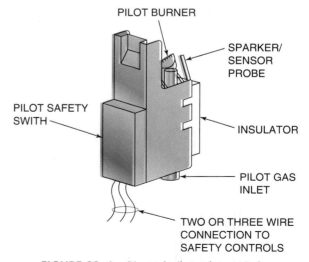

PILOT BURNER

SPARKER/
SENSOR
PROBE

PILOT SAFETY
SWITH

INSULATOR

PILOT GAS
INLET

TWO OR THREE WIRE
CONNECTION TO
SAFETY CONTROLS

FIGURE 22–6   Bimetal pilot safety switch.

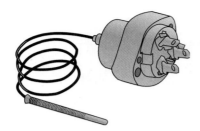

FIGURE 22–7 Mercury-bulb pilot safety switch.

TROUBLESHOOTING THE G60 SERIES CONTROLS

Troubleshooting of the G60 module is done by process of elimination. All the other causes that might explain the symptom are explored first. If no other cause can be found for the symptom, then the controller box must be replaced. The following paragraphs describe the possible external causes for each symptom. If none of the external problems are shown to exist, then the ignition controller must be replaced.

Symptom: No Spark

Check to see if the ignition controller is being supplied with 24 V (or 115 V). If it is not, check the room thermostat and all safety controls that can interrupt the voltage to the controller. Check the transformer to make sure that 24 V is available from the transformer. If proper voltage is available at the controller box, check to see that the high-voltage cable is securely connected at both ends, and that it is not brittle, burned, or cracked. Check ohms from one end of the cable to the other. The resistance should be close to zero. If the cable is good, check to see that the spark gap is 0.1.

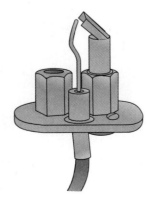

FIGURE 22–8 Flame rectification sensor.

FIGURE 22–9 G-60 electronic ignition module. (Courtesy of Johnson Controls, Inc.)

Symptom: Spark Is Present, Pilot Flame Will Not Light

It is not necessary to check voltage to the box—it is sparking. Check to see if there is voltage at the pilot valve. If there is not, then the box has failed. If there is voltage available, try to light the pilot flame with a match. If the pilot gas lights, then the spark is positioned poorly and must be repositioned until it lights the pilot gas. If the pilot will not light with a match, loosen the aluminum pilot gas supply line to see if pilot gas is available. If you can hear gas escaping from the pilot gas line, you probably have an obstructed pilot gas orifice in the pilot burner that must be removed and cleaned. If there is no gas pressure available in the pilot gas supply line, check to see if there is gas pressure available at the inlet to the gas valve. If gas is available at the valve inlet, then the pilot valve is not opening and must be replaced.

Symptom: Pilot Flame Present, Main Flame Will Not Light

We know that the thermostat has closed and that power is being applied to the G60 box because the pilot is lit. There are several possibilities that could cause the main flame to not light.

1. The sensing wire has failed.
2. The sensor is positioned badly.
3. The sensing circuit is fine, but the box is not putting out voltage to the main valve because the box has failed.

4. The box is sending voltage to the main valve, but the main valve has failed.

The easiest place to start checking is between the MV and the GR terminals on the box. If voltage is present, check the main valve coil for continuity. It has probably failed. If there is no voltage between the MV and GR terminals, then the sensing circuit must be checked to see if it is telling the box that the pilot flame has been established. The sensor wire must be located so that it is in the pilot flame. The continuity of the sensor wire can be checked with an ohmmeter. There must also be good continuity to ground. Check the ohms between the GR connection and the casing. It should be zero.

HONEYWELL AND ROBERTSHAW CONTROL BOXES

The Honeywell and Robertshaw boxes (Figure 22–10) are similar in operation to the Johnson Controls G60 box. The differences are as follows.

1. On some models, there is no sensor wire. The same wire is used to send the spark and to sense the pilot flame. This eliminates one step of troubleshooting, because if the pilot flame has lit, you don't need to check the sensor wire for continuity.

2. Instead of the valves being wired to the ground connection, a terminal called MV/PV is used. The pilot valve is wired between the PV and PV/MV terminals, and the main valve is wired between the MV and PV/MV terminals.

Except for the minor differences noted, the troubleshooting of the Honeywell and Robertshaw boxes is the same as for the G60 box.

PICK-HOLD SYSTEM

Furnaces manufactured by Carrier, Bryant, Day/Night, and Payne have their own method of accomplishing the intermittent pilot sequence. The pick-hold gas valve is unique. It has five terminals. It is not interchangeable with any other type of furnace control system. It consists of two solenoid coils that operate the pilot valve, and a small heater (called a heat motor) that opens the main valve (after a time delay of approximately 10 seconds after it is energized). The PICK coil is designed so that when it is energized, it picks up the pilot valve off its seat. The HOLD coil, when energized, is not strong enough to open the pilot valve by itself, but if the pilot valve has already been opened (by the PICK coil), the HOLD coil is strong enough to hold the pilot valve open. The PICK coil performs the same function as your thumb when you manually open the pilot valve in a standing pilot system. The HOLD coil performs the same function as the thermocouple in the standing pilot system.

The pick-hold system uses a sparker as shown in Figure 22–11. When 24 V are applied to the terminals, the box will put out a 10,000-V spark

(a) Honeywell

(b) Robertshaw

FIGURE 22–10 Honeywell and Robertshaw electronic ignition control boxes.

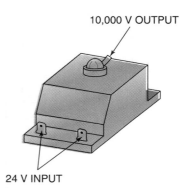

FIGURE 22–11 Pick-hold sparker.

between the sparker wire and the pilot burner hood. If the pilot does not light, the sparker will keep going for as long as it is supplied with 24 V. Therefore, this is a non-100% shut-off system. If the pilot flame is established, it will be sensed through the sparker wire, and the sparking will stop.

A simplified wiring diagram for the combustion portion of the pick-hold system is shown in Figure 22–12. When the room thermostat calls for heat, it makes the connection between the R and W terminals. The HOLD coil is energized directly from the thermostat, and the PICK coil and the sparker are energized through the normally closed contacts of the bimetal pilot safety switch. Pilot gas and spark should both be present. When the pilot flame is established, the sparker turns itself off, and the bimetal element in the pilot safety switch heats up. First, the normally closed contact opens. Some time later (maybe 15–30 seconds), the normally open contact will close, energizing the main gas valve, and the main flame will ignite. The same signal that energizes the main valve also sends a signal to the fan circuit which will be discussed later.

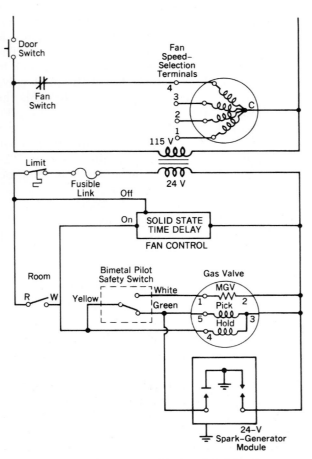

FIGURE 22–12 Simplified pick-hold wiring.

Troubleshooting: Spark Present, But No Pilot Flame

We know that the room thermostat has closed, and the normally closed contact of the bimetal pilot safety switch is closed. Potential causes are

1. No gas available to the furnace.
2. Plugged pilot gas orifice.
3. Spark positioned badly.

Use a match to try to light the pilot. If the pilot gas lights, then change the position of the sparker. If the pilot doesn't light, confirm the availability of gas to the furnace by loosening the supply gas piping. If gas is available, check the availability of pilot gas by loosening the pilot gas tubing. If no gas is available, the pilot valve has failed. If gas is available, the pilot orifice must be plugged.

Troubleshooting: No Spark, No Flame

Check across the R and W terminals to make sure that the room thermostat is closed. Then check between W and 5 on the gas valve to see if the normally closed contact in the bimetal pilot safety switch is closed. If you get 24 V across the switch (indicating it's open), it must be replaced. Sometimes a sharp rap on the switch with a screwdriver will restore it to operation, but it should still be replaced.

Troubleshooting: Pilot Flame Cycles On and Off

The HOLD coil has failed. After the pilot flame is established, the normally closed contact opens. As soon as it does, the pick coil is de-energized, and the pilot valve closes. The bimetal pilot safety switch closes, and the pilot lights again. The cycle continues. The gas valve must be replaced.

Solution If the pick coil were energized directly from the thermostat, it would work fine, except for a safety problem. Consider a situation where the furnace is operating, and there is a momentary power outage. The main valve and pilot valve would both close, and the flame would go out. When the power is restored, the pick coil will be energized, opening the pilot valve. And because the normally closed contacts of the bimetal pilot safety would still be warm, the main coil will also be energized. We would therefore have main gas being admitted to the furnace before the existence of a pilot flame has been proven.

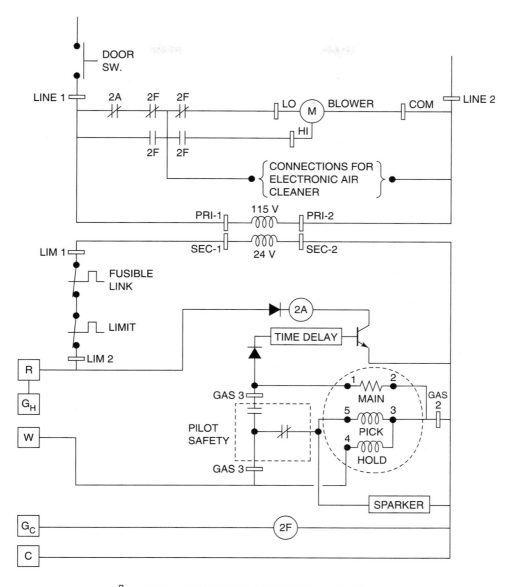

FIGURE 22–13 Pick-hold furnace.

In Figure 22–12, when power is restored, the pick coil would not be energized until after the bimetal pilot safety switch had a chance to cool. Even if the main valve opens, no main gas will flow so long as the pilot valve is closed.

Figure 22–13 shows the completed wiring diagram for the pick-hold system. The operation of the fan deserves further explanation. The control of the fan is accomplished electronically by a chip on a circuit board. There is no bonnet-sensing fan switch or mechanical time-delay switch. The time-delay control gets a constant 24-V signal from the transformer. This energizes relay 2-A coil, opening the 2-A contacts, and turning the fan off! Yes, the blower relay switches are normally closed, and a 24-V signal is required to keep the fan off. This results in a unique problem with the pick-hold furnace. If either the transformer has failed, the fusible link or the limit switch is open, or if the blower relay coil has failed, the fan will run continuously!

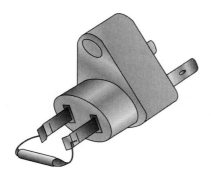

FIGURE 22–14   Fusible link.

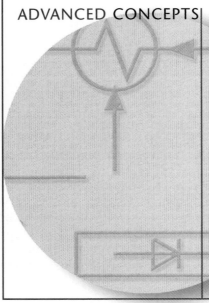

A fusible link is shown in Figure 22–14. It is used in various applications where it is desired to open a circuit if a predetermined limit temperature is reached. In the pick-hold furnace, it is located on the front of the heat exchanger and is clearly visible when the front panel of the furnace is removed. It will only reach its opening temperature if the flame, for some reason, escapes from the confines of the inside of the heat exchanger (for example, if the flue stack is blocked). In this application, it is sometimes called a **roll-out switch**. The

fusible link is also commonly found on electric heating systems where it operates as a high-temperature limit switch

For normal blower operation, the 2-A relay coil is energized immediately when the furnace is plugged in. On a call for heat, when the bimetal pilot safety operates, it energizes the MGV and, at the same time, provides 24 V to the timer circuit, which some seconds later emerges as an "off" signal, which defeats the constant 24 V to the 2-A coil. This de-energizes the 2-A coil, and the normally closed switch contacts then bring on the blower motor.

When the furnace is used for cooling as well as heating, there is an extra relay (2F). When the room thermostat makes R–G, it energizes the 2F coil, causing the blower motor to run on High speed.

FIGURE 22–15   Diode and transistor symbols.

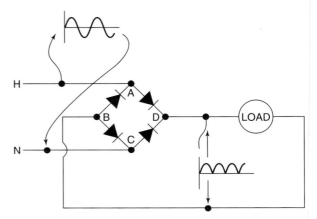

FIGURE 22–16   Rectifier.

ADVANCED CONCEPTS

The time-delay circuit includes two symbols that have been previously explained, a **diode** and a **transistor** (Figure 22–15). The service technician is normally not called on to troubleshoot down to this level.

A diode is the electronic equivalent of a mechanical check valve. It will only allow electrons to flow in one direction. In order to test a diode you can use an ohmmeter. It should show continuity across the diode in one direction, but if you reverse the meter lead locations and measure in the opposite direction, it should indicate no continuity.

A transistor is the electronic equivalent of an electric valve, which opens when an external voltage is applied. For the transistor shown, current will not flow from the collector lead to the emitter lead (like a closed valve) until there is first a small current that flows from the base lead to the emitter lead. The current from the base to the emitter can be quite small compared to the current that can flow from the collector lead to the emitter lead. "See" practical experience semiconductor devices.

Diodes are also sometimes arranged in a circuit such as shown in Figure 22–16, a **rectifier**. Even though the current in the supply voltage is continuously changing direction, the flow of electricity between terminal D and the load will always be in one direction only, to the right.

Pick-Hold with an Inducer Fan

Figure 22–17 shows the same furnace, but with the addition of an inducer fan to help pull the products of combustion through the heat exchanger. When the inducer fan is not running, the pressure switch (7 V) is in the "up" position. The pressure switch senses operation of the inducer fan and will move to the "down" position whenever the inducer fan runs.

On a call for heating, voltage is supplied through the W terminal to the 2D inducer fan motor relay. Its contacts close, energizing the inducer fan

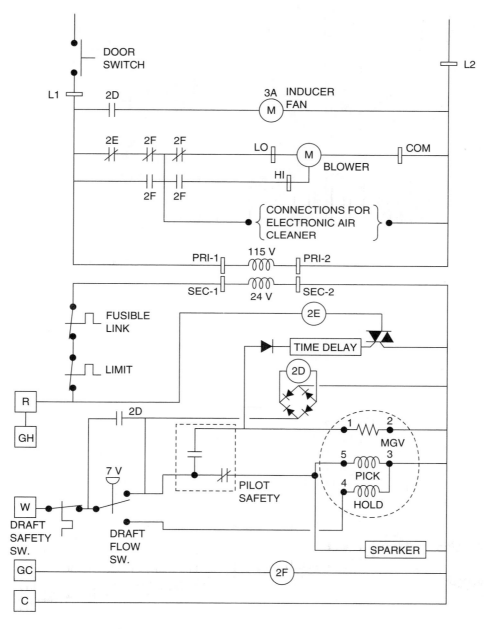

    [] = SPADE CONNECTION ON TERMINAL
    [ ] = THERMOSTAT CONNECTION ON TERMINAL BOARD
    2D = INDUCER MOTOR RELAY
    2E = BLOWER RELAY, HEATING
    2F = BLOWER RELAY, COOLING

FIGURE 22–17 Pick-hold furnace with inducer fan.

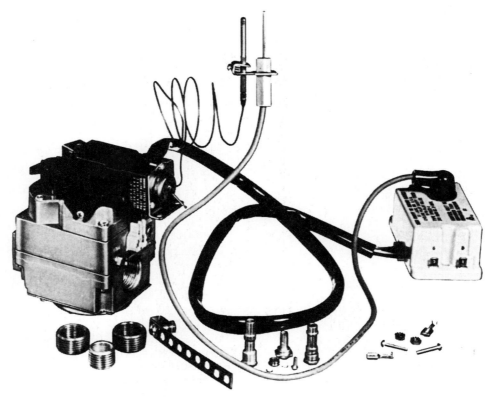

FIGURE 22–18 White-Rodgers mercury-bulb system. (Courtesy of White-Rodgers Division, Emerson Electric Company)

motor 3A. Another set of 2D contacts also close in the 24-V circuit, "sealing in" the normally closed contacts of the pressure switch. Even though the pressure switch will open as soon as it senses the operation of the inducer fan motor, the 2D coil will remain energized.

At the same time that the 2D coil is energized, the pick coil is also energized through the bimetal pilot safety switch 6H. Its operation is also "sealed in" around the 7-V pressure switch, through the operation of the 2D contacts in the 24-V circuit.

MERCURY-BULB SYSTEM

The White-Rodgers system (Figure 22–18) involves the use of a special gas valve. It has a main valve and a pilot valve (similar to other valves), but it also has an internal pressure switch (Figure 22–19) and a place to plug in the terminals of a mercury-bulb pilot safety switch. In Figure 22–20, the mercury-bulb pilot safety switch is shown as an SPDT switch in a dotted box separate from the gas valve because it is a remote switch. The pressure switch is located in

the chamber of the gas valve between the gas inlet and the gas outlet. It is a normally open switch, and it closes when the pilot valve opens.

When the mercury bulb is cool, the switch makes between terminals 4–3. When the room thermostat calls for heat, a circuit is completed from the transformer, through the room thermostat, the pilot safety switch and the pilot valve. A parallel circuit is completed through the sparker. The sparker ignites the pilot gas, and the pilot flame heats the pilot safety valve. The switch between 4–3

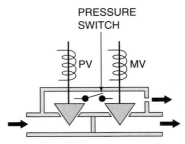

FIGURE 22–19 White-Rodgers gas valve with internal pressure switch.

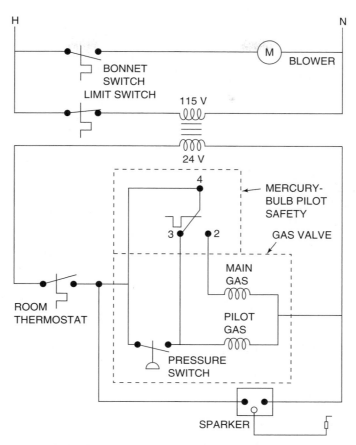

FIGURE 22–20 Wiring for White-Rodgers gas valve.

opens, and some seconds later, the switch between 4–2 closes. This completes the circuit to the main gas valve. The main gas is ignited by the pilot flame.

When the switch between 4–3 breaks, the pilot valve remains energized because the internal pressure switch closes as soon as the pilot valve opens. The pressure switch "seals-in" the circuit around terminals 4–3.

☞ If the pressure switch fails, the technician might be tempted to bypass it by simply cutting the two wires leading to it, and connecting them together. The technician might think that the pilot safety switch alone is enough protection to make sure that there is a pilot flame proved before opening the main gas. Don't do it! Bypassing the pressure switch will allow the unit to operate normally, except for one situation. Suppose that the unit is in its normal heating cycle, with the pilot valve and the main valve energized and the mercury bulb hot. Now, suppose there is a momentary power

interruption, either because of a utility interruption or an occupant who moves the room thermostat down and then back up. When the power is restored, the main gas valve, the pilot gas valve, and the sparker are all energized at the same time. It creates the unsafe situation of allowing main gas into the furnace without having first proved the existence of a pilot flame.

☞ When the pressure switch fails (not uncommon), the entire gas valve must be replaced. However, this is expensive. White-Rodgers has a "fix" available that is safe and inexpensive. Instead of bypassing the failed pressure switch, a resistor (of a specific resistance) is wired external to the valve in place of the failed pressure switch (Figure 22–21). During start-up, with the mercury bulb switch closes from 4–3, the pilot valve receives 24 V and the resistor has no effect. However, when 4–3 opens, the added resistance in series with the pilot valve reduces the current through the pilot valve circuit. There is sufficient current to

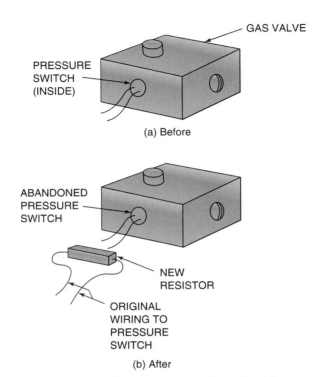

FIGURE 22–21   Repairing a gas valve with a failed pressure switch.

hold the pilot valve open, but if there is a momentary loss of power, the reduced current is insufficient to actually pull the pilot valve open from a closed position.

RETROFIT KIT

Retrofit kits are available that can be used to convert a standing pilot furnace into an electronic ignition furnace. This conversion is usually not justified unless the standing pilot gas valve has already failed. Then the incremental cost of retrofitting to electronic ignition over the cost of simply replacing the standing pilot valve can sometimes be justified. These retrofit kits may use either a flame rectification sensor or a mercury-bulb sensor. They are designed so that the sparker and sensor can be mounted using the hole in the pilot burner assembly that was used to hold the thermocouple (the thermocouple is removed when converting to electronic ignition).

LENNOX PULSE FURNACE

The Lennox system of electronic ignition is unique. It uses a pulse combustion process in which pressure pulses from the ignition process operate flapper valves inside a specifically designed gas valve. It uses a flame rectification sensor similar to other electronic ignition furnaces. A spark plug is used to provide the spark for initial ignition. A separate spark plug is used as an ignition sensor. The two spark plugs are not interchangeable due to different thread diameters. However, the control box for this furnace (Figure 22–22a and b, two different brands) has other functions as well.

On a call for heat, prior to starting the combustion sequence, this control box sends a 115-V signal to a small blower (**purge blower**) that circulates outside air through the combustion chamber (prepurge). The same blower is energized after each heating cycle (postpurge). The sequence of operation after a call for heat is as follows (times are approximate and will vary depending on the control module manufacturer).

1. Purge blower is energized.
2. At 30 seconds, the gas valve and ignition spark are energized for 5 seconds.
3. When ignition occurs and is sensed by flame rectification, the spark and purge blower are de-energized.
4. When the room thermostat is satisfied, the gas valve is de-energized and the purge blower is energized.
5. Postpurge continues for 30 seconds.

If no ignition is sensed is step 3, the purge blower continues to run. After an additional 30 seconds, step 2 is repeated (another 5-second trial for ignition). This sequence will continue to be repeated until normal ignition occurs. If no ignition is sensed after five trials, the control goes into lock-out and must be reset at the room thermostat.

A graphical representation of a successful and an unsuccessful ignition sequence is shown in Figure 22–23.

All times shown may vary slightly, depending on the manufacturer of the particular control box being used.

Figure 22–24 shows the wiring diagram for a Lennox Pulse furnace.

1. Line voltage is supplied to the unit through the door interlock switch. The door interlock switch ensures that if the door to the fan compartment is not properly closed, nothing will be energized. Door interlock switches are provided by manufacturers as a safety device and to prevent lawsuits. But they also

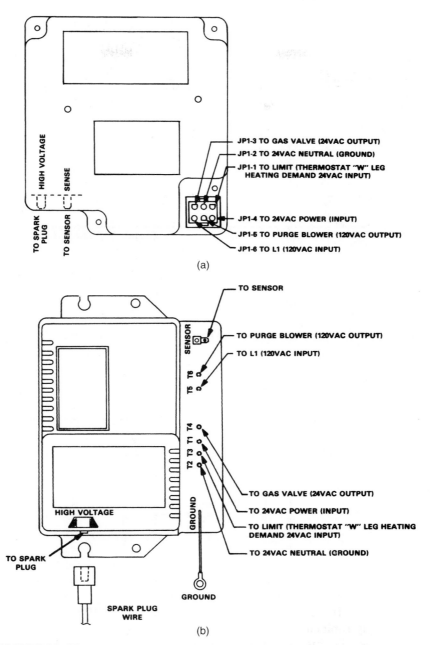

FIGURE 22–22 Lennox Pulse control boxes. (Courtesy of Lennox Industries, Inc.)

provide a steady stream of service calls from customers whose furnaces will not work after they have changed the air filters. Many owners fail to replace the access panel cover properly, and the door switch does not close.

2. The transformer provides 24-V control voltage. Note that the "common" side of the transformer is grounded at terminal T. A good ground is required for the electronic ignition primary control to operate properly.

3. The room thermostat closes, making R to W.

4. The signal from the room thermostat feeds through two pressure switches and a limit switch. The first pressure switch will open if there is excessive pressure in the flue gas exhaust pipe. The second pressure switch will open if there is excessive vacuum in the combustion air intake pipe. These conditions can occur if there is a blockage in either pipe, such as from a bird's nest.

5. The primary control is energized, and the furnace goes through its sequence of control.

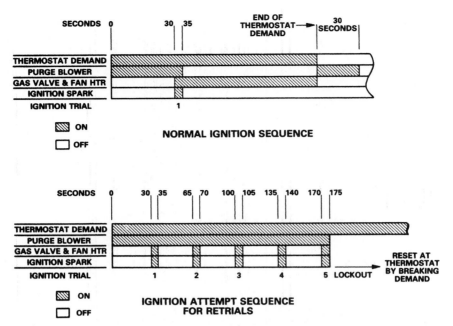

FIGURE 22–23   Timing of functions—pulse furnace. (Courtesy of Lennox Industries, Inc.)

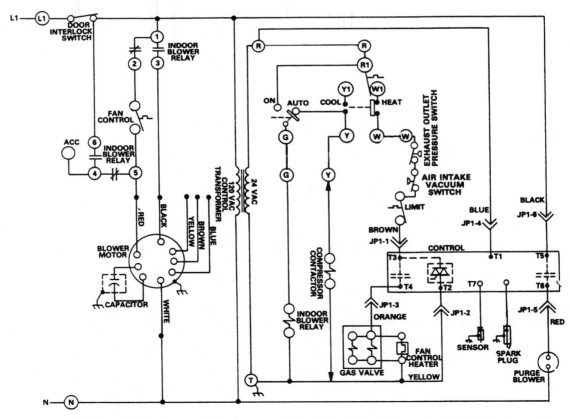

FIGURE 22–24   Lennox Pulse furnace wiring. (Courtesy of Lennox Industries, Inc.)

Note the symbols that are labelled JP1-1, JP1-2, JP1-3, etc. JP1 refers to a quick-connector, and the -1, -2, -3, and so forth, tell you which pin on the connector is being used to complete the connection. Also note that the indoor blower relay (terminals 1, 2, and 3) will operate the furnace blower motor on low speed (red lead) for heating and a higher speed (black lead) for cooling. A separate set of terminals on the indoor blower relay (4, 5, and 6) are used to provide 24 V to an accessory such as an electronic air cleaner when the fan runs on either heating or cooling.

VENT DAMPER

When a furnace (or boiler) turns off, there is still some residual heat in the heat exchanger that becomes lost through the flue stack during the furnace off-cycle. As an energy conserving measure, some furnaces use a damper in the flue stack that closes during the off-cycle. Figure 22–25 shows the typical wiring. The unique feature of the vent damper is that it is spring-loaded to open, motor-operated to close. This way, if the motor should fail, the spring will keep the damper in the open (safe) position.

When the furnace is ready for operation, 24 V are applied to terminals 1 and 2 of the vent damper,

energizing the vent damper motor, holding the damper closed. The furnace or boiler is now in standby mode. When the room thermostat calls for heat, it provides 24 V from the TH-R terminal on the pilot control to terminal 4 on the vent damper, energizing the relay in the vent damper. The relay contacts open, de-energizing the vent damper motor. The springs open the vent damper. When it is fully open, the damper pushes on the end switch (a microswitch), closing it. This completes the circuit from TH-R to TH-W on the ignition control box, and normal light-off can then proceed.

When the thermostat opens, the relay is de-energized, and the vent damper motor is energized to close the damper.

CARRIER IGC BOARD

The Carrier IGC (Integrated Gas Unit Controller) board is one of the more modern electronic control schemes used on rooftop furnaces that controls not only the ignition, but the blower motor and inducer fan motor as well. Figure 22–26 shows the terminals that are available on this control board. Also built onto the board is a combustion relay, a blower relay, a gas valve relay, and all of the fan and safety control logic.

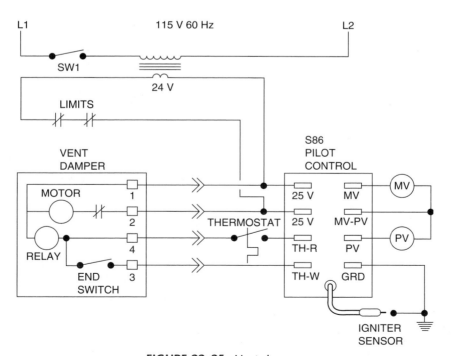

FIGURE 22–25 Vent damper.

ADVANCED CONCEPTS

The Hall Effect sensor is a magnetic device that can be mounted on the inducer fan motor. When the motor is rotating at the correct speed, a current is induced. When the Hall Effect sensor is used to prove operation of the inducer fan motor, its wiring is connected to terminals 1, 2, and 3 in the lower left corner of the IGC board.

All external components are wired to the appropriate terminals on the IGC board. The room thermostat is wired to terminals W, G, and R. A roll-out switch is wired between the two terminals RS, the limit switch is wired between the two LS terminals, and the gas valve is wired to terminals GV and GV. Then 24 V from the transformer secondary are wired between terminals R and C.

On a call for heat, the board detects continuity between R–W, energizing the on-board combustion relay, which closes a switch between L1 and the CM terminal. This energizes the inducer fan. The operation of the inducer fan can be proved in one of three ways:

1. Pressure switch
2. Centrifugal switch
3. Hall Effect sensor

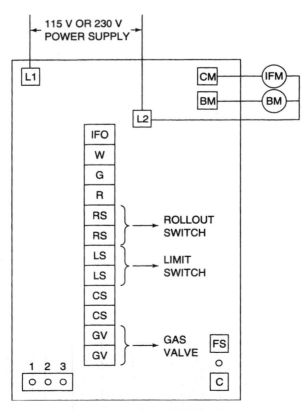

FIGURE 22–26   Carrier IGC logic board.

Once the inducer fan motor operation is proved, the board sends a 10,000 V spark through the ignition wire, and at the same time, energizes the GV terminals to supply gas. The flame rectification sensor (terminal FS) proves the flame, and 45 seconds later the BR terminal is energized, bringing on the blower motor. If the flame is not established within 5 seconds after the sparking starts, the gas valve will be de-energized, and the inducer fan motor will run for another 20 seconds to purge out the heat exchanger. Another ignition attempt follows. If, after 15 minutes of trying, no ignition is proved, the unit goes into lock-out.

Diagnostic Signals

A single LED mounted on the control board is illuminated continuously during normal operation. If it is off, there is probably no 24 V between terminals R–C. It can also give eight different error signals, as signaled by the number of flashes between pauses:

1 flash: "Fan On/Off Delay Modified" indicates that the limit switch has opened (although it may be closed now). This signal will continue until power is disconnected. You should investigate to determine why the limit switch may have opened (i.e., low air flow).

2 flashes: "Limit Switch Fault" is open now.

3 flashes: "Flame Sense Fault" is indicating a flame without the gas valve being open.

4 flashes: "4 Consecutive Limit Faults" indicates that the furnace has tried to start four times, but each time, the limit switch was open. The furnace is in lock-out.

5 flashes: "Ignition Lock-out Fault" indicates that the furnace has tried unsuccessfully for 15 minutes to light off. It is in lock-out.

6 flashes: "Induced Draft Motor Fault" indicates that there have been 24 V at the W terminal for more than 60 seconds and the control board has not gotten a signal proving that the inducer fan was running. The combustion relay remains energized, and if the board receives proof of inducer fan rotation, ignition will be attempted.

7 flashes: "Rollout Switch Fault" indicates that the rollout switch has opened, and the unit is in lock-out.

8 flashes: "Internal Control Fault" indicates that there is most likely a problem in the IGC board. If disconnecting power and reconnecting does not clear the fault, replace the IGC board.

If more than one fault exists, they will all be indicated, in sequence, separated by a three-second pause between each.

> Make sure you read the LED so that you are aware of all problems before you disconnect power, which will cancel and reset the error messages.

SUMMARY

- Many types of control devices are used in gas furnaces.
- The manufactures' service data provide the best sources of maintenance information for gas-furnace controls.

PRACTICAL EXPERIENCE

Required equipment Gas furnaces, manufacturers' maintenance data

Procedure

1. In the maintenance data, determine the system that controls the furnace.
2. Follow the circuit diagram while at the same time locating the operating and control elements of the furnace within the furnace.

REVIEW QUESTIONS

1. In Figure 22–2, which relay must be energized to provide power to the vent fan motor?
2. In Figure 22–2, when heat is called for, what voltage would you expect to measure across the RVF coil? _____ volts
3. Name the methods used with the intermittent pilot system to prove the existence of flame.
4. What method is used with the standing pilot system to prove the existence of flame?
5. In the Honeywell direct spark system, when does the spark transformer turn on?
6. What is the purpose of a purge blower?
7. With the Carrier IGC board, Figure 22–20, how long after the thermostat calls for heat does the inducer fan come on?
8. With a direct-spark system, what percentage of gas is shut off if the burner does not light?
9. With the Lenox pulse furnace, at the beginning of the heat cycle, what action must be sensed to shut off the purge blower?
10. What is the standard safety mechanism associated with vent dampers of the type shown in Figure 22–25?

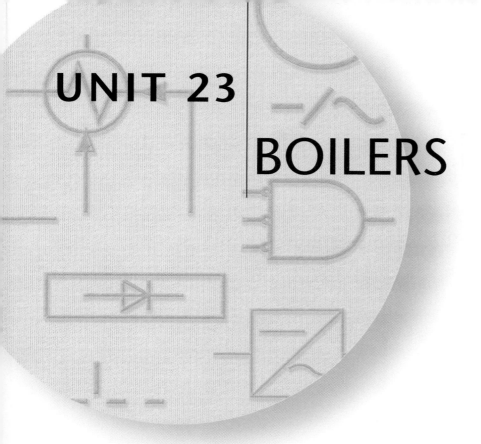

UNIT 23 BOILERS

OBJECTIVES

After completion and review of this unit, you should be familiar with the similarities and differences in the system using a boiler, as compared to standard heating systems. Included are:

- Residential hot-water boilers.
- Zone controls.
- Multiple pumps.
- Tankless heaters.
- Electronic controllers.

The control scheme for a boiler ranges from a very simple millivolt-operated gas valve on a pool heater to a very complex large boiler with control for com-bustion air fans, dampers, air-fuel mixtures, modu-lating controls, purge cycles, multistep start-up sequences, and belt-with-suspenders safety con-trols. The boiler control scheme may be thought of as two separate functions.

The first function is combustion, either gas-fired or oil-fired. This is usually controlled to maintain a constant boiler condition, either hot water temperature or steam pressure. It may be done with on-off control or with modulating con-trols. The boiler condition that is being main-tained might be constant year-round, or it may be automatically reset depending on the outside air temperature. When it is warm outside, the need for heating in a building is reduced, and the temperature of the boiler hot water may be ad-justed downward for energy savings and improved control.

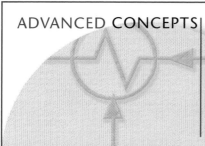

ADVANCED CONCEPTS

Sometimes the heaters that are receiving hot water are controlled with modulating valves to modulate the amount of water flowing through the heater. When the tem-perature is mild outside, the modulating valves will be operating at very close to a fully closed position. This makes control less accurate and can cause the valve seat to wear out quickly. By resetting the hot-water temperature downward, in order to pro-vide the same amount of heat, the control valve will open more, allowing more flow of the lower-temperature water.

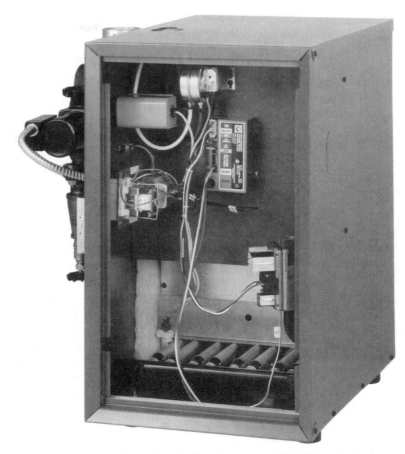

FIGURE 23–1 Hot-water boiler. (Courtesy of Weil-McLain Co.)

The second function of the control scheme is to deliver the heat to the occupied space, or in some cases, to a heat exchanger that will use the hot water to heat city water for domestic uses (sinks and showers). Often there are multiple "**zones**," such as upstairs and downstairs, or living quarters and sleeping quarters, or interior offices and perimeter offices. A zone is defined as any area that has its own thermostat that can maintain the area at a selected temperature, not dependent on what the temperatures are in the rest of the building. Zone control can be accomplished by having a thermostat adjust either the flow rate or the temperature of the hot water or steam being delivered.

RESIDENTIAL HOT-WATER BOILER

Figure 23–1 shows a boiler of the type commonly used in a residence. A simple circuit for a boiler (not the one shown in Figure 23–1) is shown in Figure 23–2. A hot-water thermostat (commonly

referred to as an **aquastat**) controls the water temperature in the boiler at a constant 180°F (sometimes as low as 160°F or as high as 200°F). When the room thermostat calls for heat, it energizes the hot water pump (Figure 23–3) through the control relay.

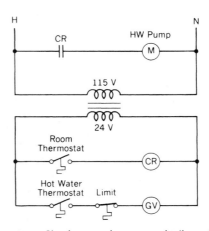

FIGURE 23–2 Single-zone hot-water boiler schematic.

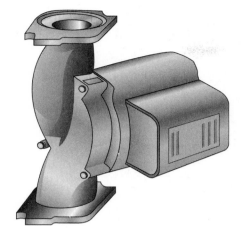

FIGURE 23–3 Hot-water pump for a residential boiler.

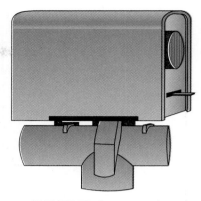

FIGURE 23–5 Zone valve.

ZONE CONTROL FOR SMALL BOILERS

Zone Control Valves

Figure 23–4 shows a boiler system that has two different zones of control, operated by two different

thermostats. It would be common to find the upstairs of a house as one zone, and the downstairs as the second zone. Each zone has its own zone valve (Figure 23–5). When the thermostat closes, it energizes a clock-type motor inside the valve that slowly opens the valve, allowing hot water to flow through the heater coil in that zone. When the valve reaches the end of its travel, an end-switch is mechanically closed. The end-switches for the zone valves are

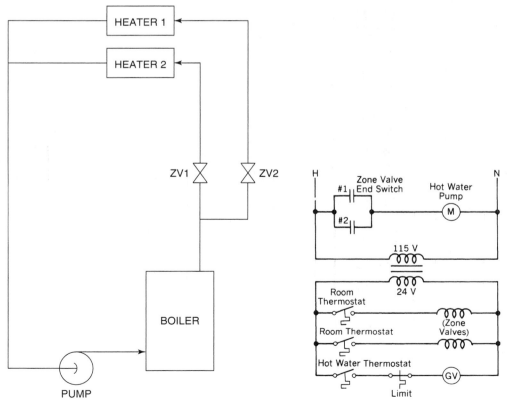

FIGURE 23–4 Boiler with zone valves.

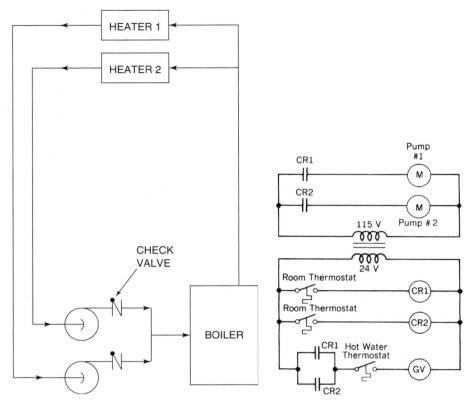

FIGURE 23–6 Boiler with multiple pumps.

wired in parallel, so that if either zone needs heat, the hot water pump will run. The pump will only shut down when both thermostats are satisfied.

The water inside the boiler is maintained at a constant temperature by an aquastat.

Multiple Pumps

Instead of each zone thermostat operating a zone valve, each zone can be provided with its own circulating pump (Figure 23–6). Each room thermostat uses a room thermostat to energize a control relay. The control relay contacts are wired in parallel so that either zone can energize the gas valve (or oil burner). In this way, during the summer months when the boiler does not need to be maintained at 180°F, it will be allowed to cool to room temperature.

 Allowing a small boiler to cool to room temperature is fine, but extreme caution must be exercised if a large boiler is to be

allowed to cool. The same applies to start-up of a large boiler. It must be done *very* slowly (maybe over a two- or three-day period) to prevent the boiler from being damaged from expansion/contraction.

BOILER WITH TANKLESS HEATER

Some boilers use an auxiliary heat exchanger located inside the boiler to supply **domestic hot water** (water for sinks, baths, and showers). Figure 23–7 shows a wiring diagram for this system. It uses a three-wire aquastat to give the domestic water zone priority over heating. For example, suppose that the demand is so great that the boiler cannot maintain 180°F. The aquastat will make R–B and break R–W, turning off the circulating pump, but letting the oil burner continue to run. Only after the aquastat reaches 180°F again will the heating system be allowed to draw heat from the boiler by turning on the circulating pump.

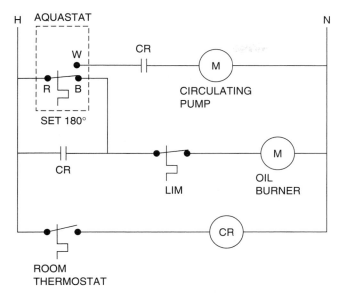

FIGURE 23–7   Boiler with tankless heater.

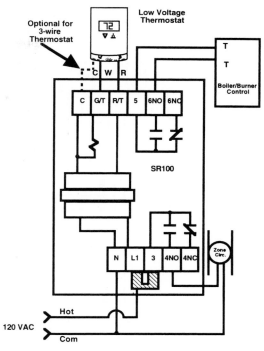

FIGURE 23–9   Connections for single-zone operation. (Courtesy of Erie Controls.)

ELECTRONIC BOILER CONTROLLER

Modern electronic controls have been marketed to replace the preceding control schemes (Figure 23–8). Figure 23–9 shows the connections for a single-zone operation where the room thermostat operates both the circulating pump and gas valve or primary controller on the boiler. This is slightly different from the scheme where the boiler is maintained at a constant temperature, as in Figure 23–2. With many modern boilers, the boiler is allowed to cool between each cycle,

because the recovery time is very quick. However, if the boiler is being used to supply domestic hot water, then a minimum temperature of 160°F must be maintained by controls outside the electronic boiler control.

The electronic boiler controllers may also be used in multiples to provide zone control (Figure 23–10) or to provide domestic water priority (Figure 23–11).

Figure 23–12 shows multiple-zone controllers that can incorporate up to six double-pole single-throw zone relays to control up to six circulators and a boiler operating control (only three zones are shown). There is a field-selectable priority for zone 1, which eliminates the need for an additional relay to provide domestic hot water priority. When priority is selected, zone 1 (which is the domestic hot water) has priority control over zones 2 through 6. If any zone is calling for heat and a demand for domestic hot water is detected, zone 1 turns on the burner and domestic hot-water circulator. Circulators for zones 2 through 6 are disabled until the domestic hot-water demand is satisfied.

Another feature that may be used with the preceding system is that it only allows the priority zone

FIGURE 23–8   Electronic boiler controller. (Courtesy of Erie Controls.)

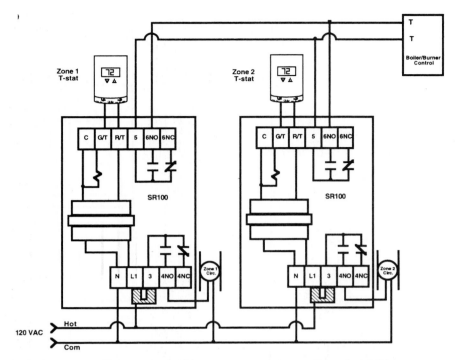

FIGURE 23–10   Zone control with water temperatures allowed to fall between cycles. (Courtesy of Erie Controls.)

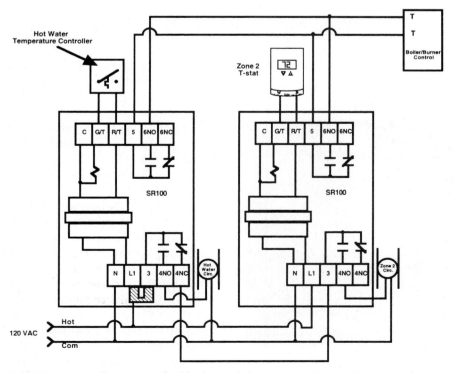

FIGURE 23–11   Zone control with domestic hot-water priority. (Courtesy of Erie Controls.)

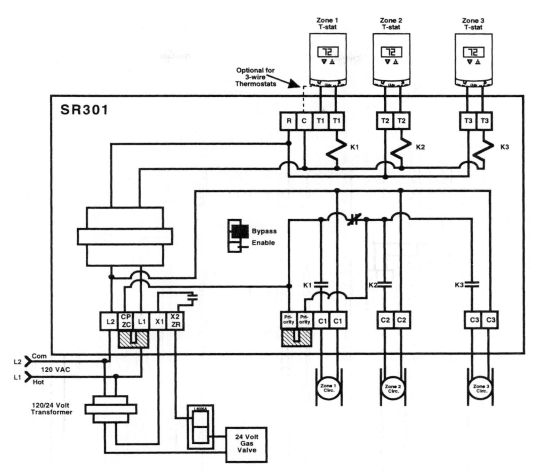

FIGURE 23–12   3-zone field-selectable priority. (Courtesy of Erie Controls.)

to take charge for a maximum of one hour. This is provided to prevent freeze-up of the heating zones if there is a malfunction in the priority zone. This feature may be enabled or bypassed with a switch on the face of the controller.

MULTIFUNCTION ELECTRONIC CONTROLLER

The controller shown in Figure 23–13 represents another step in control complexity and sophistication. It provides the following control features:

1. High- and low-limit protection.
2. Control of the circulating pump.
3. Automatic reset of the boiler temperature with outside air temperature.
4. Warm weather shutdown.
5. Thermal safety cut-off.

6. Constant operation of the circulating pump in the event of a burner failure.
7. Automatic disabling of the priority domestic hot-water zone in the event of a domestic hot-water failure.

When control becomes this complex, you often are not even given a wiring diagram that shows how it is accomplished. The most you might have is a diagram such as Figure 23–13 that shows the inputs and the outputs to the electronic "black box." For this system, the inputs include the room thermostat, the outdoor air sensor, the boiler temperature sensor, and the domestic hot-water controller. The outputs are to the circulating pumps (or zone valves) and to the status-indicating **LEDs**. LED stands for "light emitting diode." You can think of it as a small light in an electronic system. The numbers that are formed on digital displays (even including your wristwatch) may be formed using individual LEDs.

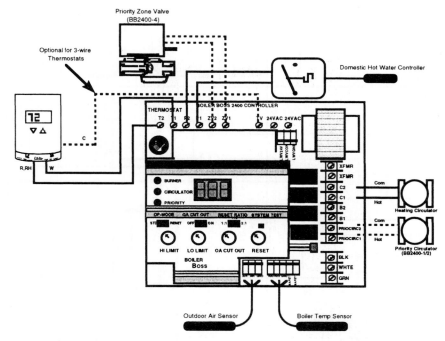

FIGURE 23–13 Complex boiler control. (Courtesy of Erie Controls.)

SUMMARY

Boiler systems use many of the same controls used in other heating systems. Exceptions are the hot-water pumps and zone valves. As with other systems, electronic controls (black boxes) are being used with greater frequency.

PRACTICAL EXPERIENCE

Redraw the wiring diagram Figure 23–9 as a ladder diagram.

REVIEW QUESTIONS

1. According to the ladder diagram Figure 23–2, the hot water pump comes on when:
   a. The hot water thermostat closes.
   b. The limit switch closes.
   c. Both the hot water thermostat and the limit switch close.
   d. When the room thermostat closes.

2. Referring to Figure 23–4, does the position of the room thermostat have direct control over the gas valve?

3. Referring to Figure 23–4, will the hot water pump come on if only one room thermostat is closed?

4. Referring to Figure 23–6, will the gas valve operate when only one room thermostat is closed?

5. Large boilers must be brought up to operating temperatures slowly.

   True _____   False _____

6. When cooling down a large boiler, it must be done slowly.

   True _____   False _____

7. According to Figure 23–7, if the water temperature is below 180 degrees, will the circulating pump operate?

8. According to Figure 23–7, if the room thermostat is closed and the water temperature is below 180 degrees, will the circulating pump operate?

9. In Figure 23–12, if the transformer marked 120/24 volt transformer (bottom left) were to open, what voltage would you expect to measure between R and C of the terminal block?

10. The controlled relay in Figure 23–2 has failed. Will this cause the boiler to trip out on high limit? Explain your answer.

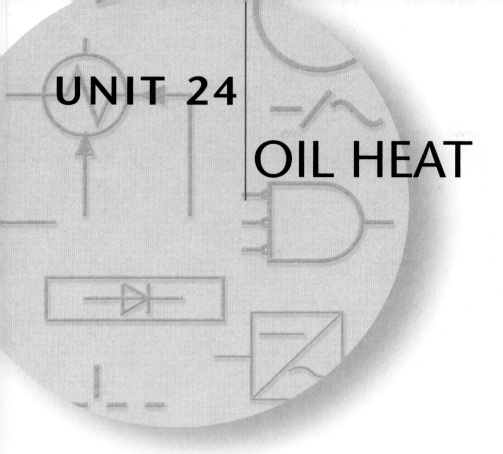

UNIT 24

OIL HEAT

OBJECTIVES

After completion and review of this unit, you should be familiar with oil fired furnaces and their components, including:

- Burner motors.
- Flame sensors.
- Oil fired heaters.

Oil is used as a fuel in both oil-fired furnaces (Figure 24–1) and oil-fired boilers (Figure 24–2). This chapter will deal only with the combustion portion of the oil-fired systems. For oil-fired furnaces, the controls used to operate the fan are the same as for gas-fired furnaces discussed in Unit nineteen. For oil-fired boilers, the control of pumps, dampers, and other accessories is the same as for gas-fired boilers, discussed in Unit 23.

OIL BURNER

Oil is far more difficult to burn than gas. If you were to have a dish of heating oil at room temperature, you would not be able to ignite it, even if you applied heat from a propane torch. The reason the oil won't easily burn is that at room temperature, there is insufficient oil vapor above the liquid to ignite. In

order to burn the oil, it must be finely atomized. This increases the total surface area of the oil and increases the rate of oil vapor formed off the surface. An oil burner (Figure 24–3) includes a pump to increase the pressure of the oil so that it can be pushed through an atomizing nozzle, a burner motor that drives the oil pump and a fan that supplies combustion air to mix with the atomized oil, a transformer that produces approximately 10,000 V, electrodes that produce a 10,000-V spark, a sensor that determines if a flame has been established, and a primary control that monitors the combustion process.

BURNER MOTOR

The burner motor (Figure 24–4) may be either split-phase or capacitor start. Residential burners will most commonly use either a 1/6- or a 1/8-hp split-phase. Manual overload protection is required, with the reset button provided on the motor housing. The motor is usually flange mounted to the burner housing. The combustion air blower is fastened directly to the motor shaft. The fuel oil pump may be either directly coupled to the opposite end of the motor shaft, or it may be belt driven. The motor is usually 3450 rpm when the pump is belt driven, and 1725 rpm where the pump is direct driven.

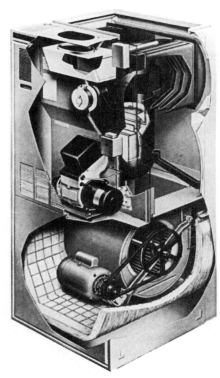

FIGURE 24–1   Oil furnace. (Courtesy of Lennox Industries, Inc.)

FIGURE 24–2   Oil-fired boiler. (Courtesy of Mestek, Inc.)

PRIMARY CONTROL

The controller for the oil burner is an electronics "black box" called a **primary control**. One type is mounted on the flue gas stack and is sometimes called a **stack control** (Figure 24–5). Another type of primary control is mounted on the oil burner (Figure 24–6). It differs from the stack control in that it uses a remote sensing device to sense whether combustion is taking place, rather than the

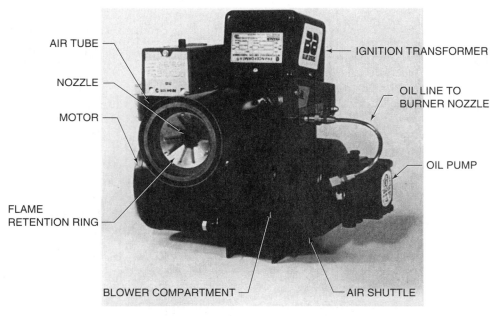

AIR TUBE

NOZZLE

MOTOR

FLAME
RETENTION RING

BLOWER COMPARTMENT

IGNITION TRANSFORMER

OIL LINE TO
BURNER NOZZLE

OIL PUMP

AIR SHUTTLE

FIGURE 24–3   Oil burner. (Courtesy of Wayne Combustion Systems.)

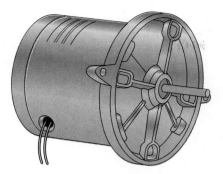

FIGURE 24–4   Flange-mounted oil burner motor.

integral temperature sensor on the stack controller that senses stack temperature.

The function of the primary control is

1. On a call for heat from the thermostat, start the burner motor and energize the ignition transformer.

2. When the oil flame has been established, some models will turn off the ignition transformer (intermittent ignition). On other models, the transformer remains energized,

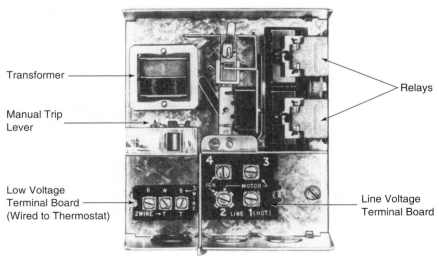

Transformer

Manual Trip Lever

Low Voltage Terminal Board (Wired to Thermostat)

Relays

Line Voltage Terminal Board

FIGURE 24–5   Stack controller. (Courtesy of Honeywell, Inc.)

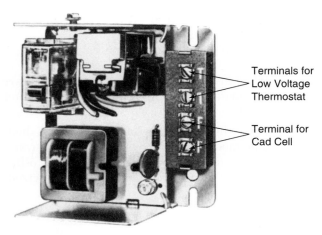

FIGURE 24–6 Burner-mounted primary control. (Courtesy of Honeywell, Inc.)

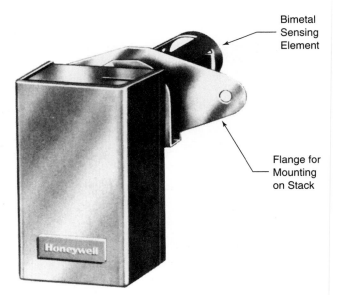

FIGURE 24–7 Stack switch. (Courtesy of Honeywell, Inc.)

and the spark continues to fire throughout the call for heat.

3. If the oil flame is not established within a specified time (usually around one minute), shut down the burner motor. If unsuccessful, the control will go into lock-out. The reset button on the primary control must be reset before a new trial for ignition may proceed.

4. When the thermostat is satisfied, the primary control turns off the burner motor and the ignition transformer (for constant ignition models).

Some primary controls provide nonrecycling control. They will attempt to restart a burner immediately upon a loss of flame. Ignition attempt will continue until control locks out. Others are recycling controllers. They will shut down the burner immediately on loss of flame, then attempt to restart the burner once before locking out.

FLAME SENSING METHODS

There are several ways in which the primary control determines whether the trial for ignition has been successful. The stack control has its built-in temperature sensing element that is inserted into the flue stack when the controller is mounted. For remote-mounted primary controllers in residential and small commercial use, the combustion is sensed either with a remote temperature sensing switch mounted on the stack (**stack switch**, Figure 24–7), or a cadmium sulfide cell (**cad cell**, Figure 24–8). The cad cell is a light-sensitive device that is mounted

where it can "see" the oil flame. A cad cell has a very high resistance when it senses darkness, but its resistance becomes very low when it sees light. It therefore behaves as if it were a switch in the primary control circuit, open when there is no flame and closed when there is flame. If the burner flame is properly adjusted, the cad cell resistance will be in the range of 300 to 1000 ohms when the burner is operating. If the resistance is higher, either the cad cell needs cleaning or reaiming, or the burner flame may need adjustment. The two wires that emerge from the cad cell will be connected to the terminals F–F on the primary control (flame) or the S–S terminals (sensor).

The stack detector switch (Figure 24–7) is either a two-wire or three-wire device. On the two-wire model, the switch is normally closed. When the sensor detects heat, the switch opens. On the three-wire model, there is an SPDT switch. In the cold

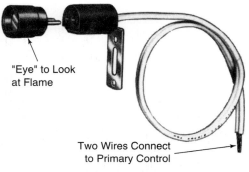

FIGURE 24–8 Cad cell.

starting position, the switch between terminals R and B is closed. On a temperature rise, R–B will open, and R–W will close.

WIRING DIAGRAMS— PRIMARY CONTROL

The wiring diagrams that follow all show the internal wiring of the primary control. Some are just electromechanical, whereas others have electronic devices such as thermistors, diodes, or triacs. It is not necessary for the service technician to understand exactly what goes on inside the primary control. As with other "black box" controllers, you simply need to know the required inputs (thermostat signal and flame signal) that will produce a desired output (a voltage output to the burner motor and transformer). The general sequence of operation is as follows. When the burner is started, a bimetal operated safety switch (inside the primary control) starts to heat. This switch will open and shut down the burner unless an oil flame is established. If the flame is detected by the flame sensor, the circuit to the safety switch element is opened, the safety switch will stop heating, and the burner will be allowed to continue to run.

Figure 24–9 illustrates the typical internal wiring for a primary control using a low-voltage

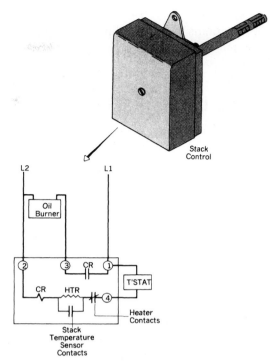

FIGURE 24–10   Wiring of stack-mounted primary control with line-voltage room thermostat.

thermostat. When the thermostat calls for heating, a circuit is completed through the 1K relay coil, safety switch heater, and the normally closed contacts of 2K1. The 1K contacts complete the line-voltage circuit through the ignition transformer, the burner motor, and the oil solenoid valve. If a flame is established and sensed by the cad cell, the 2K relay coil will become energized. This will create a short around the safety switch heater, causing it to stop heating. If the cad cell circuit is not completed within the predetermined time, the safety switch heater will open the 1K coil circuit.

Figure 24–10 shows the internal wiring diagram for a stack controller using a line-voltage thermostat.

If a cad cell system burner locks out on safety after flame is established, the problem could be the primary control or the cad cell. Push the manual reset button on the primary control to attempt ignition. Quickly, jumper the S–S (or F–F) terminals. The burner should continue to run. If it does, the primary control is good. If the control works with the jumper in place but fails when the cad cell is reconnected, the cad cell has either failed, or become covered

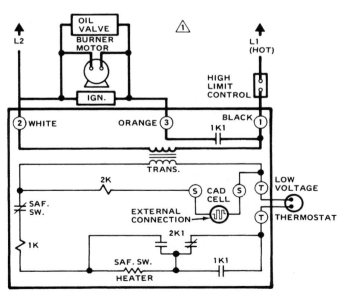

⚠ **PROVIDE OVERLOAD PROTECTION AND DISCONNECT MEANS AS REQUIRED.**

FIGURE 24–9   Wiring of primary control with low-voltage room thermostat. (Courtesy of Honeywell, Inc.)

with soot. [NOTE: do not jumper the S–S (or F–F) terminals prior to calling for heat. The primary control will think there is already a flame, and it will not allow ignition to happen.]

OIL-FIRED UNIT HEATER

Figure 24–11 shows the interconnection diagram for a unit heater that has an oil-fired burner. The sequence of operation is similar to that of the gas-fired furnaces discussed in earlier units. When the room thermostat closes, the primary control causes the burner motor and the ignition transformer to start. Assuming that the flame detector properly senses the flame, the heat exchanger will begin to warm. When the heat exchanger reaches approximately 130°F, the fan switch will close, energizing the fan motor. If for any reason the heat exchanger temperature exceeds approximately 200°F, the safety limit switch will open, shutting down the entire system.

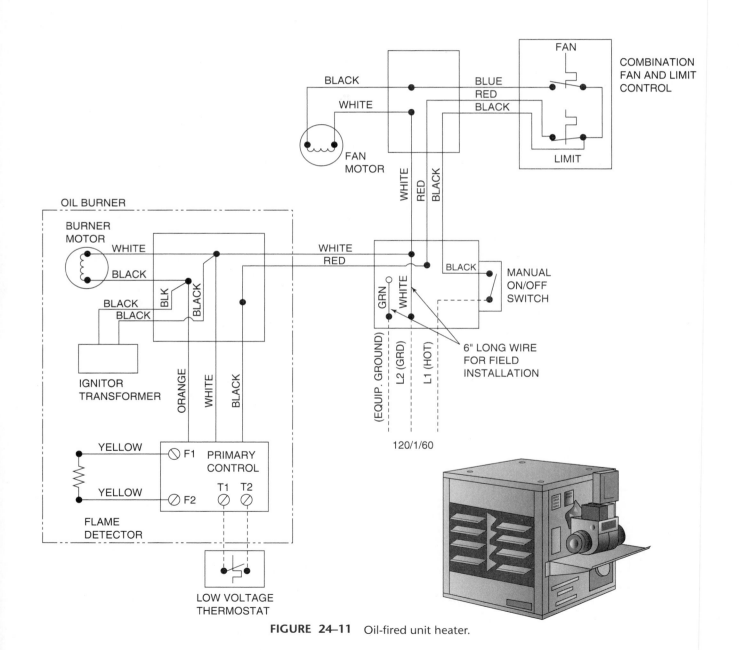

FIGURE 24–11 Oil-fired unit heater.

SUMMARY

The external controls for oil heaters for the most part are the same as with gas heat. Special devices are needed to deliver oil to the burner area in a condition suitable for combustion.

PRACTICAL EXPERIENCE

1. Redraw the diagrams in Figures 24–11 and 24–12 as ladder diagrams.

REVIEW QUESTIONS

1. In the circuit of Figure 24–9, if the coil of relay 2K were to become open, what would be the effect in circuit operation?

2. In the circuit of Figure 24–9, if the face of the CAD cell were to become blackened what would be the effect on circuit operation?

3. Referring to Figure 24–11, will the burning motor operate when the FAN switch is open?

4. Referring to Figure 24–11, what complaint might you get from a customer if the fan relay contacts should become welded in the closed position?

5. At what speed does a direct-drive burner motor usually operate?

6. If the limit in Figure 24–11, were to open, what voltage would you measure between the black and white wires of the primary control?

7. In Figure 24–9 the room temperature is above the selected temperature. What voltage would you expect to measure between terminal one (1) (black) and three (3) (orange)?

8. In Figure 24–11, if the limit switch is open, will the burner motor operate?

9. In Figure 24–11, if the limit switch is open and the fan switch is closed, will the fan motor operate?

10. In Figure 24–12, will the burner motor ever operate without the fan motor operating?

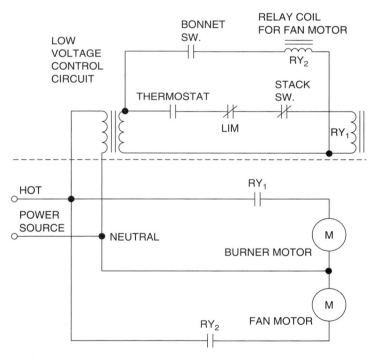

FIGURE 24–12 Exercise.

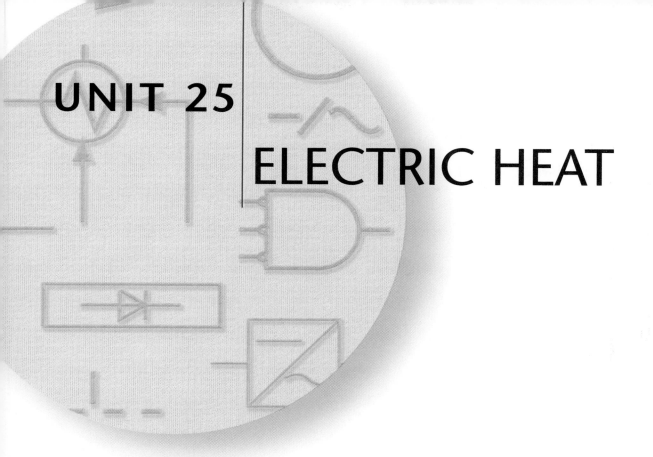

UNIT 25
ELECTRIC HEAT

OBJECTIVES

After completion and review of this unit, you should be familiar with electric heat systems, their components, controls, and safety devices, including:

- Heating elements.
- Limit fuses.
- Sequencers.

RESISTANCE HEATING

When an electric current is passed through a resistor, the electrical energy is converted into heat energy. The principle of **resistance heating** is used in toasters, electric ovens, electric water heaters, and electric furnaces and heaters. All the electrical input is converted into heat, at the rate of 3415 Btu/hr for each kW of input. It is 100% efficient because of this complete conversion, but it is very expensive to operate because the energy input in the form of electricity is three to four times as expensive as energy in the form of fossil fuel. Resistance heat may be designed either to stand alone (without a fan), as a resistance heater with a fan (electric furnace or self-contained electric heater), or as an insert into the ductwork of a system that already has a fan (duct heater).

ELECTRIC FURNACE

Figure 25–1 shows a cutaway view of an upflow electric furnace. Instead of a heat exchanger, the blower sends the room air through an electric heating element that operates on 230 V. When the thermostat calls for heat, it energizes both the heating element and the blower relay.

Figure 25–2 shows the heating element. It is a long spiral-wound wire, supported by insulators. It is made of a nickel and chromium alloy commonly called **Nichrome**. A furnace will use several of these elements, each drawing between 3 and 8 kW. The furnace control panel will have a 20- to 40-amp cartridge fuse on each leg of each heater. Therefore, an electric furnace with three heating elements would have six cartridge fuses.

In addition to the large fuses that protect the furnace wiring against a shorted heating element, there is also a thermal cutout and a limit switch mounted on the heater element. The limit switch senses the air temperature around the heating element and will de-energize the element if the limit set point is reached (usually between 140°F to 190°F).

The **thermal fuse** will also open if it detects an abnormally high temperature around the heating element. The primary cause of a potentially too-high air temperature is restricted or lost airflow

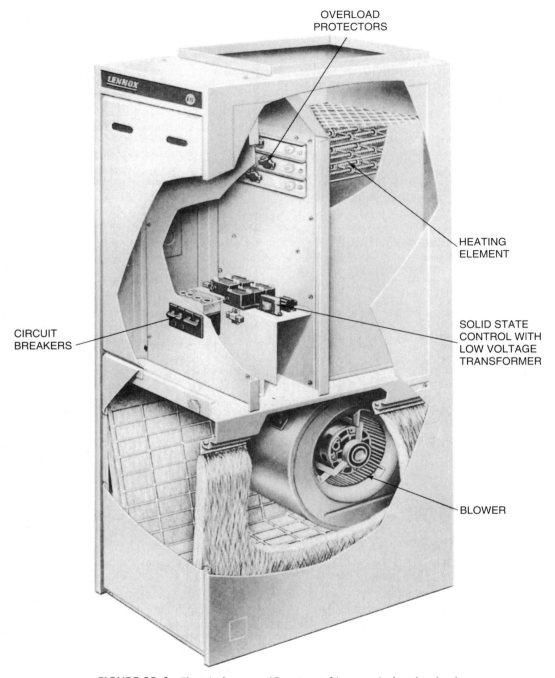

OVERLOAD
PROTECTORS

HEATING
ELEMENT

SOLID STATE
CONTROL WITH
LOW VOLTAGE
TRANSFORMER

CIRCUIT
BREAKERS

BLOWER

FIGURE 25–1 Electric furnace. (Courtesy of Lennox Industries, Inc.)

across the heating element. The heating element will continue to produce the same amount of heat, regardless of the airflow. If the airflow is too low, it will get too hot. The resistance heating element will also get hotter than normal, and if left unchecked, can get hot enough to burn out the heating element.

Most electric furnaces (as well some other types of units) have a door latch that is operated by a solenoid coil that is energized whenever power is applied to the unit. There is nothing more embarrassing to a service technician than not being able to open the front panel, as the customer watches. The

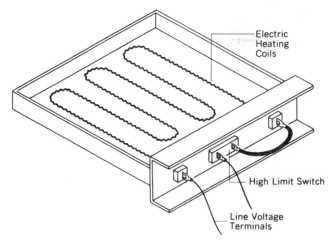

FIGURE 25–2 Electric heating element.

only way to release the latch is to turn the power off to the unit from the circuit breaker or disconnect. After the door is opened, the power may be switched on again.

DUCT HEATER

A **duct heater** is an electric heater element that is installed in a run of duct. The element slides inside the duct, and an attached control compartment

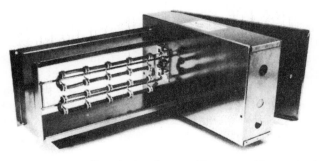

FIGURE 25–3 Electric duct heater. (Courtesy of Tutco, Inc.)

hangs outside the duct (Figure 25–3). Duct heaters require a minimum airflow across them in order to keep the temperature of the elements below an acceptable level. Turns in the ductwork can cause uneven airflow distribution, which can result in a portion of the electric element not getting sufficient airflow. This can be a difficult-to-diagnose reason for elements burning out. Sometimes a sail switch is used in the duct to prove that airflow is present before the electric heating element is allowed to operate.

PACKAGED COOLING WITH ELECTRIC HEAT

Figure 25–4 shows a wiring diagram for a conventional air conditioning system with an electric resistance heater. The low-voltage thermostat connected to the unit's terminal board is designed for electric heat. On a call for heat, it makes R–W, energizing the heat sequencer. The heat sequencer has a pilot-duty heating element inside that causes a bimetal line-duty switch to close. When that switch closes, it energizes the heating element and the fan motor (through the normally closed contacts of the blower relay). On other heaters, a specially designed electric-heat thermostat is used. On a call for heat, it makes both R–W and R–G. The R–W circuit operates only the heater, whereas the R–G circuit energizes the blower relay.

SEQUENCING THE HEATING ELEMENTS

When there are two or more heating elements, they are sometimes staged or sequenced to come on one at a time. There are two ways that this can be accomplished:

1. Two-stage thermostat
2. Heat sequencers

Figure 25–5 and Figure 25–6 show the comparison of a two-stage thermostat with that of a single-stage thermostat controlling two stages of heat. The single-stage thermostat uses a heat sequencer.

With the two-stage thermostat Figure 25–5, the thermostat switch feeding W1 closes at a temperature two to three degrees higher than W2. If the selected temperature is well above the room temperature,

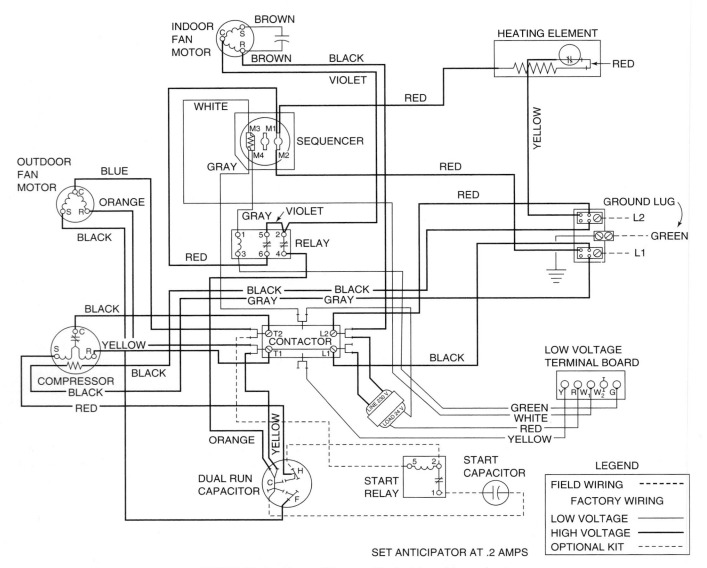

FIGURE 25–4 Air conditioner with electric resistance heat.

both relays will energize. As the room heats and approaches the selected temperature, W2 will open, and heat relay #2 will de-energize. If the room temperature can be maintained with heat element #1 alone, it will do so. If not, the temperature will drop two to three degrees below the selected temperature, and W2 will close energizing heat relay #2, putting heat element 2 back into the system.

In Figure 25–6 a single-stage thermostat is used to control two heating elements. Control is obtained through the heat sequencer. The sequencer consists of a heating element controlling two sets of contacts. The contacts close at two different temperatures so there is a time delay between the closing of contacts set A and contacts set B.

If the selected temperature is above the room temperature, R will mate to W supplying power to the heat sequencer. After a short delay, contacts A will close completing a circuit to element #1. The room starts to heat. Two possibilities for the next action now exist. If the single heat element heats the room fast enough, the thermostat will open, stopping the heating until the room cools and the thermostat calls for heat again.

If the room does not heat fast enough, power to the heat sequencer is maintained. Contacts B close and heating element 2 receives power. The room will heat until the thermostat opens. NOTE: In some equipment sequencers are used simply to keep all the resistive heating elements from being

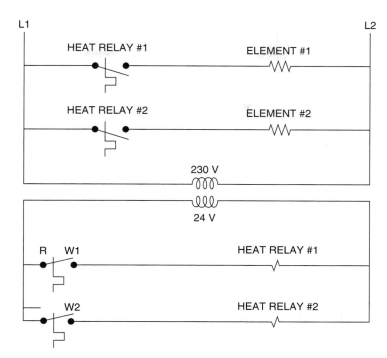

FIGURE 25–5   Electric heat with two-stage thermostat.

connected to the line at once. Remember the resist-ance goes up as the element heats, so the current decreases as the element heats. This is particularly true where the power company is charging for peak demand.

BLOWER CONTROL

Another method of operating the blower on an elec-tric heat system is by using a switch that senses when current is flowing through the electric heating

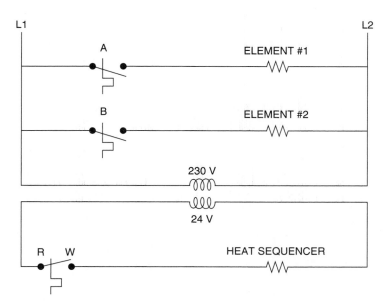

FIGURE 25–6   Two stages of heat controlled by a single-stage thermostat.

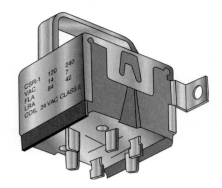

FIGURE 25–7 Current-sensing fan switch.

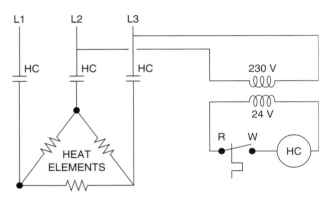

FIGURE 25–8 Three-phase electric heat.

FIGURE 25–9 SCR.

element (Figure 25–7). The wiring to the element is passed through the current-sensing loop of the blower control. Whenever the element is energized, the current-sensing loop will cause a set of normally open contacts in the control to close, energizing the fan motor.

THREE-PHASE ELECTRIC HEAT

All electric heat is single phase, but electric heat is used on three-phase systems. Figure 25–8 shows how three single-phase heaters can be applied to a three-phase system. This diagram also illustrates the use of a contactor (HC = heating contactor) to energize the heaters.

SCR CONTROLS

An **SCR (silicon-controlled rectifier)** is an electronic device that can be used to regulate the flow of electricity to an electric heater. The SCR can turn the electric heating element on and off many times

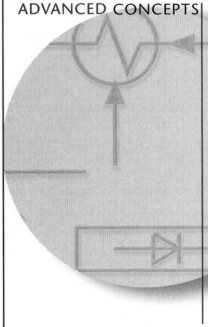

ADVANCED CONCEPTS

When connected in an AC circuit, the SCR acts as a rectifier, and its output is DC. The symbol for an SCR is shown in Figure 25–9. When the current flowing through the gate is high enough (as modulated by a temperature-sensing device), it will turn the SCR on, allowing current to flow through the anode-cathode path and through an electric heating element. Figure 25–10 shows how the first portion of the AC waveform is "chopped off," because the rising voltage is not high enough to cause the gate to open the current flow to the heater. When the voltage gets high enough to cause the turn-on current to flow through the gate, the remaining portion of the waveform is allowed to flow through the heater. When the downside of the applied voltage waveform signal drops to zero, the SCR turns off again, until the turn-on current is reached once again. By adjusting the turn-on current setting, the amount of the waveform that can be allowed to pass through the heater can be adjusted. By connecting two SCRs in tandem, one for each direction of current flow, it becomes easy to control AC current flow. Each half-cycle is "chopped" at low-power settings, and the full AC sine wave can be passed through when full power is required.

Some devices use a power **triac**, which is essentially the same as two back-to-back SCRs in a single semiconductor package. Higher-power capability can be achieved by using two discrete SCRs as described earlier. Often the user is unaware of the difference, since most manufactured power controllers are prepackaged by the manufacturer.

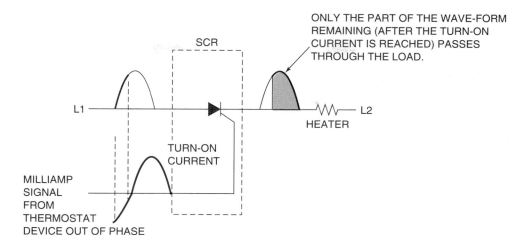

FIGURE 25–10   "Chopped off" wave form to control heater capacity.

each second. By modulating the time on and the time off, the SCR acts as a modulating control for the heating element.

SUMMARY

Using resistive elements is an effective though expensive method of obtaining heat. Control of electric heat is simple and direct.

PRACTICAL EXPERIENCE

Review electric heating equipment manufacturers' literature. Pay particular attention to suggested methods of control. Look for the use of two-stage thermostats as well as sequencers.

REVIEW QUESTIONS

1. Referring to Figure 25–5, what might you expect as a customer complaint if the coil of heat relay 1 were to open?

2. Referring to Figure 25–6, what would you expect as a customer complaint if the heat sequence A contacts failed to close while the B contacts operated normally?

3. With reference to Figure 25–4, the fan motor is operating. What voltage would you expect to measure from the indoor fan motor (R) to ground?

4. In Figure 25–4, with cool selected what voltage would you expect to measure from the sequencer (M1) to ground?

5. In Figure 25–4, when does the device called out as "Relay" energize?

   a. Whenever the indoor fan is to operate.

   b. During heat.

   c. During cool.

   d. When fan is selected.

6. With the two-stage thermostat used in Figure 25–5, what is the temperature differential between the actuation of W-1 and W-2?

7. In Figure 25–4, what type of motor is the indoor fan motor?

8. In Figure 25–4, what type of motor is the outdoor motor?

9. In Figure 25–4, what type of motor is the compressor motor?

10. Referring to Figure 25–8, if the fuse controlling L2 were to open, how much heat would be developed in the heater elements compared with normal operation?

   a. 100 percent

   b. 75 percent

   c. 66 percent

   d. 33 percent

UNIT 26

ICE MAKERS

OBJECTIVES

After completion and review of this unit, you should be familiar with different types of ice-making machines, their components, and the circuits that control their operation. New components include:

- Augers.
- Water level controls.
- Ice cubers.
- Ice thickness probes.
- Fill and dump controls.
- Purge systems.

There are a great number of ice makers and ice maker manufacturers. There is much less uniformity among those manufacturers about how they control operation, as compared to air conditioners or refrigeration units. Understanding and troubleshooting ice machines is probably the most complex challenge that will be encountered by many technicians. But the task can be simplified if you know what tasks must be accomplished by the ice machine controls.

TYPES OF ICE MAKERS

There are two general types of ice makers, ice flakers and ice cubers. An ice flaker (Figure 26–1) con-

sists of an evaporator formed into the shape of a cylinder in which a level of water is maintained inside by the use of a mechanical float control. The evaporator temperature is 30°F–40°F below freezing, so ice forms on the inside surface of the evaporator. An auger inside the evaporator rotates slowly and continuously, so that as the ice forms, it is scraped off the inside surface and pushed up and out the evaporator to a storage bin.

The ice flaker is a continuous operation. It continues until the bin is full, at which point a **bin thermostat** or **bin switch** turns off the refrigeration system. The bin switch usually is a thermostat that opens on a drop in temperature. The sensing bulb is located near the top of the bin. When the ice touches the probe, the switch opens.

The most common complexity that is added to the ice flaker circuit is a safety circuit that prevents the accumulation of ice in the evaporator. This is important, because if the auger becomes frozen in place, the auger motor will exert enough force on the output shaft of the gear box to snap it or break the internal gears. If the compressor is allowed to run without the auger, the entire water contents of the evaporator may become frozen solid. When water freezes, it expands, and the result will be a cracked evaporator. The protection of the ice flaker generally takes two forms:

1. Auger delay—When the bin switch signals that the bin is full, it turns off the compressor.

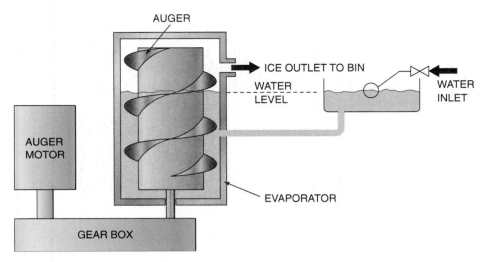

FIGURE 26–1   Ice flaker.

But the auger motor is allowed to run for a short while longer to clear all of the ice that has already been formed in the evaporator.

2. Proof of auger rotation—Two methods are used that check for proper auger operation before the compressor is allowed to start. One uses a centrifugal switch on the auger motor. This switch is wired into the compressor circuit, and if the auger motor doesn't run, the compressor won't start. The second method is to sense the auger motor amps. If the amps get too high, a switch will open, de-energizing the compressor motor circuit.

The ice cuber is far more complex than the ice flaker. Instead of being a continuous process like the ice flaker, it is a batch process. That means that it goes through different cycles of operation to make a batch of ice, and when a batch has been completed, the sequence of operation repeats. One cycle consists of a "freeze" cycle and a **"harvest" cycle**. During the "freeze" cycle, a pump circulates water over a cold evaporator that has been formed into a shape that will make cubes (Figure 26–2). As the water circulates over the evaporator, the ice cubes form. After the cubes are fully formed (usually 15 to 25 minutes), some method is used to sense that they are ready. That activates the "harvest" cycle in which the pump motor stops, the condenser fan motor stops, and a hot gas solenoid valve is energized, circulating hot gas through the evaporator. Some units energize a "dump valve" in which the unfrozen water is drained, and replaced with new water. Some units have an inlet water solenoid valve to accomplish the same purpose.

When the ice falls off the evaporator into the storage bin, the unit is returned to the "freeze" cycle. The operation of most modern cubers can incorporate many other features and are controlled through a "black box" control board. There is no way for the service technician to know all the functions of each manufacturer's black box controller, unless the service manual is available. Many of the features of these black boxes are explained later in this chapter.

Both ice flakers and ice cubers have a selector switch that has positions for Off, Ice, and Clean. In the Clean position, only the water pump is energized. This allows the technician to dump in some ice machine cleaner and circulate it through the entire water system.

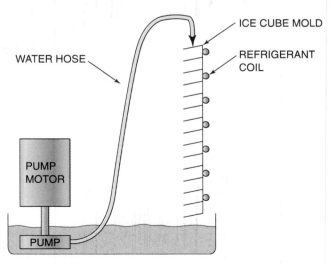

FIGURE 26–2   Ice cuber.

MANITOWOC J MODEL CUBER

Sequence of Operation

Figure 26–3 shows the ladder diagram. The control board has relays (switches labeled 3, 1, 2, 4), connections for an ice thickness probe, a discharge line thermistor, a low-voltage plug, and outputs to the hot gas solenoid, the dump solenoid, water pump, and contactor coil. When the bin switch first closes, relay switches 3 and 1 and 2 close, starting the water pump and the water dump solenoid for 45 seconds to purge old water from the ice machine. The hot gas valve is also energized during the water purge, but nothing flows through it because the compressor has not yet started. At the end of 45 seconds, switches 3 and 1 open, switch 2 remains closed, and 4 closes. The compressor and condenser fan motor start. The dump solenoid and water pump stop. Five seconds later, switch 2 opens and the hot gas solenoid is de-energized. For the next 30 seconds, the evaporator prechills. Then switch 1 closes again and the water pump starts. This is the "freeze" cycle.

The "freeze" cycle continues until the cubes are fully formed. This is sensed by a probe that consists of two metal fingers that hang in front of the evaporator. As the ice thickness builds, the circulating water moves closer and closer to the fingers until contact is made (the newer units use a one-finger probe). If the contact with water is maintained for seven seconds, the control board will initiate the "harvest" cycle. The water dump valve energizes for 45 seconds to purge the water from the trough, and the hot-gas valve opens, diverting hot refrigerant gas into the evaporator. After 45 seconds, the water pump and dump valve de-energize, but the hot-gas valve remains open (the newest control boards have an adjustable water purge time.) "Harvest" continues until the cubes slide out of the tray, in one sheet, moving a water curtain away from the evaporator,

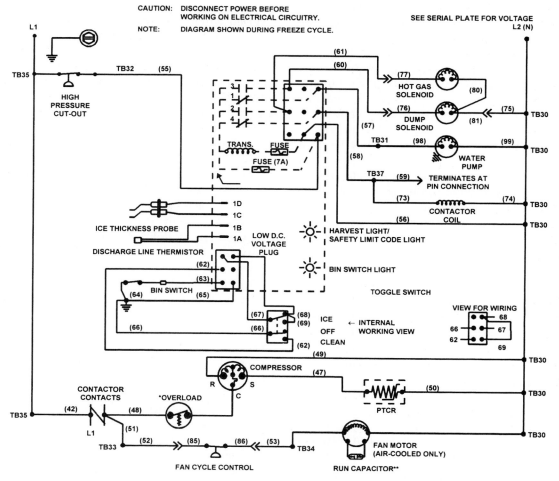

FIGURE 26–3 Manitowoc J model cuber. (Courtesy of Manitowoc Ice, Inc.)

which, in turn, trips a microswitch that acts as the bin switch as the cubes fall into the storage bin. This momentary opening and reclosing of the microswitch (bin switch) terminates the "harvest" cycle and returns the machine to the "freeze" cycle. If the ice storage is full enough that the last ice made can't fall all the way into the bin, the bin switch will remain open, and the ice machine will turn off until enough ice is removed to allow the sheet of new cubes to drop clear of the water curtain.

Safety Features

Once the unit has shut off, it will remain off for at least three minutes, even if the microswitch is closed. This is to allow the compressor PTC device to cool before a restart is attempted.

Once the unit has started, the control board locks the machine in the "freeze" cycle for six minutes. Even if water contacts the ice thickness probe during these six minutes, the ice machine will stay in the "freeze" cycle to prevent short cycling (earlier models did not have this feature). However, to allow the service technician to initiate a "harvest" cycle without delay, this feature is not used on the first cycle after moving the toggle switch to Off and then back to Ice.

If the water curtain has been removed during the "freeze" cycle, the "freeze" cycle will continue normally. However, if the water curtain remains off during the "harvest" cycle, there will be no way to trip the microswitch when the ice sheet falls. As a safety, any time the "harvest" cycle time reaches 3.5 minutes and the bin switch has not closed, the ice machine will stop as though the bin were full.

In addition to standard safety controls such as the high-pressure cut-out, the control board has built-in safety limits. Safety limit #1 will shut down the system if the freeze time exceeds 60 minutes for three consecutive "freeze" cycles. Safety limit #2 will shut down the machine if the harvest time exceeds 3.5 minutes for three consecutive "harvest" cycles. Safety limit #3 will trip if the compressor discharge temperature falls below 85°F for three consecutive "harvest" cycles. Safety limit #4 will trip if the compressor discharge temperature exceeds 255°F for 15 continuous seconds.

Status Indicators

The control board has LEDs to indicate the status of the machine. The Bin Switch LED (green) is on when the bin switch (water curtain) is closed and off when the bin switch is open.

The Harvest/Safety limit LED (red) primary function is to turn on as water contacts the ice thickness probe during the "freeze" cycle and will remain on throughout the entire "harvest" cycle. Prior to the "harvest" cycle, the light will flicker as water splashes on the probes. The secondary function of this light is to continuously flash when the ice machine is shut off on a safety limit and to indicate which safety limit shut off the machine.

The Clean LED (yellow, when provided) is on whenever a cleaning cycle is in progress.

Troubleshooting

Thermistors, which send temperature information to the control board by changing resistance, can fail. You can check its operation by removing it from the circuit and measuring its resistance at various temperatures. In an ice/water bath (32°F), the resistance should be 283 K to 377 K ohms. At 70°F–80°F, the resistance should be 93 K to 119 K ohms. In boiling water (212°F) the resistance should be 6.2 K to 7.3 K ohms. If the discharge line thermistor fails closed, the ice machine would stop on safety limit #4 15 seconds after contact #4 on the control board closes (15 seconds after the compressor starts). If the discharge line thermistor fails open, the ice machine would start and run through two normal "freeze" and "harvest" sequences. During the third "harvest" sequence, the ice machine would stop on safety limit #3.

The Ice-Off-Clean switch may fail. Because of the electronics involved, it is recommended by the manufacturer that a voltmeter not be used to check the toggle switch. Instead, with the power off, isolate the toggle switch. Then, using ohms, check the switches for continuity in all three toggle switch positions. The switches between terminals 66–62 and between terminals 67–69 should be closed when the toggle switch is set to Clean. The switch between terminals 67–68 should be closed when the toggle switch is set for Ice.

The ice thickness probe (two-finger model) can fail if water hardness creates a bridge across the two probes, across the plastic insulator that separates them. On older model machines without the minimum freeze time feature, this would cause short cycling. The quickest check for this problem is to simply disconnect one of the probe wires from the control board. If the short-cycling stops, replace the probe. On the newer models, if the ice machine cycles into harvest before water contact with the probe, disconnect the ice thickness probe from the control board. Bypass the minimum freeze time feature by moving the Ice-Off-Clean switch to Off and back to Ice. Wait about 1.5 minutes for water to begin flowing over the evaporator. If the harvest light stays off and the ice machine remains in the "freeze" sequence, the ice thickness probe was causing the malfunction. If the harvest light comes on, and 6–10 seconds later, the ice machine cycles from "freeze" to "harvest," the control board is causing the malfunction.

If the ice machine does not cycle into "harvest" when water contacts the ice thickness probe, place a jumper wire across the ice thickness probe terminals at the control board (Figure 26–4a). If the "harvest" light comes on, and 6–10 seconds later harvest is initiated, replace the ice thickness probe. (On a single-finger probe, instead of jumping across the two thickness probe connections, jumper from the single-thickness probe connection to ground as in Figure 26–4b)

If the ice machine will not run at all, verify that voltage is available, the HPC is not open, the transformer and fuses at the control board are okay, the bin switch functions properly, the Ice-Off-Clean toggle switch is making contact from terminals 67–68, and the DC voltage is properly grounded. If no problem is found, replace the control board.

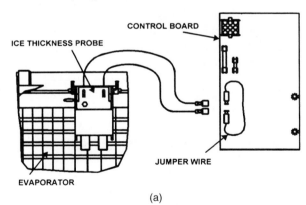

(a)

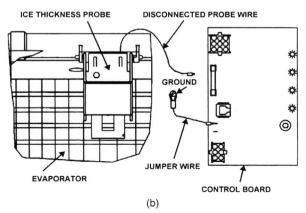

(b)

FIGURE 26–4 Bypassing the ice-thickness probe. (Courtesy of Manitowoc Ice, Inc.)

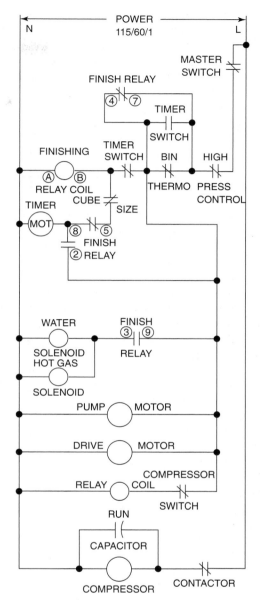

FIGURE 26–5 Scotsman AC30 cuber.

Scotsman AC30 Cuber Figure 26–5 shows the ladder diagram for a Scotsman AC30 model cuber. Ice maker wiring diagrams sometimes use a different convention. Instead of showing switches in their true "normal" position, they will show the switches in the actual position during the "freeze" cycle. Figure 26–5 is actually shown in the "finishing freeze" portion of the "freeze" cycle, as explained following. The unique features of this ice machine are

1. The drive motor operates a rotating water distribution tube similar to that which would be found in the bottom of a dishwasher. The

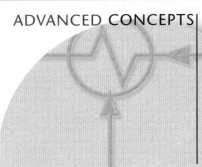

ADVANCED **CONCEPTS**

Instead of simply having the cube size thermostat begin the "harvest cycle," a timed "finish-freeze" cycle is used after the cube size thermostat closes. This is done because during the "freeze" cycle, the evaporator temperature (sensed by the cube-size thermostat) drops fairly rapidly through the first three-quarters of the cube formation. After that, the rate of temperature change flattens out considerably. If we attempted to use evaporator temperature to determine when the cubes were fully formed, we would find that small variations in the temperature at which the evaporator temperature started the "harvest" cycle would produce large variations in actual cube size.

water distribution tube sprays water upward, into an upside-down ice cube evaporator mold. During "harvest," the cubes drop down onto a table, where the rotating water distribution tube sweeps them out into a chute that leads to the ice storage bin.

2. The "harvest" is controlled through the operation of a timer motor that operates an SPDT timer switch.

3. The "harvest" is initiated by a cube-size thermostat. It is actually a thermostat that senses evaporator temperature. As the cube size builds, the evaporator becomes more and more insulated from the 32°F circulating water. The reduced load on the evaporator results in a continuously dropping evaporator temperature as the cube size grows.

4. There is a **finishing relay**. Its function is to allow the completion of the batch of cubes that have been started and has entered into the "harvest" cycle, even if the bin switch should open during this time.

Sequence of Operation

During the "freeze" cycle, the manual master switch and compressor switch are both closed, as is the high-pressure control and the bin thermostat. The normally open timer switch is actually closed, and the normally closed timer switch is open. The finishing relay is simply a three-pole double-throw control relay. The finishing relay coil is energized, causing the contacts to be closed between terminals 4–7 and 5–8. The contacts between terminals 2–8 and 3–9 are open. The master switch causes the pump motor and the drive motor to run. With the compressor switch Off, this is the cleaning mode of operation. With the compressor switch On, the compressor contactor coil is energized and the single pole contactor closes, bringing on the compressor.

The "freeze" cycle continues until the cube size is large enough to close the cube-size thermostat (not shown as a thermostat in the ladder diagram, but a thermostat none the less). This energizes the timer motor, beginning the "finish freeze" cycle. After six minutes, the timer switches operate. The normally open timer switch closes, sealing in around the bin thermostat. Now, it won't matter if the bin thermostat opens. The cycle will complete itself.

The normally closed timer switch opens, de-energizing the finishing relay coil and starting the "harvest" by energizing the water and hot-gas solenoid valves through finishing relay contacts 3–9. The timer motor continues to run, energized through finishing relay contacts 2–8. This "harvest" cycle continues for three minutes, until the timer motor returns its switches to the "freeze" cycle positions.

ICE-O-MATIC C SERIES CUBER (600 LB/DAY)

The four diagrams in Figure 26–6a–d show the step-by-step sequence for an Ice-O-Matic C-61 cuber manufactured in the 1980s.

Sequence of Operation

During the first step of the "freeze" cycle, the bin switch is closed and the selector switch is in the Ice position. Power is supplied to the contactor coil, bringing on the compressor. The water pump is supplied power through the selector switch and the normally closed contacts of the cam switch. The fan motors are supplied power through the normally closed contacts of the control relay.

Step two begins when enough ice has formed on the evaporator to lower the suction pressure sufficiently to close the low-pressure control (note that

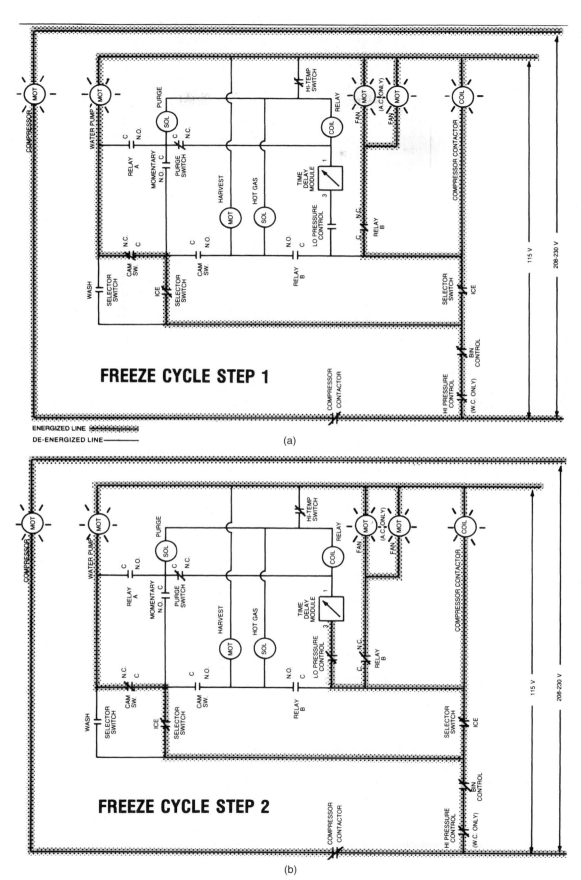

FIGURE 26–6 Ice-O-Matic C series cuber. (Courtesy of Ice-O-Matic.)

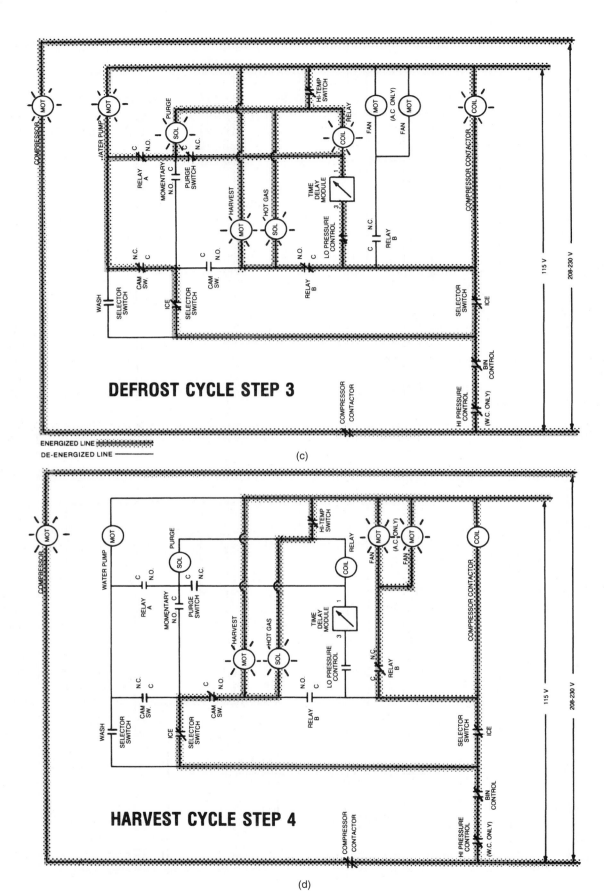

ENERGIZED LINE ▓▓▓▓▓▓▓
DE-ENERGIZED LINE ──────

(c)

DEFROST CYCLE STEP 3

HARVEST CYCLE STEP 4

(d)

FIGURE 26–6 (Continued)

this is a low-pressure cut-in, and not a low-pressure cut-out). This provides power to the time-delay module. By adjusting the set point of the time-delay module to longer or shorter, the bridge thickness (the ice bridge connecting each cube to the adjacent cubes) may be adjusted. The proper bridge thickness is thick enough to hold the entire ice sheet together so that it releases in one piece, but thin enough so that the individual cubes break apart when they fall into the storage bin. Generally, a 1/8″ bridge is about right.

After the power has been applied to the time-delay module for the set amount of time, it passes through (Step 3). This energizes the relay coil, the purge solenoid, the water pump (through the "A" normally open relay contacts), the hot-gas solenoid and harvest motor (through the "B" normally open relay contacts). At the same time, the normally closed relay contacts "B" open, shutting off the condenser fan motors. The harvest motor has two functions. First, it pushes a probe (through a slip clutch) against the back of the ice sheet to help push the sheet of cubes out of the evaporator. Second, it drives a cam that operates an SPDT cam switch. Within a few seconds after the harvest motor is energized, it will rotate enough to operate the SPDT switch.

In Step 4, the harvest motor has operated the cam switch, the low-side pressure has built up due to the hot gas, opening the low-pressure control, and de-energizing the relay coil. The harvest motor and hot-gas solenoid remain energized through the normally open contacts of the cam switch. Initially, the clutch between the harvest motor and the probe will be slipping, because the probe is unable to push the ice sheet out of the evaporator. This mode will continue, until the ice "releases" from the evaporator. Then, the probe will once again be able to move. As the probe pushes the ice out and returns to its original position, the cam switches return to their normal positions, and the "freeze" cycle begins anew.

ICE-O-MATIC C SERIES CUBER (1200 LB/DAY)

Figure 26–7a–d shows a diagram similar to the previous wiring, but with some added complexity. The differences are

1. There are two evaporators, each with its own harvest motor. The two ice sheets will not release at the same identical moment. The unit must remain in the "harvest" cycle until both ice sheets have been harvested.

2. The compressor operates on 230 V, whereas everything else operates on 115 V.

3. There is a finishing relay, which does not allow the unit to shut down if the bin switch opens during the "harvest" cycle.

Sequence of Operation

In Step 1, the machine is in the "freeze" cycle. The compressor, condenser fans, and water pump are running. Relay 3 is energized and will remain so as long as the high-temperature safety switch does not trip. When a sufficient thickness of ice has been formed, the low-pressure cut-in closes.

In Step 2, instead of the low pressure cut-in energizing the timing module directly, it energizes relay R-4. Relay R-4A contacts then energize the timing module, and R-4B contacts close in parallel around the bin control. If the bin control opens after this point in time, it will have no effect so long as R-4B contacts are closed (seal-in circuit) through the end of the "harvest" cycle.

Step 3 begins when the timing module allows current to pass, starting the hot-gas cycle. This energizes relay R-1 and the water purge solenoid. The R-1 normally open "B" contacts close, energizing relay R-2 coil and harvest motor #1. The relay R-2 contacts energize both the hot-gas solenoid and harvest motor #2. When harvest motor #2 moves the cam slightly, the normally open contacts of cam switch 2 close, energizing relay R-5, whose contacts seal in again around the bin control.

Shortly after the hot gas starts, the low-side pressure rises and the low-pressure cut-in opens, de-energizing relay R-4, the purge solenoid, and the water pump. Harvest motor #2 will continue to turn until harvest motor #1 returns its cam switch to the normal position. Harvest motor #1 will always stop turning before harvest motor #2. The hot-gas valve and relay coil R-5 will remain energized as long as either cam switch is in the not-normal position. As soon as both cam switches return to their normal positions, the "freeze" cycle will start over, unless the bin switch has opened during "harvest." In that event, the machine will stop at the completion of the "harvest" cycle.

HOSHIZAKI MODEL KM CUBER

Figure 26–8 is typical of the Hoshizaki cubers. As with the other modern ice makers, a "black box" controller board is used to control the "freeze" and "harvest" cycles. Figure 26–9 shows the layout of

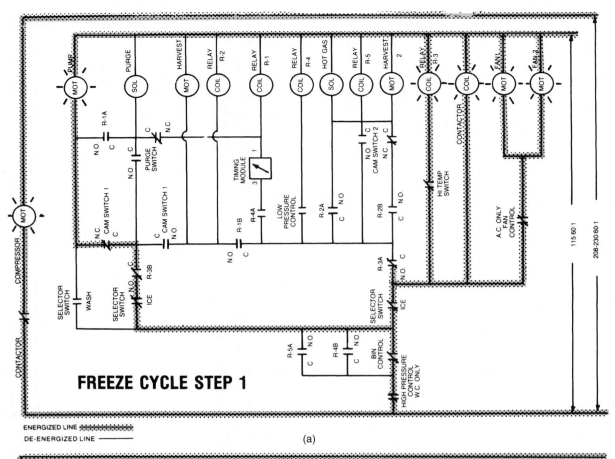

ENERGIZED LINE ░░░░░░░░░░
DE-ENERGIZED LINE ─────────

(a)

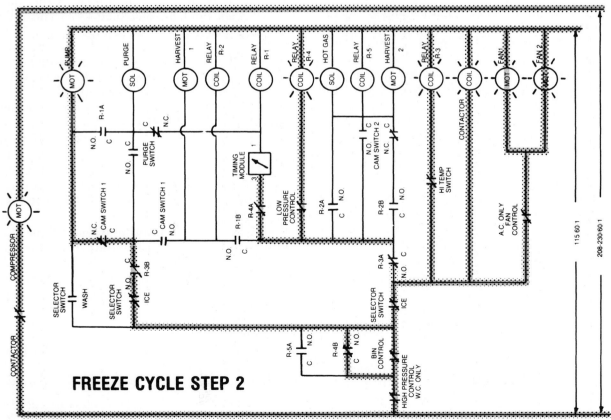

(b)

FIGURE 26–7 Ice cuber with two evaporators. (Courtesy of Ice-O-Matic.)

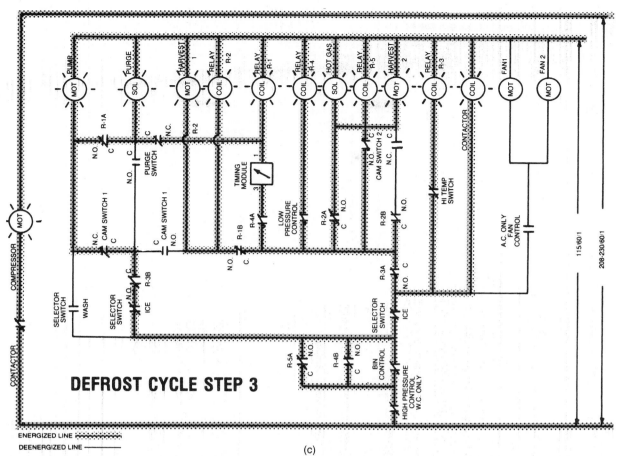

DEFROST CYCLE STEP 3

ENERGIZED LINE ·······
DEENERGIZED LINE ———

(c)

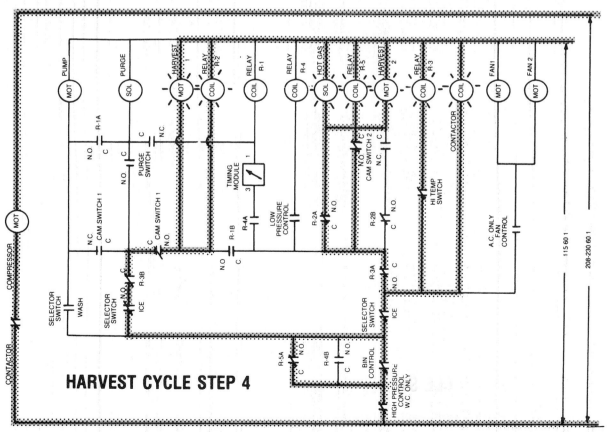

HARVEST CYCLE STEP 4

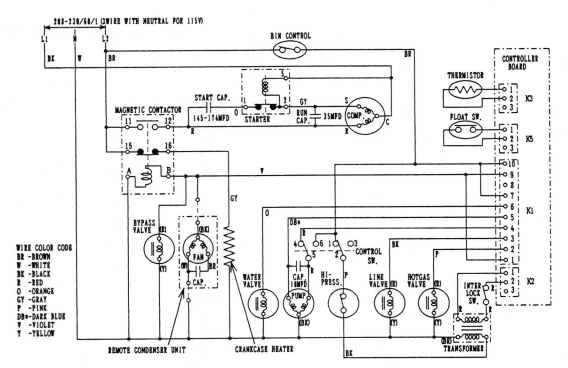

FIGURE 26–8 Hoshizaki KM cuber. (Courtesy of Hoshizaki America, Inc.)

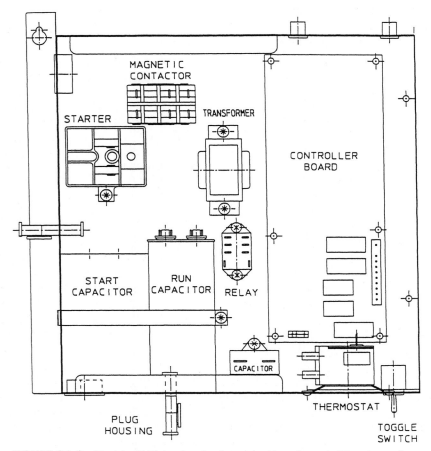

FIGURE 26–9 Hoshizaki KM cuber basic control box layout. (Courtesy of Hoshizaki, America Inc.)

the control box for this cuber. One unique feature of this sequence is that in between the "freeze" cycle and the "harvest" cycle, there is a pump-out cycle when the water pump is used to help clean out any deposits in the water system.

Sequence of Operation

The unit always starts in the 1-minute fill cycle shown in Figure 26–10a. When power is applied to the unit, the water valve is energized and the fill period begins. After one minute, the controller board checks for a closed float switch, indicating that sufficient water has been supplied to the unit water sump. If the float switch has not closed, the water valve will remain energized through additional 1-minute cycles until the float switch does close.

In Figure 26–10b, the unit actually goes through a short "harvest" cycle prior to making any ice. The compressor starts, the hot-gas valve opens, the water valve remains open, and "harvest" begins. The evaporator will warm quickly, because there is no ice on it. When the suction line thermistor reaches 48°F (average 2 minutes), the "harvest" is turned over to the adjustable control board defrost timer. This adjustment can vary the defrost timer from 1 to 3 minutes.

Figure 26–10c shows the "freeze" cycle. After the timer terminates the "harvest" cycle, the hot-gas and water valves close, and ice production starts. For the first 5 minutes, the controller board will not accept a signal from the float switch. This 5-minute minimum "freeze" cycle acts as a short cycle protection. As ice builds up on the evaporator, the water level in the sump lowers. When the float switch opens, the "freeze" cycle is over.

In Figure 26–10d, the float switch has opened, and the unit begins its "harvest" cycle. The hot-gas valve opens and the compressor continues to run. The drain timer starts counting the 10/20 second pump out. The water pump stops for 2 seconds and reverses, taking water from the bottom of the sump and forcing pressure against the check valve seat allowing water to go through the check valve and down the drain. At the same time, water flows through the small tube to power flush the float switch. When the drain timer stops counting, the pump-out is complete. Pump-out always occurs on the second harvest after start-up. Some control boards allow for adjustment for pump-out to occur every cycle, or every second, fifth, or tenth cycle from this point.

After the pump-out, the unit goes into its normal "harvest" cycle. The water valve opens to allow water to assist the "harvest." When the evaporator warms to 48°F, the control board receives the thermistor signal and starts the defrost timer. The water valve is open during "harvest" (defrost) for a maximum of 6 minutes, or the length of the "harvest," whichever is shorter. When the defrost timer completes its count down, the defrost cycle is complete and the next "freeze" cycle begins. The unit continues through the "freeze," pump-out, and normal "harvest" cycles until the bin control senses ice and shuts the unit down.

Troubleshooting Tips

The thermistor can be checked using the following temperature/resistance table:

Sensor temp. (°F)	Resistance (K ohms)
0	14.4
10	10.6
32	6.0
50	3.9
70	2.5
90	1.6

If the thermistor is open, the unit will go through a 20-minute "harvest" cycle. If the thermistor is shorted (zero ohms), the unit locks out on manual reset high temperature safety. If the evaporator reaches 127°F, the thermistor resistance will drop to 500 ohms, shutting down the unit on manual reset. To reset, turn the power off and back on. This is the only manual reset safety on this unit.

KOLD-DRAFT CUBER

The Kold-Draft ice cuber is unique in many respects. The ice is formed in an evaporator that has the cube openings on the bottom side. There is a water distribution plate that fits up under the evaporator. Water is pumped through the distribution plate and out the top of the distribution plate through holes. Each hole squirts water into a different compartment of the ice cube mold.

As the ice cubes form, the space between the ice and the water plate diminishes. When the cubes are fully formed, the ice is sufficiently close to the water plate holes that it causes an increase in the output pressure of the pump that supplies the water to the water plate. This increased pressure causes some of the water to shoot over a weir and into a drain.

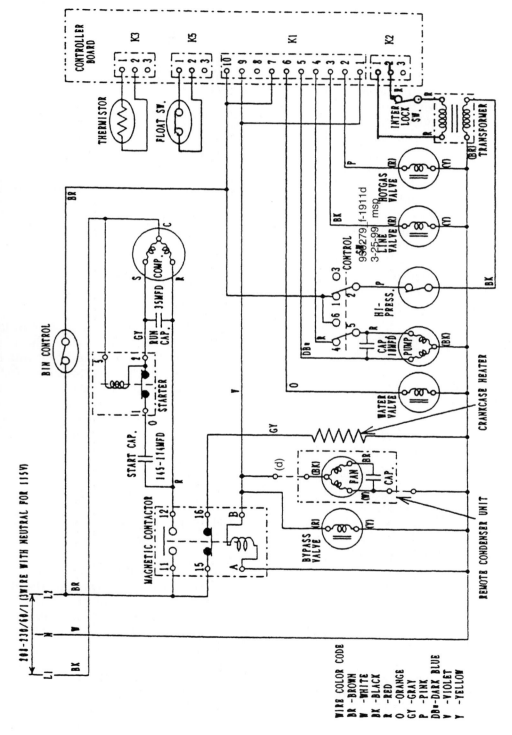

ONE MINUTE FILL CYCLE

TO INITIAL HARVEST

FIGURE 26–10   Hoshizaki sequence of operation. (Courtesy of Hoshizaki America, Inc.)

(a)

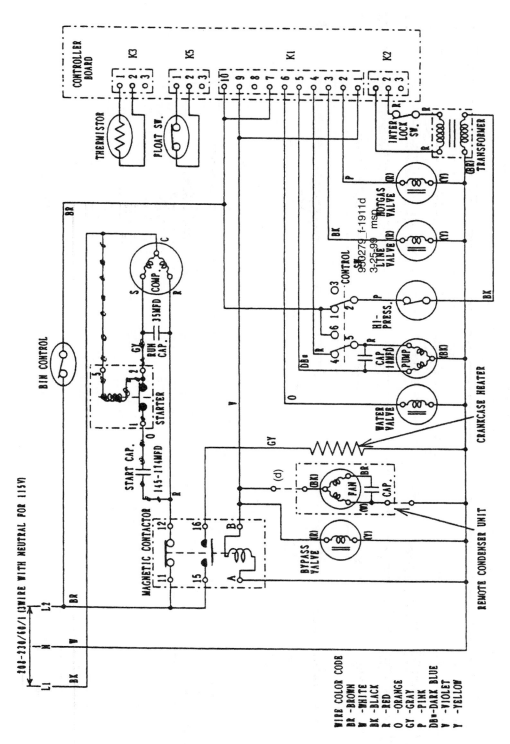

INITIAL HARVEST / NORMAL HARVEST

TO FREEZE CYCLE

FIGURE 26–10 (Continued)

(b)

FREEZE CYCLE

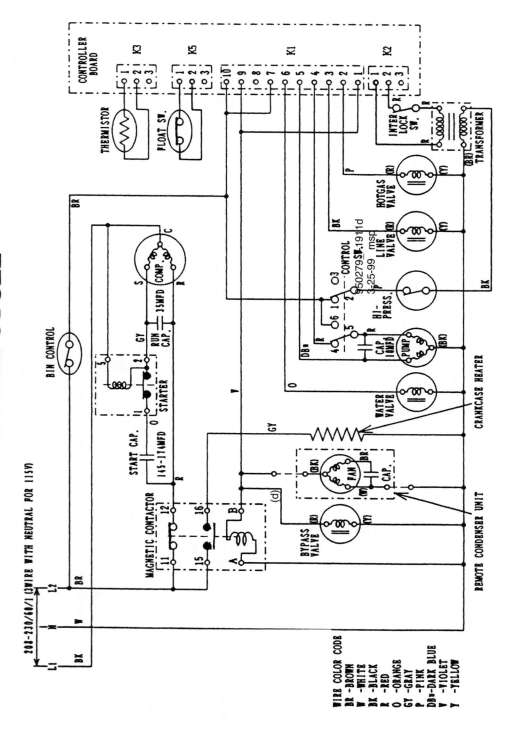

(c)

FIGURE 26-10 (Continued)

TO PUMP-OUT AFTER 1ST FREEZE (MAY SKIP TO NORMAL HARVEST DEPENDING ON BOARD ADJUSTMENT)

10/20 SECOND PUMP-OUT CYCLE

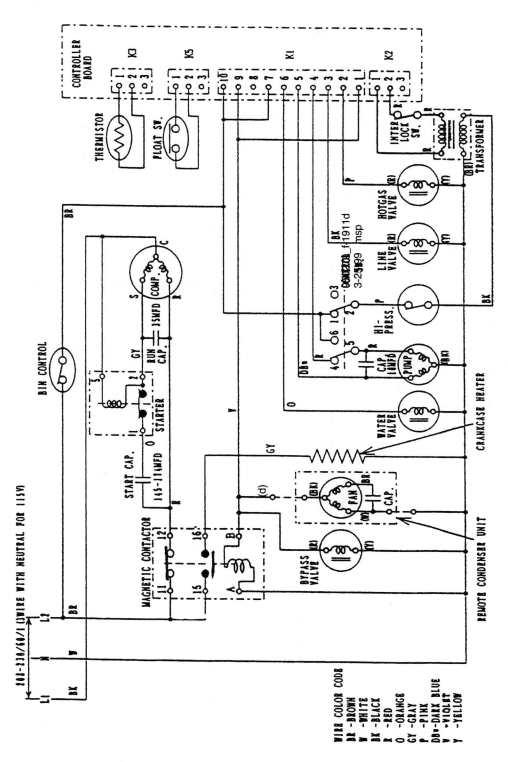

TO NORMAL HARVEST CYCLE ⟶

FIGURE 26-10 (Continued)

(d)

The level of water in the system falls quickly. A water level switch is used to sense when it is time to begin the "harvest" cycle.

During "harvest," an actuator motor pivots the water plate out of the way. Hot gas is introduced into the evaporator (in the usual fashion). The cubes fall out by gravity.

Sequence of Operation

The wiring diagrams are shown in Figure 26–11a–g. The condensing unit runs continuously, so long as the bin is not full.

Figure 26–11a   The sequence begins by filling the water sump. A water level controller energizes the water valve, and the pump and defrost snap action toggle switch operates the water pump. The pump and defrost snap action toggle switch is in the Up position when the water plate is up, next to the evaporator. Later, when the water plate moves down, away from the evaporator, it will kick the pump and defrost snap action toggle switch into the Down position. If the incoming water is below 40 to 45°F during this fill process, the cold water control will energize the defrost valve to prevent freezing until the fill is complete. When there is sufficient water in the sump, the water level control opens, de-energizing the water valve. The water level switch actually consists of a tube, with a thermistor probe near the top and another near the bottom. The thermistor probes are connected to a circuit board, which interprets the thermistor probe signals to determine the water level in the sump.

Figure 26–11b   As the "freeze" cycle continues, the temperature of the evaporator drops. When it falls to 20°F, the actuator control (operated by another thermistor that senses evaporator temperature) moves to the Cold position. This does not affect the "freeze" cycle. It simply is preparing for the "harvest" cycle.

Figure 26–11c   When the cubes are almost fully formed, the interference with the water supply holes in the water plate cause the pump discharge to increase. This pushes water over a weir to a drain. The water level in the sump drops quickly. When it reaches the lower thermistor probe in the level sensing assembly, the water level control switch switches to Low, energizing both the defrost valve and the actuator motor. The actuator motor is a reversible motor. In this mode, it lowers the water plate away from the evaporator.

Figure 26–11d   As the water plate moves away from the evaporator, it trips the pump and defrost snap action toggle switch to the Down position. This provides a parallel switch to operate the defrost valve (in addition to the evaporator-temperature-sensing actuator control). It also de-energizes the water pump.

Figure 26–11e   When the water plate opens fully, the actuator toggle lever on the actuator motor shaft pushes the actuator toggle switch (to the left on the wiring diagram), stopping the actuator motor and completing a third circuit to the defrost valve. As this cycle continues, the evaporator warms, and the cubes fall out.

Figure 26–11f   With the ice out of the evaporator, the evaporator thermistor warms, causing the actuator control to switch from 2–1 to 2–3. This causes the actuator motor to operate in the opposite direction, raising the water plate back toward the evaporator. The cold water control will warm up at about this time and switch to the warm side, keeping the defrost valve energized.

Figure 26–11g   With the water plate almost closed, it pushes up the pump and defrost toggle switch, starting the water pump. When the water plate reaches the fully closed position, the actuator toggle switch is pushed to the right, stopping the actuator motor with the water plate up. This, then, is the same as Figure 26–11a. The cycle repeats until the bin control is satisfied, and then everything turns off.

HOSHIZAKI MODEL F OR DCM ICE FLAKER

As you will recall, flakers do not have separate "freeze" and "harvest" cycles. They make ice and harvest continuously, at the same time, until the bin is full. This flaker also incorporates, on some models, a periodic flush cycle.

Sequence of Operation

The interconnection diagram and control box layout are given in Figures 26–12 and 26–13. On initial start-up, there will be a Fill cycle before the gear motor starts (Figure 26–14a). With the Flush switch and Ice switch closed, power is supplied to the inlet water valve. The unit will not start until the reservoir is full and both floats on the dual float switch are closed (Figure 26–14b). The operation

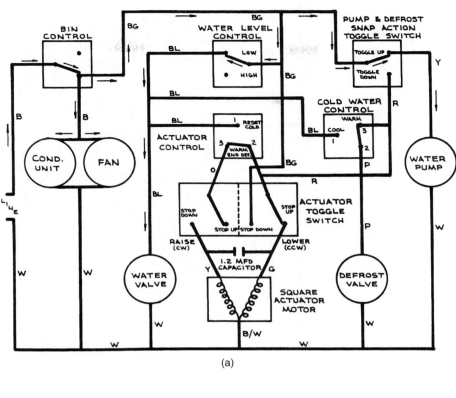

(a)

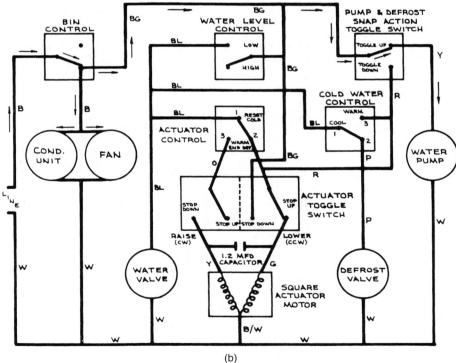

(b)

FIGURE 26–11   Kold-Draft sequence of operation. (Courtesy of Kold-Draft®.)

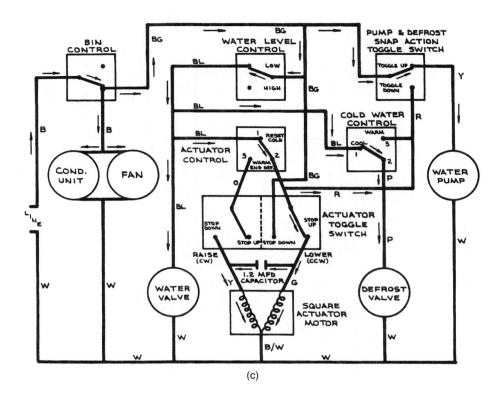

(c)

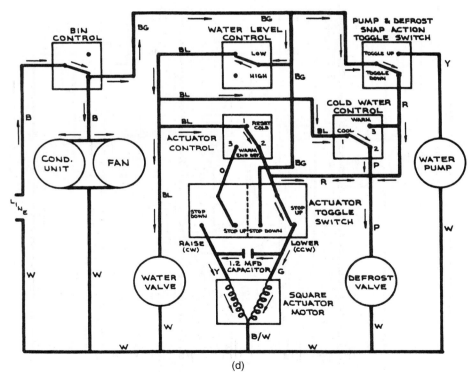

(d)

FIGURE 26–11   (Continued)

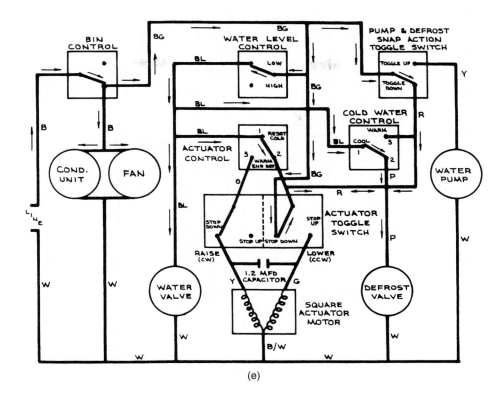

(e)

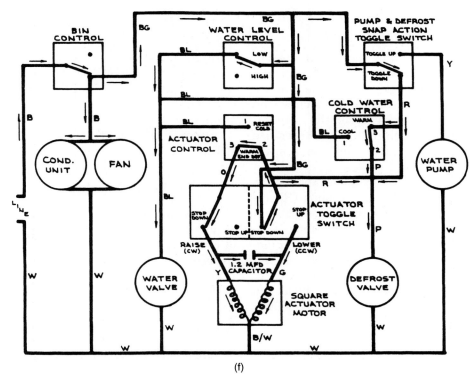

(f)

FIGURE 26–11 (Continued)

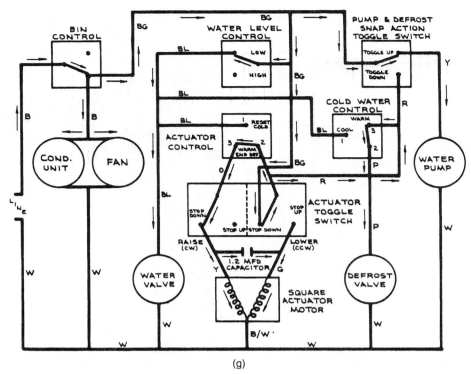

(g)

FIGURE 26–11 (*Continued*)

is then turned over to the bin control. If the bin control is closed (calling for ice), the gear motor and condenser fan motor are energized. One minute later, the compressor starts (Figure 26–14c). As the refrigeration system cools the water in the evaporator, ice will start to form within 2 to 5 minutes. Ice production will continue until the bin control opens.

When the bin switch opens, on some models, the entire unit shuts down within six seconds (Figure 26–14d). On other models, 90 seconds after the bin control switch opens, the compressor stops. One minute later, the gear motor and condenser fan motor stop. This sequence is accomplished through a series of timers within the solid-state timer board. Its purpose is to provide a period of time to clean the ice out of the evaporator to prevent the auger from freezing in place during the Off cycle.

Periodic Flush Cycle

On some of the larger flakers, a periodic flush cycle is included to prevent the buildup of mineral deposits from the water system. A 12-hour timer will cycle the unit down and open the flush valve which allows the complete water system to drain

(Figure 26–14e). The unit will remain off for 15 minutes, which allows any ice remaining in the evaporator to melt and flush the evaporator walls and mechanical seal. The inlet water valve is not energized during this flush period. The unit will automatically restart after 15 minutes on the flush timer.

Circuit Protect Relay

This flaker, as does many others, uses a 230-V compressor, whereas all the other loads operate on 115 V. However, this flaker uses an interesting circuit protection relay to protect the 115-V circuit from a careless technician miswiring the power supply (Figure 26–14f). If either of the power leads (L1 or L2) is mistakenly connected to the terminal that should have the neutral wire, 230 V would be applied to the 115-V portion of the circuit, causing significant damage. However, here, if 230 V is applied to the coil in the circuit protect relay, the switch will move from terminal 1 to terminal 3. No part of the circuit is connected to terminal 3, so no part of the 115-V circuit will "see" the improper 230 V. When the proper 115 V is applied to the circuit protect relay coil, the switch remains connected to terminal 1. Once the unit

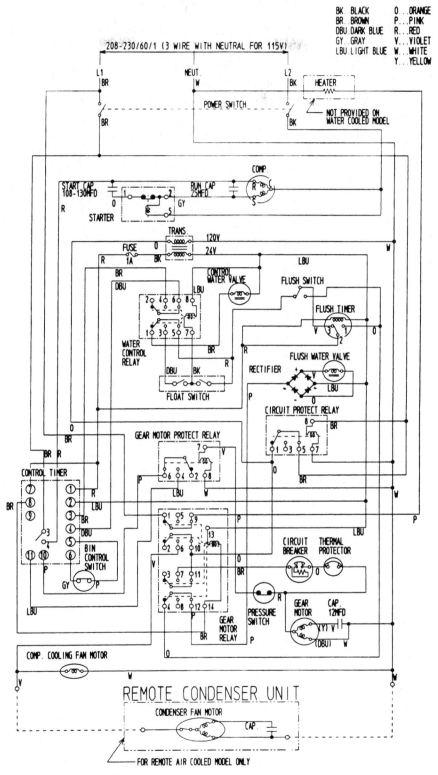

FIGURE 26–12 Hoshizaki interconnection diagram (Courtesy of Hoshizaki America, Inc.)

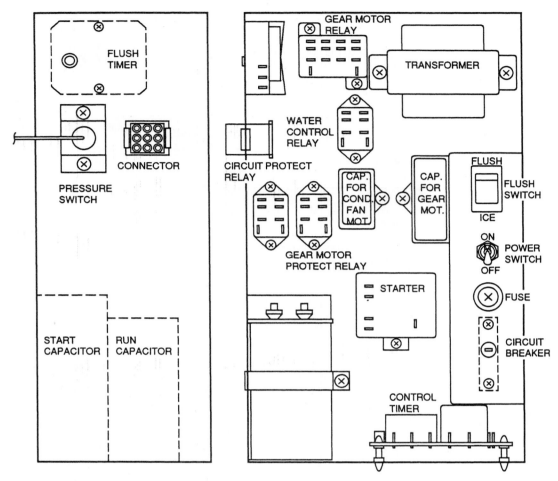

FIGURE 26–13 Hoshizaki flaker control box layout. (Courtesy of Hoshizaki America, Inc.)

has been properly wired, the circuit protect relay has no further function.

ROSS-TEMP MODEL RF FLAKER

The flaker in Figure 26–16 operates as follows. When the bin switch closes, the power relay and the gear motor are energized, and the compressor runs. After the delay thermostat senses a cold evaporator, it switches from the RD-BK position to the BL-BK position. This bypasses the bin switch, but it has no effect until the end of the cycle when the bin is full. The gear motor and the compressor motor are both started through current relays.

When the bin is full, the bin thermostat opens, de-energizing the power relay. But the gear motor continues to run through the delay thermostat. This allows the auger to clear out any accumulated ice in the evaporator. After a few seconds, the evaporator warms, and the delay thermostat switches back to the RD-BK position, and the gear motor stops.

The interlock relay has no function, until something goes wrong. The current through the gear motor is also passed through a current-sensing element in the gear motor protector (between terminals 1–2). If the current through the gear motor becomes higher than normal, the current-sensing element in the gear motor protector will close the switch between terminals 2–3, energizing the coil in the interlock relay. (This high current might be caused by binding of the auger bearings or lack of lubrication in the gear box.) This operates two switches in the interlock relay. The normally closed switch (terminals 1–2) open, de-energizing the power relay and the gear motor. The normally open switch (terminals 4–6)

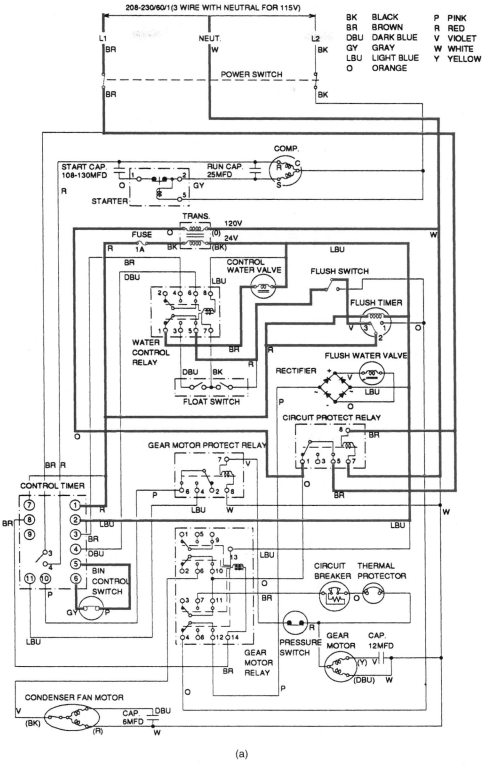

(a)

FIGURE 26–14 Hoshizaki F or DCM flaker sequence of operation. (Courtesy of Hoshizaki America, Inc.)

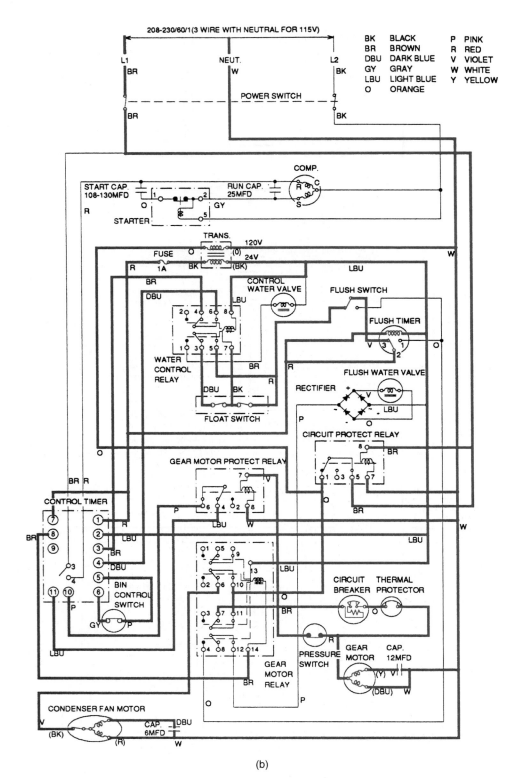

(b)

FIGURE 26–14 (*Continued*)

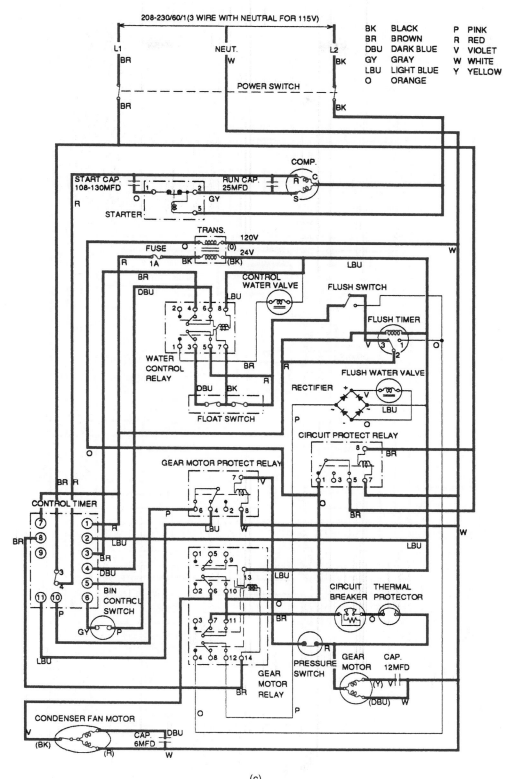

(c)

FIGURE 26–14 (*Continued*)

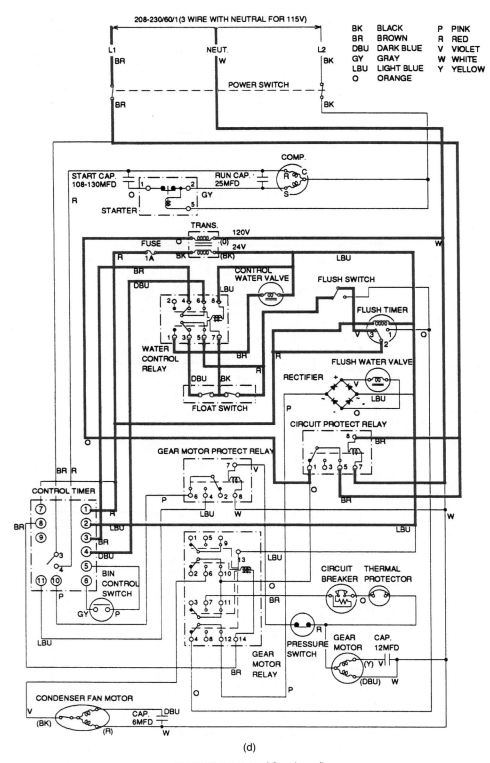

(d)

FIGURE 26–14 (Continued)

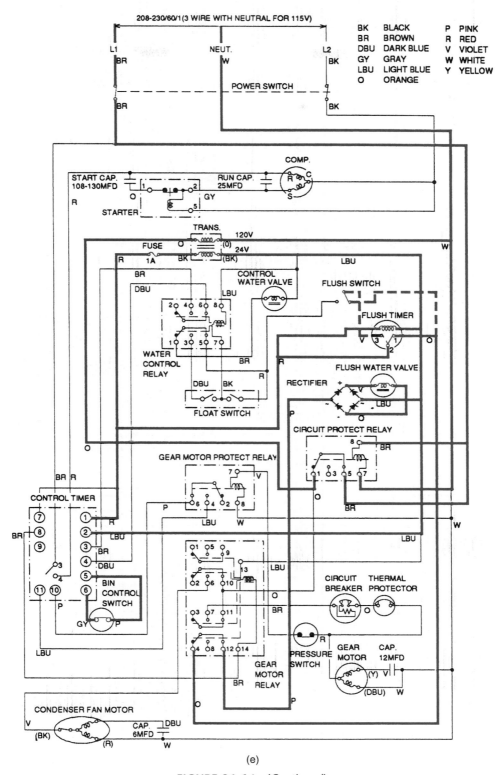

(e)

FIGURE 26–14 (*Continued*)

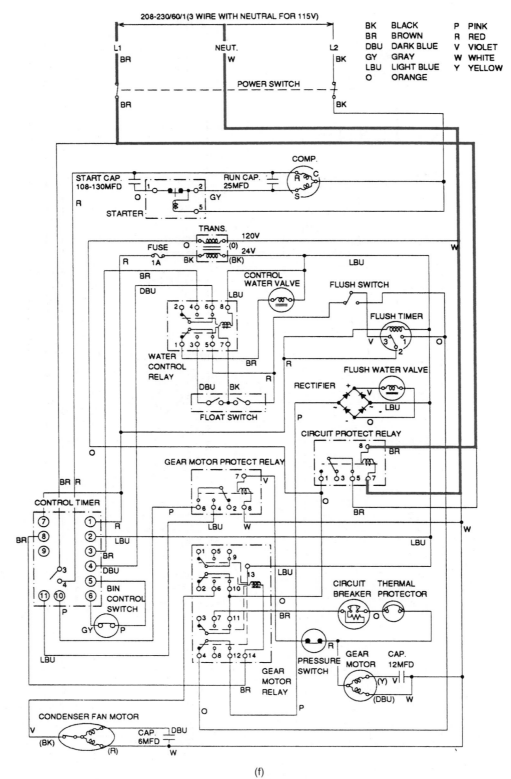

(f)

FIGURE 26–14 (*Continued*)

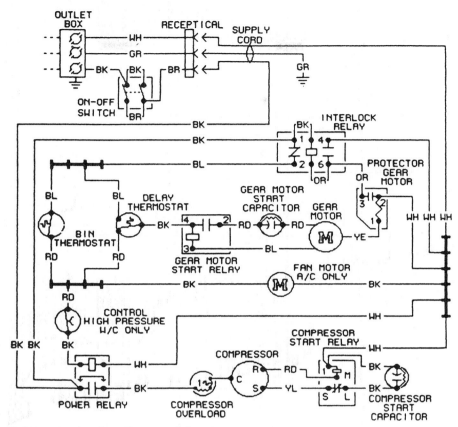

FIGURE 26–15 Ross Temp RF flaker. (Courtesy of Ross Temp, a division of IMI Cornelius, Inc.)

closes, locking in the interlock relay coil. Even if the gear motor protector now returns to its normal position, the interlock relay will remain energized, and the power relay will remain de-energized. The only way to return from this lock-out position is to turn the unit off at the On-Off switch.

SUMMARY

Ice makers are varied in their construction and operation. The circuit diagram is the key to understanding the individual system. Where electronic control boards are used, it is necessary that the technician recognize required input and outputs.

PRACTICAL EXPERIENCE

1. Figure 26–16 shows an interconnection diagram for the C-61 ice maker. Use a highlighter or other method to indicate which wires are at L1 voltage and a different color to indicate which wires are at L2 voltage during the "freeze" cycle.

2. Redraw the diagram in Figure 26–16 as a ladder diagram. Show all the wire colors. You may combine the gear motor, its start relay, and capacitor into a single circle in your diagram. The same applies for the compressor motor.

REVIEW QUESTIONS

1. Referring to Figure 26–7, in step 1, why will relay R1 energize after the compressor contact?

2. Referring to Figure 26–7, in step 2, relay R4 is energized. One set of contacts closes the circuit to the Timing Module. What does the other set of contacts do?

3. Referring to Figure 26–7, what stops the system after the ice bin fills?

4. Referring to Figure 26–16, what type motor is the compressor motor?

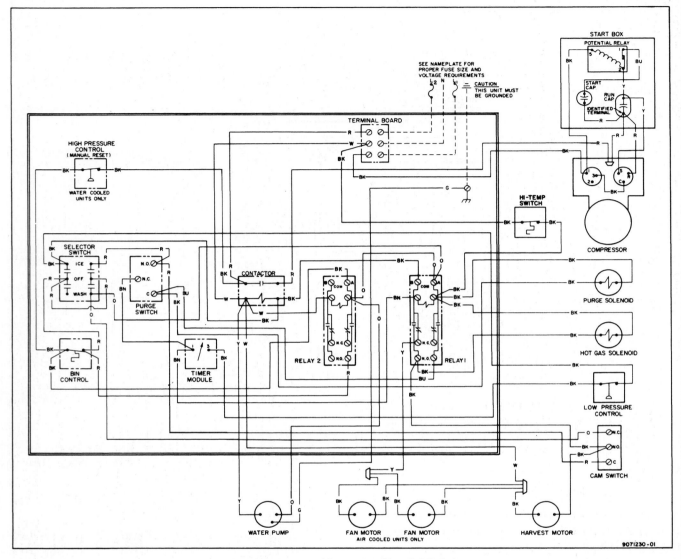

FIGURE 26–16   (Courtesy of Ice-O-Matic.)

5. Referring to Figure 26–16, the bin control contacts have a set of contacts in parallel with them. What relay provides these contacts?

6. Referring to Figure 26–16, what do these contacts (#5) provide for?

7. Referring to Figure 26–5, which relay provides for the completion of the cycle after the bin is full?

8. Referring to Figure 26–5, is the normal position of the relay contacts 4 and 7 closed as shown?

9. Referring to Figure 26–11, what happens when the bin control switch is actuated?

10. Referring to Figure 26–11, what two conditions must be met for the water valve to be energized?

UNIT 27

TROUBLESHOOTING STRATEGIES

OBJECTIVE

After completion and review of this unit, you should be familiar with:

- Troubleshooting procedures for air-conditioning and refrigeration systems.

Problems relating to an air-conditioning or refrigeration system malfunction can be divided into two categories: electrical and mechanical. It is estimated that about 80% of all service calls are because of electrical problems. Additionally, many mechanical problems become evident with the indication of an associated electrical problem.

FACTORS TO CONSIDER

Practical troubleshooting consists of relating the knowledge that the technician has to the problem to be solved. It is impractical to attempt to cover the infinite variety of possible troubles in a textbook. It is suggested that the technician concentrate on the basic theory of electricity, manufacturers' bulletins, controls, and system schematics of equipment.

Most air-conditioning and refrigeration systems consist of series circuits in which individual controls are used to interrupt or complete electrical paths whenever a basic limitation is exceeded. The control operates to impose the limit on the circuit. It is the technician's responsibility to become familiar with the different control devices used in the related systems. Again, the manufacturers' bulletins, tech notes, and component information sheets provide basic information on the operation of controls. It is most important that the technician become familiar with electrical troubleshooting procedures. The following factors must be considered:

1. **Profit** Actual time spent locating a defective part is just as important as (perhaps more important than) the time spent making the repair or replacement. For example, if the problem is a defective fuse or circuit breaker, minimal time should be spent in determining the problem. The customer cannot afford—nor can the service company charge for—an excessive number of hours spent troubleshooting.

2. **Procedure** It is important that a systematic procedure for troubleshooting be developed by the technician. By following a systematic procedure, the technician should be able to locate any problem in a minimum amount of time. For example, a defective major component (such as a compressor) should be detected almost as rapidly as a minor component (such as a fuse).

SYSTEMATIC TROUBLESHOOTING

Electrical troubleshooting of most air-conditioning systems may usually be divided into two subsystems: low voltage and high voltage. The low-voltage control circuit is usually 24 volts, supplied by a step-down transformer. The high-voltage power circuit is usually 220 volts, although some are 110, 115, 120, 208, 230, and even 440 volts. Steps involved are as follows.

1. **Localize**   When a problem exists in a system, the first procedure is to localize the problem. If the system contains sections, such as in a split air-conditioning system, the problem can usually be rapidly localized to a section, for example, the evaporator or condenser section.

2. **Isolate**   When troubleshooting, the technician should isolate the problem to a particular group or subsystem. Isolation of the problem usually leads to a more rapid discovery of the defective component.

3. **Locate**   After the problem area is isolated to a specific section or system, the actual fault can be quickly located by following the suspect circuit to an open, short, or defective component. The final location of a single defect on an open or shorted component is not necessarily the end of the troubleshooting procedure. Often the defective part is merely an indication of a problem in another area. For example, a blown fuse or an open circuit breaker may indicate a shorted system component.

Conclude the Troubleshooting Procedure

The conclusion of a problem is not accomplished until all defects relating to the problem are discovered and corrected. For example, an open high-pressure control indicates excessive high-side system pressure. If the cause of the excessive pressure is not corrected, such as a defective condenser fan motor, resetting the pressure control will be only a short-term cure for the complaint.

CUSTOMER COMPLAINTS

Often the information from the customer provides a lead to the trouble in a system. For example, when called on to troubleshoot and repair a home air conditioner, 3-ton split system, the customer might comment: "No matter what I do with the thermostat, nothing happens. The air-conditioning unit doesn't make a sound." This is an indication that there could be a problem with either the high-voltage (220-volt) source or the low-voltage (24-volt) control. When the thermostat mode selections are varied, the sound of the high-voltage contactor energizing and de-energizing in the unit can usually be heard. If there is no high voltage, there may not be low voltage. Without low-voltage control, the contactor will not operate.

On another service call, the customer's comment may be: "When I adjust the thermostat, I can hear a click in the air conditioner, but the unit doesn't come on." This is an indication that 24-volt control power is available. It is possible that one side of the 220-volt source might be missing, provided that the step-down transformer is 110 volts to 24 volts. Of course, any number of other troubles are possible, but at least there is an indication that power is available.

Do not assume the customer's indication of the trouble is completely accurate. Good customer relations require that you listen to the comments. Then act according to established troubleshooting procedures.

LINE VOLTAGE

It is good practice to measure and record the line voltage present at the terminals of the power disconnect when on an air-conditioning service call. The line voltage greatly affects the operation of air-conditioning units. It should always be within the limits set by the manufacturer. The voltage should be recorded when on the service call for reference on a possible return call. Any change in normal voltage (higher or lower) is a possible trouble source. Also, be sure that the meter is accurate by checking it against a known value.

TROUBLESHOOTING RULES

There are only two rules for troubleshooting. They are simple to state, and they are always true:

1. *If you measure a voltage across a switch, the switch is open.*
2. *If you measure a correct voltage across a load and the load doesn't work, the load has failed.*

Some comments are in order regarding the application of these two rules.

1. With digital meters, voltage readings that we have always considered to be zero when using analog meters will indicate some very small voltage. A very small voltage reading across a switch could indicate either a meter inaccuracy or a very slight resistance across the contacts of a closed switch. If you read a voltage of less than one or two volts, assume that it is zero volts.

2. Rule #1 does *not* say that if you read zero volts across a switch, the switch is closed. There are many situations in which you might read zero volts across an open switch.

3. Rule #2 indicates when the load has failed. This only means that the problem is with the load, and not with any other part of the circuit. The "fix" might be as simple as turning a knob or resetting an overload switch on the load. It does not necessarily mean that the load must be replaced.

CHECK THE EASY ITEMS FIRST

Always look for the easy fix first. Whatever electrical components there are that might explain the symptom that you observe, go first to the devices that are most accessible and easiest to check. If a 230-V 1ϕ unit is dead, check for any fuses or disconnect switches at the unit. Check to see that the unit is supplied with the proper incoming voltage. However, if the condenser fans are running and the compressor is not, don't bother checking the fuses no matter how easy it is. It is a waste of time, because the running condenser fans tell you that the fuses are good.

The trick to effective troubleshooting is to zero in as quickly as possible by testing the components that are most likely to be the cause of the problem. This chapter presents several of the techniques that are second nature to experienced technicians and some that they may even be helpful to experienced troubleshooters.

TROUBLESHOOTING TOOLS

If power may still be applied to the malfunctioning unit, it is easier troubleshoot using a voltmeter than an ohmmeter. Disconnecting components is not normally necessary in order to make voltage measure-

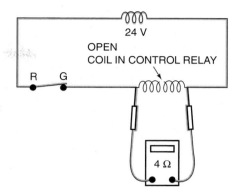

FIGURE 27–1   An erroneous ohm measurement.

ments. Disconnecting components is necessary to make resistance measurements.

To demonstrate why that is necessary, refer to Figure 27–1. Suppose the blower motor does not operate. You think that the coil in the control relay may have failed, and you want to use your ohmmeter to see if there is continuity across the coil. If you simply place your ohmmeter probes across the coil contacts, you may get fooled. The coil you are trying to measure may actually be open, but if you don't disconnect it from the circuit before measuring it, you will measure continuity through the transformer secondary, and you might mistakenly conclude that the control relay coil has continuity.

So it is necessary to disconnect a device from the circuit before attempting to measure its resistance. But when you remove the wiring from the device, you will be moving, removing, pushing, and jostling wires, with several potential undesirable results:

1. There may be an intermittent problem that resolves when you move the wires. Then you will not be able to find it. The problem will most likely reoccur sometime later, and you will have a call-back.

2. Some wiring is old, and its insulation is brittle. If you start moving wires around, the insulation may become damaged. You may wind up needing to replace some or all of the wiring.

3. Whenever you remove wires, you need to identify them with tape so that you won't forget on which terminals they get reconnected. This is extra work and is not required if you troubleshoot with voltage.

When you use your voltmeter to troubleshoot, you will either find a switch that is open (you measured voltage across it) or a load that has failed (you measured proper voltage across it, and it doesn't work). You will be able to do this without moving any wires and without changing the circuit in any way. You may then remove the device, and double-check it with ohms if you like.

CHECK THE SWITCHES AND PUSH THE BUTTONS

Sometimes the wiring for the unit you are troubleshooting is either very complicated or the wiring diagram is missing (or both). It is not always necessary to figure out everything about how a system is supposed to work before you can find the problem.

Sometimes there is a variety of pressure switches, temperature switches, or other operating or safety controls that are easily visible and accessible. You may have no idea which switches may or may not be wired in series with the nonworking conditioner as in Figure 27–2. The low-voltage room thermostat controls a compressor and condenser from a single contactor CC and a blower motor from a control relay CR.

Suppose that the condenser fan was running, but the compressor was not. The logical first step would be to go directly to the compressor, because that is the only device not working. It would not make sense to check the contactor, because we know that its switches are closing because we can observe

that the condenser fan is running. It would not make sense to check the transformer, or the thermostat, or the fuses in the power supply for the same reason. It would not make sense to check the blower relay, or the blower motor, or the condenser fan motor, because none of these components could fail in a way that would cause the compressor to not run.

If neither the compressor nor condenser were running, but the blower motor was, your troubleshooting approach would be different. Although it is certainly possible that both of these loads could have failed, it is extremely unlikely, and your initial approach should be to proceed as if there is only a single device that has failed. What devices in this circuit, if they failed, could cause all the problems you observe? It could be a failed compressor contactor or a failed room thermostat (not making R-Y). But the problem could not be a failed transformer or a blown fuse in the power supply (the blower motor is running).

EXAMPLE

Suppose the wiring for the preceding problem were slightly more complicated, as shown in Figure 27–3. There is a high-pressure cut-out (HPC) and a low-pressure cut-out(LPC) and a thermal overload (OL) in the compressor circuit. There is a fan-cycling switch in the condenser fan circuit. If the condenser fan and blower are both running but the compressor is not, what devices would be logical to check first?

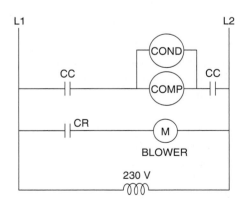

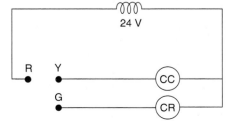

FIGURE 27–2 Rooftop air conditioner.

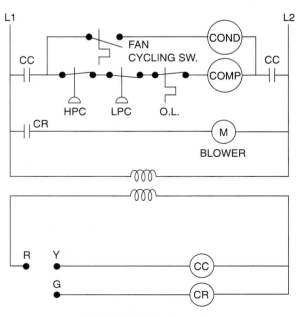

FIGURE 27–3 Example.

Solution

Compressor, HPC, LPC, and OL. No other single device could have failed that would explain the symptom that has been observed. ■

JUMPER WIRES

When used properly, a jumper wire (a wire with an alligator clip at each end) is a valuable troubleshooting tool. Used improperly, a jumper wire can cause damage to an electrical circuit and, just as importantly, can cause embarrassment to you and your company when you create a short circuit and make all the lights go out in your customer's place of business. The most important rule for use of a jumper wire in the field is this:

 Never place a jumper wire across a load. Only place a jumper wire across a switch.

The reason for this rule is illustrated in Figure 27–4. Suppose that the switch has failed. If you place a temporary jumper wire across the switch (Figure 27–4a), it will complete a circuit through the load, and the load should run. However, suppose the switch is still good, and the load has failed. If you place your jumper wire across the load (Figure 27–4b), it will create a short circuit. Aside from blowing a fuse or a circuit breaker, the high current flow through the switch could easily ruin the switch because it was never designed to carry so much current. When you place a jumper wire across a switch, you will not get into that type of trouble. You are only making contact between two electrical points that normally get connected together when the switch closes normally.

There are two other precautions to observe when using a jumper wire.

1. Make sure that your jumper wire is sufficiently heavy gauge to carry the current. If it is as heavy as the wiring that is attached to the switch you're jumping, that is sufficient.

2. If the alligator clips on the ends of your jumper wire are not insulated, make sure the power is off when you attach them, and make sure that they are securely fastened and will not move and create unwanted contact when you turn the power back on.

3. If you jumper across a switch in a live circuit, there will be a spark created when you attach the second alligator clip. That occurs because you are completing a circuit. Don't get startled and drop the alligator clip.

4. Do not place a jumper wire across a switch that carries a high current. For example, do not jumper across the switches of a contactor that is supplying current to a compressor.

Figure 27–5 shows an alternative way of jumping across a switch when you don't have or can't

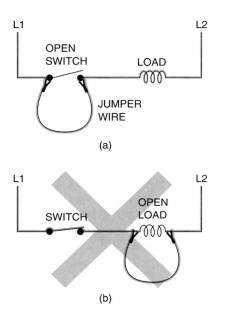

(a)

(b)

FIGURE 27–4 Never place a jumper across a load.

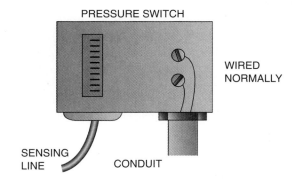

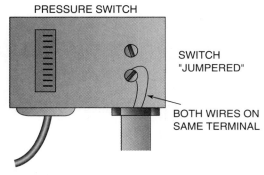

FIGURE 27–5 Jumpering a switch without a jumper wire.

use a jumper wire. The wire(s) on one terminal can be moved onto the other terminal. This creates the same effect as if the switch had closed.

Why Jumper Wires?

A jumper wire is not to be used to find the problem. You use your voltmeter for that. But after you have found an open switch with your voltmeter, you should jumper across the switch to make sure that everything will work properly after you replace the switch. If you don't do that extra step, the following sad story can happen to you.

You find that the compressor in Figure 27–6 won't operate because the LPC is open. You tell the customer that it will cost $80 to replace it, and the customer approves. After replacing the LPC, you find that the compressor runs, but it doesn't cool. Further investigation reveals that there is no charge in the unit. You find the leak and advise the customer that it will cost a lot more than $80 to fix the leak and evacuate and recharge the unit. Your customer will be upset, because if had you given him or her all of that information before you replaced the pressure switch, they might have decided to replace the unit instead of repairing it. At the very least, you will appear to be incompetent if you quote a price to repair a unit, and then go back to the customer to say that you were only kidding.

The same story can happen in a number of different ways. You can replace a high-pressure switch only to find that your condenser fan has failed. You can replace a low-pressure switch only to find that there is a restricted metering device. You can replace a thermostat only to find that the load that it is controlling has shorted, and the new thermostat fails for the same reason as soon as it is installed.

Jumper Wire Caution

The last situation described (a shorted load) can cause a problem when you use a jumper wire. Assume that the circuit in Figure 27–2 has a shorted coil in the compressor contactor. The high current through the thermostat caused it to fail open. When you arrive, you discover the open thermostat, but not the shorted coil. You place your jumper wire across the R–Y terminals of the thermostat, and you recreate the short circuit. If you do not remove the jumper wire immediately, you can burn out the transformer. Therefore, as a general rule, if you place the jumper wire and it doesn't produce the anticipated result immediately, turn the power off (or remove the jumper) and look for another problem.

KEEP CHANGING THE QUESTION

Figure 27–7 shows a compressor that is energized by a contactor, which, in turn, is energized by other controls. The compressor does not run. This troubleshooting strategy involves starting with the disabled load and asking "Why isn't the load running?"

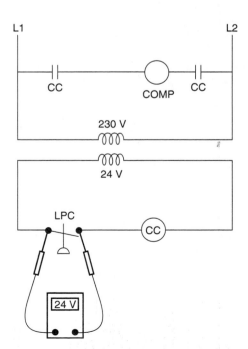

FIGURE 27–6   The LPC is open, but is that the problem?

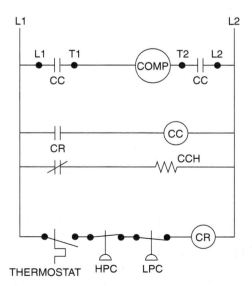

FIGURE 27–7   Keep changing the question.

There are only two possible answers—either the load is not getting energized, or if the load is getting energized, then it must have failed. Measure the voltage across the load. If it's 230 V, you're done. If it's zero, the new question becomes, "Why isn't the compressor getting energized?"

The only possible answers are that either the compressor contactor switch(es) is (are) not closing, or there is no power available to the circuit. Measure voltage across L1–T1. If it's zero, you need to check the incoming voltage between L1 and L2. If you read voltage across the switch, it's open, and the new question becomes, "Why is the contactor switch open?"

Again, there are only two possible answers—either the contactor coil is not getting energized, or if the contactor coil is getting energized, then the contactor must have failed. If the contactor coil is not getting energized, the new question becomes, "Why isn't the contactor coil getting energized?"

You may find that the contactor coil is not getting energized because the normally open contacts from CR are open. Then you ask, "Why are the contacts open?" You may find that the contacts are open because the CR coil is not getting energized, and then you ask, "Why is the CR coil not getting energized?" Eventually, you will get to a point where there are no more questions. For example, you find that the LPC is open because the unit has leaked out all its refrigerant. At that point, your troubleshooting ends, and the repair can go forward.

CUTTING THE PROBLEM IN HALF

If you had to guess a secret number between 1 and 16, random guessing would usually get the right answer in about 8 guesses. However, if you could guess in the middle of the possible numbers each time, and you were told whether the actual number was higher or lower, you would guess the right answer in no more than 4 guesses. This is the principle involved with cutting the electrical troubleshooting problem in half.

In a complex wiring diagram, there can be dozens of potential culprits. Figure 27–8 shows a contactor in series with eight switches. These may be pressure switches, temperature switches, safety switches, contacts from other relays. You have read zero volts across the compressor contactor coil, and you suspect that one of the eight switches is open. If you take hopscotch readings as shown, you may wind up having to follow the wiring and identify eight different devices and take 16 different readings before you get to the actual culprit, switch #8. Why not take your first reading at switch #4 or switch #5? If you get 230 V, then you know all the switches ahead of it are closed, and the problem is to the right of the measured switch. If you get 0 V, the open switch is one of the ones to the left.

Another way of cutting the problem in half is to go directly to the contactor for the motor that is not running. Manually depress the contactor switches using a *well-insulated* screwdriver. If the motor still does not run, the problem is in the power wiring

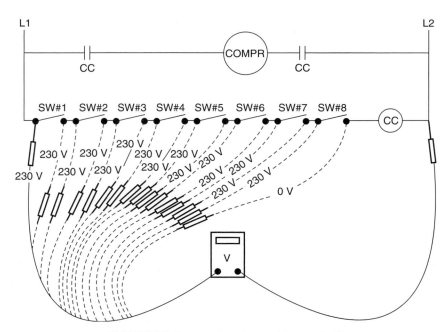

FIGURE 27–8 Cutting the problem in half.

part of the circuit. If it does run, then the problem is in the control voltage part of the circuit.

☞ When the customer or the maintenance manager is watching you work, this is a particularly good strategy. It is very impressive to the customer when the service technician can make the system run within just a few seconds after removing the access panels!

☞ Don't hold the contactor switches closed for more than a few seconds. The problem might be that the contactor coil is not getting energized because of an open safety control. By holding the contactor switches closed, you are bypassing the safety control and subjecting the system to whatever hazard the safety control was protecting it against.

IS THE COMPRESSOR WARM?

You have come upon a system whose compressor is not running. You test the compressor motor windings. The readings between the pins are infinite, infinite, and 12Ω. This indicates that the windings are continuous, but the internal overload is open (Figure 27–9). On a hermetic compressor, the overload cannot be replaced or bypassed. If it does not reclose, the compressor must be replaced.

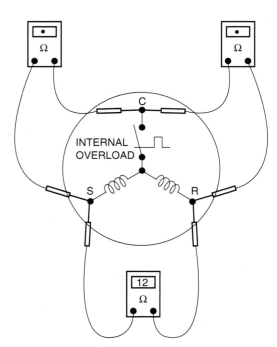

FIGURE 27–9 Resistance readings indicate that the windings are OK but the internal overload is open.

Feel the compressor. If it is not warm, replace it. However, if it is at all warm, that indicates that the compressor tried to start not too long ago, and it went out on the thermal overload. It may take an hour or more for the motor windings to cool sufficiently to make the overload reclose. Lock out the unit, and go to another job. Come back when the compressor is at ambient temperature. If the overload still has not closed when the compressor is cold, then the compressor must be replaced.

TURN THE UNIT OFF, AND THEN ON

If you come upon a unit that has nothing running, but voltage is available coming into the unit, the unit may be in a lock-out condition. Before even trying to determine if there is a lock-out relay, simply turn the unit off at the disconnect, and then back on. If the unit starts up, but then stops again, there is a good possibility that there is a lock-out relay. You may be able to troubleshoot during the short period when the unit runs, before going into lock-out. Otherwise, you may need to bypass the lock-out system by disconnecting a wire from the lock-out relay.

MAKE DECISIONS FROM NONZERO VOLTAGE READINGS

You always need to be aware of the possibility that your voltmeter is either not functioning properly or set incorrectly. For example, test leads do not last forever. Sometimes, a test lead on your meter can appear to be perfectly normal, but inside the insulation, the wire has broken. If that happens, every voltage reading you take will be zero. Make sure that your meter actually reads a nonzero voltage before you get involved in each troubleshooting project. Then, using the first two rules of troubleshooting, you will eventually get to a switch or a nonfunctioning load that has a voltage across it, and you will have found the problem.

CYCLING—TIMING

Sometimes, you come upon a unit that has a component cycling. That is, it is turning on, then turning off, and repeating the cycle continuously. For example, you might have a compressor that cycles on for five seconds, off for 30 seconds, on for five seconds, and so on.

Whenever a load is cycling, the problem is most likely not at the load. There is a switch that is opening and closing to cause the load to be energized intermittently. Sometimes you can get a clue where to look by simply observing the length of the cycles.

If a compressor shuts down within just a few seconds of starting, and then restarts within a minute, check the low-pressure cut-out. There is a good chance that the system is low on charge, and the low-side pressure reaches the cut-out setting very quickly. When the compressor turns off, the low-side pressure rises to the cut-in pressure, and the cycle repeats.

Sometimes a motor cycles, but over a much longer time period. For example, it might stay off for two or three minutes or more before restarting. This sounds like a thermal overload is opening and closing. When an overload must cool down before reclosing, it takes much longer to react than a pressure switch.

Cycling on fan motors can be fast or slow. For example, a condensing fan motor that cycles quickly might have a pressure switch with the differential set too close. But a condensing fan motor that cycles over a longer period might have a faulty run capacitor. A run capacitor makes a motor operate more efficiently. When it has failed, the motor draws too many amps, and the internal thermal overload opens. When it cools, the cycle repeats.

When a load cycles, make sure that you take your voltage readings across switches while the motor is de-energized. Otherwise, you might go right past the offending switch if you happen to measure across it while it is closed.

TWO FAULTS

Occasionally you will encounter a situation where two devices in the same circuit are open. For example, in Figure 27–10, suppose that both of the limit switches are open. If you are using an analog meter, testing voltage across either of the limit switches will give you a zero reading, not because there is not a voltage drop from one side of the switch to the other, but because of limitations on how the meter works. The wire between the two switches is an "island." That is, it is not connected to either side of the transformer or any other part of the circuit. When one probe of the voltmeter is touched to the island, it is almost as if that probe is not touching the circuit at all.

If you are using a digital meter, the result will be somewhat different. Digital meters are more sensitive than analog meters, and reading across either switch will give you a "funky" reading. Usually, when you measure voltages, you expect a reading equal to the voltage available (230 V, 115 V, 24 V, etc.) or a zero reading. But in this case, the digital

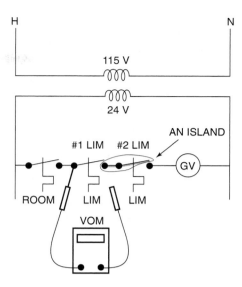

ANALOG METER = 0 VOLTS
DIGITAL METER = "FUNKY" READING
FIGURE 27–10 Two faults.

meter will give you a reading that is more than zero, but less than the available 24 V.

So what do you do if you have 24 V available at the transformer secondary, but you cannot find any device in the circuit that has a 24 V drop? The answer is to take hopscotch voltage readings. Park one probe on the right side of the wiring diagram as shown in Figure 27–11. Starting with reading #1, there are 24 V available. Keep moving one probe

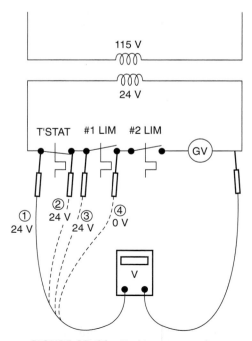

FIGURE 27–11 Parking one probe.

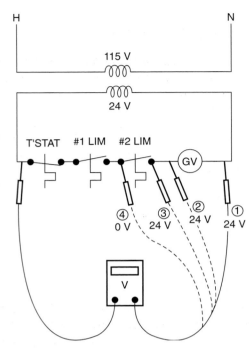

FIGURE 27–12 Parking the other probe.

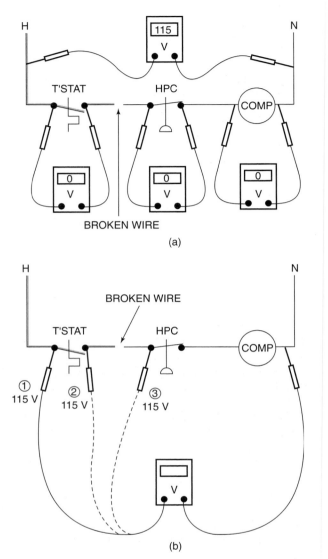

FIGURE 27–13 Diagnosing a broken wire.

until you lose voltage. In this illustration, you will lose voltage between the third and the fourth readings, indicating that the first limit switch is open. Then, you can approach the problem again from the other direction as in Figure 27–12. With one probe parked on the left side of the wiring diagram, move the right probe. You will lose voltage between the third and fourth readings, indicating that the second limit switch is also open.

BROKEN WIRE

A broken wire presents a situation similar to the preceding one. You have voltage available, but there is not voltage at the load, and the voltage readings across each of the switches are zero (Figure 27–13a). Start hopscotching (Figure 27–13b). Reading #1 tells you that there is voltage available. Reading #2 tells you that the thermostat is closed, because it is reading the same electrical pressure as the wire coming into the thermostat. Reading #3 tells you that between the connection leaving the thermostat and the connection entering the HPC, you have lost electrical pressure. This reveals that there is a broken wire (or a poor connection) between the thermostat and the HPC.

 Some technicians will park a probe on a ground, instead of the right side of the diagram, which happens to be Neutral in

this case. This is alright for 115-V circuits only. Ground is at the same zero electrical pressure as Neutral. However, this will not work on a 230-V circuit. See Figure 27–14. Suppose the thermostat is open. Readings on either side of the thermostat will give 115 V. The technician would not realize that reading #1 is actually reading L1-Ground, and reading #2 is actually reading L2-Ground.

FINDING A GROUND

Finding a ground is one of the trickiest troubleshooting problems. A ground exists when an electrical conductor touches a metal casing. What makes it tricky to troubleshoot is that if there is a ground

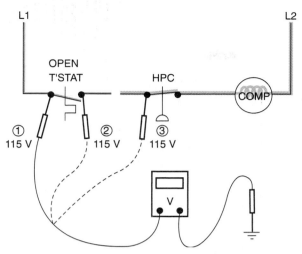

FIGURE 27–14   You can't use ground to troubleshoot a 230 V circuit.

fault in one device, every point that is electrically connected to that device will also appear to be grounded. See Figure 27–15. If the control relay coil is grounded, then measuring from M1 to ground will indicate continuity, even though there is nothing wrong with M1.

The ground condition could be caused by a switch or a load that has failed, or it could be a wire whose insulation has worn or been cut, and the conductor inside is touching the casing. The way to find it is to select one place where you are measuring continuity to ground. Then, start disconnecting devices, one by one, rechecking for the ground condition each time until the ground condition disappears. Then you will have isolated the ground to the

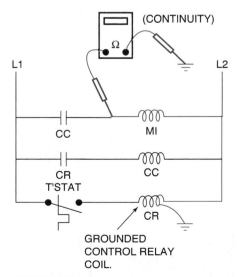

FIGURE 27–15   Finding a ground fault.

part of the circuit that you just disconnected, and you can start looking for a ground condition there.

👉 When disconnecting more than one or two wires, use masking tape on each wire, and identify where each goes. It's not hard to lose track of where everything goes, and figuring it out is a potentially time-consuming chore that can easily be avoided by proper wire marking. You can simply assign a number to each wire and make a sketch for yourself showing which numbered wires attach to which terminals.

WHICH ELECTRICAL MEASUREMENTS CAN YOU TAKE TO THE BANK?

You may be familiar with the terms false-negative or false-positive when referring to medical testing. These terms refer to medical tests where, for example, if the test shows positive, you can be sure that you have the condition tested for, but if the test shows negative, it is likely (but not sure) that you don't have the condition.

A similar situation arises when taking electrical measurements for troubleshooting. Sometimes, your reading is one that gives you a sure answer. Other times, your answer is just a maybe. Following are some examples.

Infinite Resistance or Zero Volts

Any reading of infinite resistance or zero volts, by itself, cannot be relied on with certainty. Either one may be an incorrect reading if your meter is not set on the correct scale, an internal fuse in your meter has opened or if your meter has a test lead that is broken inside the insulation. Check all these factors by touching your probes together when using the ohm scale or measuring a place where you know there is voltage present when using the volt scale. Note that none of these factors could be present if you had a noninfinite resistance reading or a nonzero voltage reading.

👉 Some meters have very limited ohm ranges that are unsuitable for HVAC/R troubleshooting. One high-tech-looking digital meter only has an ohm scale that ranges from 0–2000 ohms. Resistances over 2,000 ohms will display as infinite resistance. This is unsuitable because a clock motor (for example) might have a resistance of 3,000 ohms. If you checked the motor windings with this meter, you might mistakenly conclude that the motor winding is open.

Other inexpensive analog meters might have several ohm scales, but the lowest is the 1 K scale. On these meters, you would not be able to distinguish between zero ohms and just a few ohms. Therefore, you would not be able to tell if a motor winding was good or if it was shorted.

Compressor Windings

If you measure an open winding (infinite resistance), you can be sure that the compressor has failed. But beware of internal overloads that may be open. If you are not aware, an open internal overload can fool you into thinking that you have measured an open winding.

If you measure a ground condition (less than infinite resistance) between any compressor pin and the compressor casing (with all external wiring disconnected) the compressor has failed. But beware—the new, very accurate digital ohmmeter can read out millions of ohms, whereas older meters would display millions of ohms as infinite ohms. If your ohmmeter shows millions of ohms when checking for a ground condition, you can take this as an infinite resistance reading.

Mechanical Devices

Many devices such as motors, solenoid valves, contactors, and relays use electricity to produce some mechanical motion. Electrical testing of these devices can tell you for certain that the device has failed. However, even if all of the electrical readings show that the device is good, that's no guarantee. Even though these devices may check out okay electrically, they can still have failed mechanically. Compressors can seize, bearings can fail, and any moving mechanical device can be stuck.

Shortcut Readings

You have learned that when taking an ohm reading across a device, you must remove the wiring from at least one side of the device to obtain a reliable reading. But moving wiring is time consuming, and on old wires with crumbling insulation or brittle conductors, you may be getting yourself into an unwanted rewiring job. Some technicians will measure ohms *without* removing the wiring. If it reads infinite ohms, you can be sure that the device is open. If it reads continuity, then *that* reading cannot be believed until one side of the device is disconnected.

FAILED TRANSFORMERS

If, while troubleshooting, you discover a zero voltage reading on the secondary of a transformer, and a correct voltage reading on the primary, you can be sure that the transformer has failed and must be replaced. But every once in a while, when you replace the transformer, it will immediately fail also. If this happens, you must locate a short within the rest of the control wiring that is causing the transformer to fail.

Some technicians may measure the resistance of the transformer windings and find that it is the primary winding that has opened. They will then mistakenly assume that there is a problem on the primary side of the circuit. However, there is nothing that can go wrong on the primary side of the circuit that would cause the transformer to burn out. The problem *must* be on the secondary. A short on the secondary side will draw more amps from the transformer secondary than the transformer can supply. The secondary winding will, in turn, draw more current from the primary winding. Even though both transformer windings are carrying higher-than-design amps, it is the primary that is more likely to fail first because the primary wire is thinner than the secondary wire.

In order to locate the short, disconnect the secondary portion of the circuit from the transformer. Using an ohmmeter, measure what the transformer secondary is "seeing" (Figure 27–16). If there is a

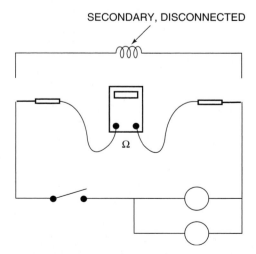

SECONDARY, DISCONNECTED

FIGURE 27–16 Testing the control circuit for a short.

short, the ohmmeter will read zero ohms. Disconnect devices from the circuit until the short disappears. When the shorted device or wiring has been repaired, then the transformer can be replaced.

☞ If the transformer you are replacing is particularly expensive, you should perform this check before replacing it. But with most small systems, the transformer costs less than $10. Many technicians will save time by not performing this check. If once each 20 times the replacement transformer also fails, the cost of the extra transformer is probably worth having avoided spending the extra time on the other 19 times.

☞ Technicians have been known to replace a part such as a transformer, only to find that the new one is also failed. They say to themselves, "How do you like that! This new replacement part was defective, right out of the box." They then get another replacement part from their truck, and after installing it, find that it also has failed. Some will say to themselves, "Isn't that something! What are the chances that I would have two new replacement parts that were defective right out of the box?" Don't fall into this trap. The chances of a new part being defective are very small and certainly much smaller than some technicians would have you believe. The chances of it happening twice on the same unit are small enough to say it's almost impossible. If you find a new replacement part has failed in the same way as the part you are replacing, you need to look for something in the system that is causing the part to fail.

SUMMARY

In troubleshooting, the proper procedure is localize, isolate, and locate.

- Listen carefully to the customer's complaint for clues to the trouble area.
- Make a good visual inspection of the equipment for obvious troubles.
- Review the schematics and wiring diagrams for clues to probable trouble spots.
- Localize, isolate, and locate the trouble.
- Return the device to proper working order in as short a period of time as possible. Do not, however, sacrifice time for accuracy and completeness.

- Practice safe habits. Remember that getting too friendly with electricity can be a shocking experience.

PRACTICAL EXPERIENCE

Required equipment A malfunctioning air-conditioning system

Procedure

1. Use the standard troubleshooting procedure as follows:
   a. Localize the problem.
   b. Isolate the problem.
   c. Locate the problem.
2. Replace the malfunctioning devices and return the unit to normal operation.

Exercise Suppose you find a voltage reading of zero volts on the secondary, and 120 V on the primary. You know that the transformer has failed, but do you know which winding has failed? Explain your answer.

Exercise Explain, using wiring diagrams as examples, why, when the wiring has not been removed from a device, an infinite ohm reading can be believed, but a continuity reading cannot.

REVIEW QUESTIONS

1. In the circuit of Figure 27–2, the compressor is not running. The voltage measured between L1 and L2 is correct at 220 volts. The voltage measured along the contact coil is 24 volts. This proves:
   a. The contactor coil is open.
   b. The contactor coil is shorted.
   c. The transformer is open.
   d. The thermostat contacts are closed. Explain your answer.
2. In the circuit of Figure 27–2, the correct voltage 220 volts is available between L1 and L2. The compressor is not running. When the contactor is mechanically depressed with an insulated screwdriver, the compressor remains off. This proves:
   a. The contactor coil is not being actuated.
   b. The compressor is open or shorted.

c. Something is wrong with either the compressor or the contractor.

d. None of the above.
Explain your answer.

3. The customer complaint is, "Nothing operates." If the circuit is Figure 27–3, your first step would be to:

a. Remove the access panel from the condensing unit.

b. Check the line voltage at the fuse box.

c. Check the thermostat setting.

d. Check the low voltage.

4. In the circuit of Figure 27–3, with cool selected and the temperature well above the selected temperature, what voltage would you expect to measure?

a. Between L1 and L2 _____ volts.

b. Across the compressor contactor coil _____ volts.

c. From the connection of HPC and LPC to L1 _____ volts.

d. From the connection of HPC and LPC to L2 _____ volts.

5. In the circuit of Figure 27–7, the thermostat called for cooling, but the system is not functioning properly. The following voltages are measured. Which measurement points to the problem?

a. L1 to T1 zero volts

b. L1 to T2 zero volts

c. Across the compressor contactor coil 220 volts

d. L2 to T2 220 volts

e. T1 to T2 zero volts

6. Given the voltage measurements of Problem 5, which component is bad, and what is the problem?

7. Referring to Figure 27–8, the voltage measured between L1 and L2 is 220 volts. The voltage measured from the left side of SW5 to L2 is zero volts. This indicates:

a. SW5 is open.

b. SW4 is open.

c. SW1 is open.

d. The line between L1 and SW5 is open.

8. Referring to Figure 27–7, the thermostat is closed, but the compressor is not operating. The compressor contactor is actuated with an insulated screwdriver, and the compressor starts to run. This proves:

a. The compressor contactor is bad.

b. The compressor contactor relay is bad.

c. The compressor is bad.

d. There is voltage between L1 and L2.

9. Referring to Figure 27–7, the thermostat is open. What voltage would you expect to measure from the junction of PHC and LPC to L2?

10. Referring to Figure 27–7, the thermostat is open. What voltage would you expect to measure from the junction of HPC and LPC to L1?

UNIT 28

MISCELLANEOUS DEVICES AND ACCESSORIES

OBJECTIVES

After completion and review of this unit, you should be familiar with some of the special devices used in air-conditioning and heating systems. Included are:

- Mechanical timers.
- Electronic timers.
- Cold controls.
- Electronic modules.
- Humidifiers.

TIMERS AND TIME DELAYS

Figure 28–1 shows a mechanical timer that has been used for many years to turn off heating or air conditioning in applications where the space would be unoccupied for many hours at a time (such as in an office building). The normally open contacts can be wired in series with the low-voltage side of the transformer (or the line-voltage side for that matter) so that whenever the timer switch opens, the heating or air-conditioning system will be inoperative. It is a seven-day timer, so the large wheel rotates once each week. The trippers on the outside of the wheel may be placed by the service technician for any ON time or OFF time each day. There are 24-hour timers available as well, but they are only useful if the same schedule is to be used every day. The seven-day timer can provide different schedules for each day, which is important if the space will be unoccupied during part of the day.

There is a fixed pointer, which should be pointing to the actual time of day when you are setting the trippers. Then, as each ON tripper passes a trip point, the normally open switch closes. At the end of the occupied period, the OFF tripper will pass the same trip point, returning the normally open switch to its open position. There is a manual override lever so that the unit may be turned on or off, regardless of the position of the trippers.

One of the problems with these timers is that they accumulate power interruption time. Suppose an office building experiences six 15-minute power outages over a two-year period. At the end of two years, the timer clock is an hour and a half behind the actual time. Your customer will call you in the morning, and by the time you get there, the unit will be running fine. Go check to see if there is a time clock tucked away someplace.

Suppose the customer tells you that his building will be occupied from 8:00 a.m. to 5:00 p.m. each day. Do not set the timer trippers for those times. If you do, and it happens to be summertime when you are setting

379

FIGURE 28–1   Mechanical timer. (Courtesy of Maple Chase Co.)

the clock, then when daylight savings time ends and nobody bothers to reset the time clock, the heating will then shut off at 4:00 p.m. just as winter is approaching. If it happens to be wintertime when you are setting the clock, when everybody "springs ahead" for daylight savings time, they will be getting to the office an hour before the air conditioning becomes operable.

Electronic Delay-On-Make, Delay-On-Break Timers

Figure 28–2 shows a "black box" timer. It is a significant departure from the electromechanical timers that use a clock motor to drive a series of gears or cams to operate a switch. It is simply a two-wire device. When a voltage is applied to one terminal, some time later, that same voltage comes out the other terminal as if an internal switch closed. Remarkably, some of these timers will operate over a dramatic range of input voltage (18–240 V), and the input voltage may be AC or DC. Some models have a fixed time delay, whereas others have a knob that can be used to adjust the length of the time delay.

A timer may be a Delay-On-Make timer or a Delay-On-Break timer. For the circuit shown, when the thermostat closes, a Delay-On-Make timer will allow 24 V to the control relay coil five minutes (for example) later. If a Delay-On-Break timer is used, when the thermostat closes, it will provide 24 V to the control relay coil immediately, except if the thermostat has been open for less than five minutes. If the thermostat has only been open for two minutes, then the Delay-On-Break timer will time out another three minutes before it allows the voltage to pass to the control relay coil. The Delay-On-Break timer is actually superior for this application (antirecycle timer) because it doesn't delay the operation of the condensing unit unless there has been an off-cycle of less than five minutes. The Delay-On-Make will provide a time delay even if the condensing unit has not run for two hours. The Delay-On-Make timer would be more appropriate for staging the starting of multiple units to reduce peak power requirements.

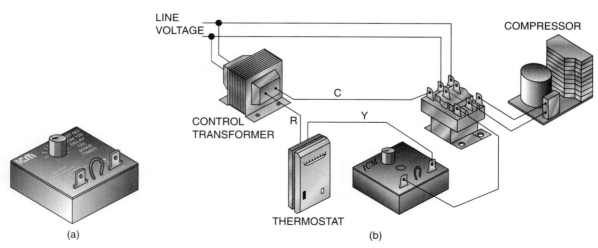

FIGURE 28–2   Delay-on-make timer.

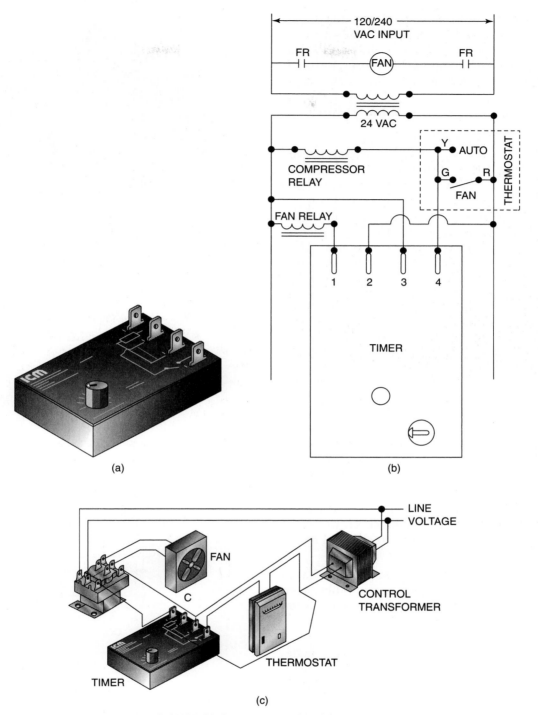

(a)

(b)

(c)

FIGURE 28–3 Postpurge fan delay timer.

A less-sophisticated model of this electronic timer uses four terminals. Voltage is applied to the two input terminals, completing a circuit. Minutes later, the voltage is available at the two output terminals to supply voltage to the contactor or relay coil.

Electronic Post-Purge Fan Delay Timer

Figure 28–3 shows a timer that provides an OFF delay, rather than the ON delay provided by the timers described earlier. When the room thermostat calls for cooling, it makes R–G, providing power to terminal 4 of the timer. As soon as power is applied to terminal 4, the power on terminal 2 (which is always available) will be applied to the fan relay coil. There is no delay during start-up. However, when the thermostat is satisfied, and it opens R–G, the power on terminal 2 will continue to be applied to the fan relay coil. The fan will run during the time delay period, and then it will shut down when power is no longer provided from terminal 1.

ELECTRONIC MONITORS

An electronic monitor can be used for a number of different functions:

1. If the available line voltage is not within acceptable limits, the monitor will prevent the compressor contactor from becoming energized.

2. If, on a three-phase system, there is an unacceptable voltage imbalance between the phases, the compressor contactor will not be energized.

3. If the contactor wants to short cycle based on input from the operating or safety controls, the monitor will provide a delay on break period.

Figure 28–4 shows a line monitor for a single-phase system. Figure 28–5 shows a monitor for a three-phase system.

COLD CONTROLS

The term **cold control** (Figure 28–6) is jargon in the refrigeration industry for line-voltage thermostat in a small refrigerator. There are two general types:

1. Cold controls that sense coil temperature
2. Cold controls that sense box temperature

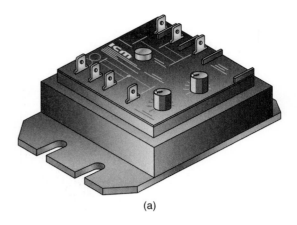

(a)

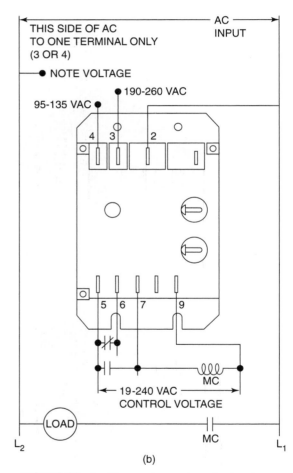

(b)

FIGURE 28–4   Electronic monitor-single-phase.

The difference is important, because although the two types may look identical, they are not interchangeable. The difference lies in their cut-in and cut-out temperatures. For example, a cold control that senses box temperature for a 38°F box might have a cut-in temperature of 39°F and a cut-out of 37°F. However, for a cold control that senses coil temperature, when the

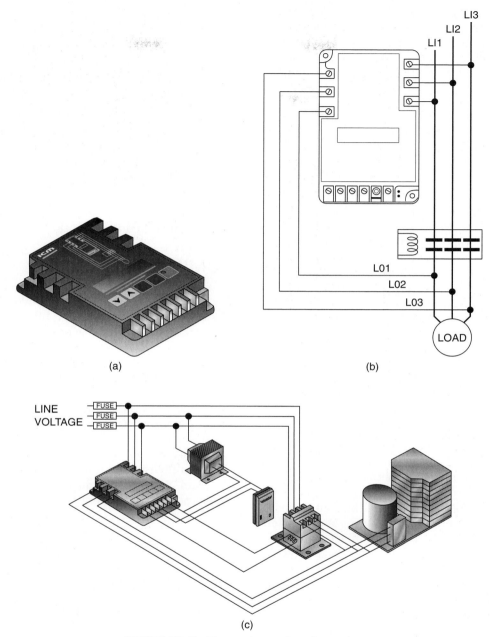

(a)

(b)

(c)

FIGURE 28–5   Electronic monitor-three-phase.

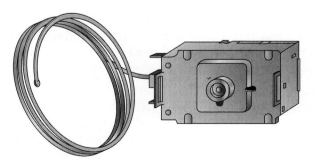

FIGURE 28–6   Cold control.

compressor is running, the coil temperature will be much colder than the box. This type of cold control for the same 38°F box might have a cut-in setting of 39°F, but a cut-out of 0°F to −15°F, depending on the design of the coil operating temperature. A cold control that senses coil temperature may be replaced with a box-temperature-sensing cold control if the sensing element can be placed where it senses box temperature. However, the reverse is not true, because you would not know the actual design coil temperature of the unit you are servicing.

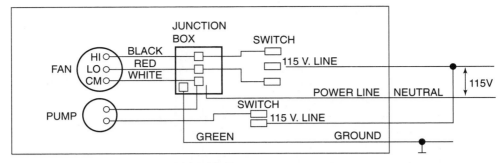

FIGURE 28–7   Evaporative cooler.

When either of the preceding two types of cold control is adjusted by the customer, it lowers both the cut-in and cut-out temperature. This presents a risk of the customer setting the thermostat for a too-low temperature—so low that the ice that forms on the evaporator coil during the "freeze" cycle does not completely melt during the "off" cycle. There is another type of cold control for refrigerators called a constant cut-in that remedies this problem. Regardless of how the customer adjusts the control, the cut-in setting remains fixed at 38°F to 39°F. This ensures that all ice will be melted off the coil before the compressor is allowed to restart.

EVAPORATIVE COOLERS

Evaporative cooling is a low-operating-cost alternative to air conditioning. It works by passing outside air over a wet pad. As the air passes through the pad, it is cooled (but also humidified) by the evaporation of the water from the pad. The cool, moist air is then supplied to the occupied space. Figure 28–7 shows a wiring diagram for an evaporative cooler. The switch on the 115-V line (unlabeled Hot) supplies power to either the black or the red wire on the motor that drives the fan that draws the outside air through the pad and delivers it to the room. The common side of the fan motor is attached, through the junction box, to the other side of the power line (the unlabeled neutral line). A separate 115-V supply comes into a different switch and supplies power to the recirculating pump motor. The other side of the pump motor is connected to the neutral through the junction box.

The Honeywell R8183 control panel (Figure 28–8) is a packaged control scheme that can be used to control the operation of an evaporative cooler. It contains three relays (fan on/off, fan high/low, and pump on/off) and a 20 VA transformer. The pump relay operates the water pump and provides wetting of the evaporator pad. Two fan relays control fan operation and fan speed. All R8183 functions are remotely controlled through the thermostat sub-base.

HUMIDIFIERS

Figure 28–9 shows a typical residential humidifier. On this type of system, the furnace motor pushes air through the furnace heat exchanger, where it is warmed and delivered to the room. But a small portion of the supply air is diverted to flow to the humidifier, where it passes over a wetted pad and then back to the return air side of the furnace. A small motor in the humidifier causes the wetted pad to rotate through a pan of water, where the pad is continuously wetted (the level of water is maintained by means of a water supply through a mechanical float valve). Figure 28–10 shows the wiring diagram to control this system. The furnace fan runs under two different circumstances:

1. The normally open switch on the blower relay is closed. The furnace fan runs on high speed. Or,
2. The blower relay is not energized, but the furnace is running and the bonnet switch (labeled as fan switch) is closed.

There is a humidistat in the room that only closes when the room humidity is lower than the set point of the humidistat. When the humidistat closes, it causes the humidifier motor to operate (in this case, it is a 24-V motor.) When the humidistat is satisfied, the wetted pad stops rotating, and it dries out, causing the humidification to stop.

CONDENSATE PUMPS

Normally, the condensate that forms on the evaporator coil of an air conditioner is allowed to drain by gravity. However, sometimes, a gravity drain is not

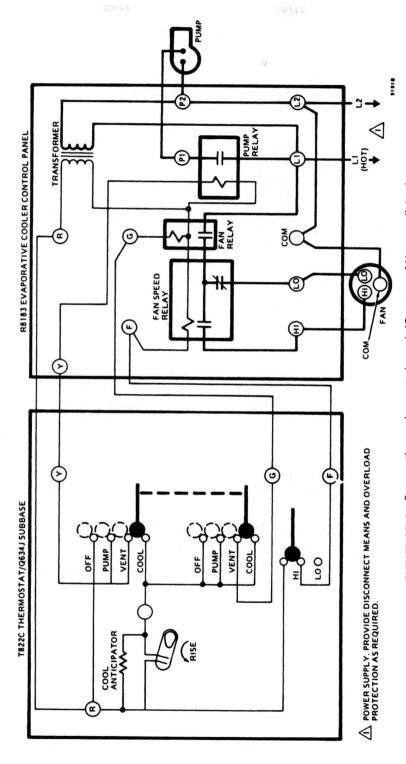

FIGURE 28-8 Evaporative cooler control panel. (Courtesy of Honeywell, Inc.)

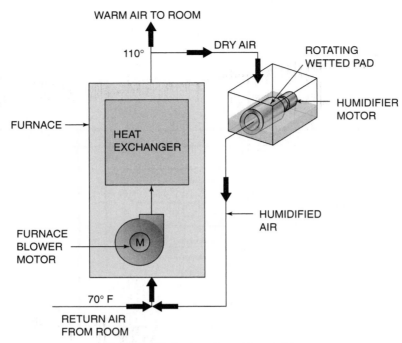

FIGURE 28–9 Residential humidifier.

possible. For example, if air conditioning is being added to a central heating system, and the furnace is located in a closet area in the middle of the house, you cannot install a gravity drain to the outside. In that event, the condensate is allowed to drain into a condensate pumping unit (Figure 28–11). This unit has a level control that operates a pump motor. When the tank is almost full, the level switch closes, the pump motor starts, and the unit pumps the water out of the tank. When the tank is emptied, the level switch opens and turns off the pump.

In addition to the preceding, there is another level switch that is used as a safety switch. While the main level switch closes on a rise in level, the safety switch opens on a rise in level. The set point of the safety switch is slightly higher than the main level switch, and if everything works normally, the safety switch never operates. However, if the tank level rises above the set point of the main level switch due to a failed pump or a clogged discharge line, the safety switch will open. The safety switch is not wired to anything in the condensate pump unit (dry contacts). The technician wires the safety switch in series with the compressor contactor. So, if the level gets too high, the air conditioning will not be allowed to operate, thus preventing a flood below the evaporator.

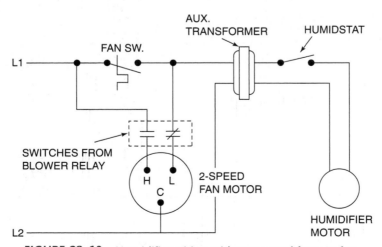

FIGURE 28–10 Humidifier wiring with two-speed furnace fan.

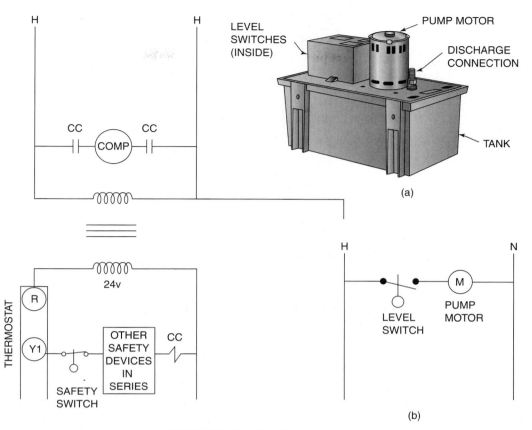

FIGURE 28–11   Condensate pump.

MODULATING DAMPER MOTORS

Figure 28–12 shows a modulating motor that can be used to control damper positions, valves, or a series of switches. Unlike a conventional motor that rotates to turn a fan blade or a compressor, this motor shaft only rotates through a limited arc (one-quarter to less than one-half of a turn). The electronic damper motor operates on a principle of a circuit called a Wheatstone bridge (Figure 28–13). In this circuit, when the switch is closed, if the resistances are all equal, the potential at C equals the potential at D. The net potential difference between C–D is zero, and there is no flow of electricity through the galvanometer (G). However, if the resistance of any one leg is changed, the bridge will become unbalanced, and there will be a current flow between C–D. This current flow is used to drive the damper motor.

Typically, the resistors that form one part of the bridge are located in a temperature sensor, and the resistance changes in response to temperature. The rotational movement of the damper motor operates a device in response to this temperature change. Internally, the rotational movement also

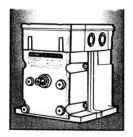

FIGURE 28–12   Modulating damper motor.

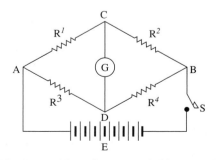

FIGURE 28–13   Wheatstone bridge circuit.

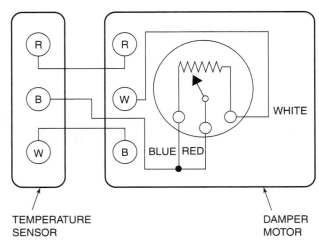

TEMPERATURE
SENSOR

DAMPER
MOTOR

FIGURE 28–14   Damper motor responds to temperature change.

changes the resistances internal to the damper motor, so that the motor reaches a new electrical balance point, but with the damper motor shaft in a new position. In this way, the damper position is proportional to the temperature input from the sensor. Figure 28–14 shows a typical wiring connection diagram for a temperature sensor and an electronic damper motor. As the sensed temperature rises, the damper motor shaft rotates in one direction. As the sensed temperature falls, the damper motor moves in the other direction.

SUMMARY

The controls and systems used in air conditioning and refrigeration are many and varied. It is the technician's responsibility to learn as many of them as possible.

PRACTICAL EXPERIENCE

Obtain manufacturers' information on mechanical and electronic timers. Become as familiar as possible with the controls and their uses as described by the manufacturers.

REVIEW QUESTIONS

1. Suppose the owner told you that the office was going to operate from 7:00 a.m. to 6:00 p.m. It's daylight savings time. At what hour should you place the ON tripper and the OFF tripper so that the heating and air conditioning will be operable for the least number of hours, and yet cover the occupied hours year-round?

2. Figures 28–9 and 28–10 refer to residential humidifier circuits used in heating applications. If the same system with an evaporator above the heat exchange were used for cooling, which fan speed would provide for dehumidifying?

3. What is the purpose of the time delay relay in Figure 28–2?

4. Referring to Figure 28–9, when the room humidity reaches the selected humidity, what action prevents the humidity from increasing above the selected value?

5. Referring to Figure 28–10, what two conditions must be met for the fan motor to operate at low speed?

6. Referring to Figure 28–11, it's a hot, humid day. The pump motor fails. What complaint would you expect from the customer?

7. The compressor motor in Figure 28–11 operates at 220 volts. At what voltage does the pump motor operate?

8. Once the pump motor starts in Figure 28–11, what shuts it off?

9. What is the purpose of a damper in an air-delivery system?

10. Referring to Figure 28–13, what condition must exist in the bridge if the damper motor is to rotate?

UNIT 29

REPAIR STRATEGIES

OBJECTIVES

After completion and review of this unit you should be familiar with repair or replace considerations involving:

- Motor bearings.
- Multispeed motors.
- Compressor/condensing units.
- Fan cycling switches.
- Overloads.

REPAIR OR REPLACE?

Whenever customers are faced with a major repair such as a compressor replacement, they will have to consider whether to go ahead with the repair or simply discard the entire unit and replace it.

What's It Worth?

Before undertaking any repair, the cost of the repair will have to be compared with what the unit or appliance will be worth after it is repaired. For example, if a 10-year-old residential refrigerator needs a new compressor, the cost of the repair will probably exceed the value of the refrigerator when you are done. It would not make sense to repair it. For less money, the customer could simply replace it with an equivalent model that is operable, or they have the option to upgrade to a new unit. For this reason, many technicians that work on relatively inexpensive units such as domestic refrigerator/freezers or window air conditioners will restrict their repairs to replacing only components that do not require opening the refrigeration system (switches, heaters, timers, thermostats, etc.). When opening the refrigeration system is required, the cost of repair is very likely to approach or even exceed the value of the appliance.

DIRTY MOTOR BEARINGS

It is common to find a motor that has failed mechanically, but is alright electrically. Often, motors in furnaces, evaporator blower-coil units, and condenser fan motors will operate for years without receiving any maintenance. Some of these motors have provisions for oiling the bearings (Figure 29–1), whereas others do not. The "permanently lubricated" motors are ones that use brass bushings instead of bearings. Brass is a rather "slippery" material when used as a bushing and requires no further lubrication. However, all bearings can become dirty after time and present resistance to the free rotation of the motor.

The rotating part of the motor (rotor or armature) is supported by either bearings or bushings

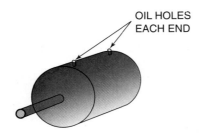

FIGURE 29–1   Oil ports on a motor.

(Figure 29–2). The symptoms of dirty motor bearings (or bushings) are as follows.

1. With the motor de-energized, the fan does not rotate freely when you give it a spin with your finger.
2. When you energize the motor, it gets warm, but it will not rotate.

When a motor has failed due to dirty bushings or bearings, you have two choices:

1. Replace the motor.
2. Attempt to repair the motor.

Replacing the Motor

Motor replacement is the safest choice for the service technician. You should explain to the customer that although the motor might be repairable, you don't recommend it because you might waste your time (and the customer's money) in an attempt that is sometimes futile. Further, even if you are successful, there is no way to know if the motor bearings (or bushings) have been damaged, or how long they will last. If you don't explain this to the customer and you simply say that the motor failed and needs to be replaced, you run the risk of incurring some bad

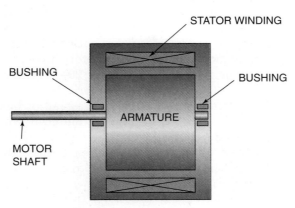

FIGURE 29–2   Motor bushings.

customer relations. Consider the following scenario. You install a new motor, and you leave the failed motor with the customer (which you should, unless you specifically ask if they want you to discard it). They have a "mechanic" friend who later inspects the motor and repairs it by simply lubricating it (see following). The customer thinks you "ripped them off" because you replaced a "perfectly good motor" when all it needed was a few drops of oil! Obviously it's better to discuss the options with the customer at the beginning to avoid this kind of misunderstanding.

Repairing the Motor

Most times (but certainly not always) a motor that has dirty bushings or bearings can be restored. This is done by lubricating it liberally with a light-weight oil such as WD-40, until it is clean and rotates freely. Remove the motor if required to get free access to it. Remove the fan blade, or whatever other device it is driving.

Try to "wiggle" the shaft. There should be very little "play" or movement. A lot of play indicates that the bushing or bearing is worn, and repair should not be attempted. Note that for all except very large motors, it is usually more cost effective to replace an entire motor rather than to replace the bushings or bearings.

With the motor de-energized, rotate the shaft by driving it with a variable-speed drill motor (Figure 29–3). While the rotor is turning, spray the WD-40

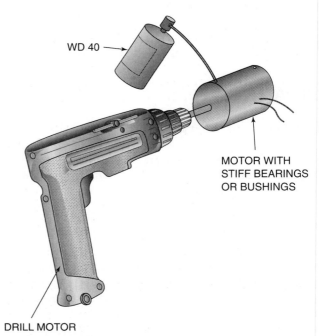

FIGURE 29–3   Restoring a stiff motor with external lubrication ports.

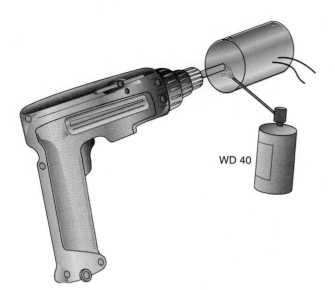

WD 40

FIGURE 29–4   Restoring a stiff "permanently lubricated" motor.

into the oil ports provided, or if no ports are provided, spray the lubricant directly into the shaft-bushing area (Figure 29–4). Some technicians will do this while the motor is energized and rotating (having started it rotating with a manual spin). This can be dangerous. Although most times this is a successful approach, there have been cases where a spark from the motor ignited the WD-40 and caused the entire can to explode.

It may take five minutes or more of rotating the motor and adding lubricant until it finally rotates freely. When (and if) the motor loosens to where it rotates freely, lubricate the motor with oil of the recommended viscosity. WD-40 is a great cleaner, but is too light to provide adequate lubrication for most motors.

Before you undertake this repair, you should explain to the customer what is wrong. Recommend that they replace the motor, but give them the option of attempting a repair. Do not place yourself in a position of recommending a repair, because if it doesn't work, the customer will only remember that it was your idea. Explain that even if the repair does work, you cannot say how long it will last. And lastly, when you make out your service ticket, write, "Recommended motor replacement. Customer declined." This way, if the motor does fail one month later, the customer will call and say, "You were right, the motor should have been replaced." If you don't write that you recommended replacement, when the motor fails, the customer will call you and say, "You just fixed this motor a month ago and now it's broken again." No amount of verbal explanation will serve your purpose as well as having the customer sign a ticket that states "Recommended motor replacement."

Even though occasionally a bushing/bearing lubrication repair will not work, sometimes taking the gamble is a good choice. This is particularly true on some specialty motors on blower-coil units inside walk-in coolers or freezers. These are sometimes small motors, but can be very expensive to replace. Another situation where repair makes especially good sense is when a replacement motor is not readily available and you must get the system running to avoid a costly loss of product.

MULTISPEED MOTORS

On residential furnaces, packaged rooftop units, and many other applications, the evaporator fan motor (sometimes referred to as the blower motor) is sometimes supplied with several speeds, some of which are unused (Figure 29–5). This is done so that the manufacturer can use one standard motor for several different size models and simply select the appropriate speed to match the heating or cooling capacity of the model on which it is installed.

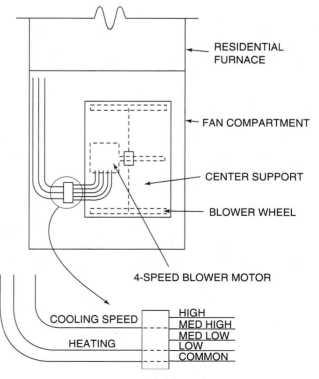

FIGURE 29–5   Multispeed motor.

Suppose you have determined that the low-speed winding has failed. Certainly, you can replace the motor to effect a repair. But you may want to try removing the wire that is energized on heating from the low-speed connection and connect it to the medium-low speed, which may still be operational.

What are the drawbacks to using an incorrect speed? Whenever you go to a higher speed, there will be more noise. Run the unit on that speed, and ask the customer if the increased noise level is acceptable. Higher fan speed will also consume more electricity, but it also improves heating or cooling efficiency, so the overall increase is usually not significant. Also, at higher speed, a cooling system will not produce as much dehumidification in the room, but this is usually acceptable in air conditioning applications. When a heating system is operated at higher speed, the discharge air temperature will be slightly lower, and that, in combination with the higher airflow, may be perceived by the occupants to be a draft. When going to a lower-than-design motor speed, all the effects are the opposite of those described. The efficiency and the capacity of the heating or cooling system will be reduced. On cooling, the lower airflow may not be sufficient, and it could cause the evaporator coil to freeze up. But this can be checked by measuring the operating pressures at the lower speed. On heating, the higher discharge air temperature and lower airflow could create problems with temperature stratification, where the difference in temperature between the air near the ceiling and the air near the floor becomes exaggerated.

Having described all the potential drawbacks, most times the effects of using a different fan speed are imperceptible to the customer, and the savings realized in not having to replace the motor will be appreciated.

REPLACE COMPRESSOR OR CONDENSING UNIT?

A condensing unit consists of a compressor, condenser, condenser fan, controls, and sometimes a receiver. When the compressor fails, it can be replaced. But you should also get pricing from the wholesaler on a new condensing unit. Although it will be more costly to the customer to replace the entire condensing unit, it may be attractive for the following reasons.

1. Replacement of the entire condensing unit will take you less time than replacement of the compressor alone. The reduction in labor cost will partially offset the increased material cost.

2. There is a risk that the compressor failed due to a restriction elsewhere in the condensing unit. You cannot know this before you replace the compressor, and you avoid this risk by replacing the entire condensing unit.

3. The newer model of condensing unit may be higher thermal efficiency than the existing unit and will offset the higher equipment cost through lower operating cost.

4. The newer model of condensing unit may be designed to operate with one of the newer "ozone friendly" refrigerants.

5. The risk of future failure of the condenser fan, start relay, contactor, capacitors, and so forth etc. is significantly reduced when they are all changed out together with a new condensing unit.

REPLACE OR BYPASS THE SWITCH?

When a switch fails in the open position, you can make the system run by placing a jumper wire across the switch. If the jumper wire makes the system run normally, the switch should be replaced. However, there is another option. The switch may sometimes be bypassed. This suggestion will alarm some people. Some technicians believe (correctly so) that the safest strategy is to leave the unit in the same condition as the manufacturer designed it. Some will be afraid to modify the system for fear of incurring legal liability. This also is a valid reason to reject bypassing a switch. But despite these concerns, with the willing and informed consent of the customer, you may wish to sometimes bypass a switch, depending on what it does. Following are some examples of where it might make sense.

Condenser Fan Cycling Switch

Suppose that you have determined that an ice machine is not making ice because the condenser fan cycling switch (Figure 29–6) has failed in the open position. You know that the fan cycling switch has been provided by the manufacturer for low-ambient operation. The manufacturer does not know whether the unit will be installed indoors or outdoors, so the fan cycling switch is installed just in

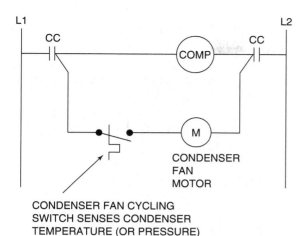

CONDENSER FAN CYCLING
SWITCH SENSES CONDENSER
TEMPERATURE (OR PRESSURE)

FIGURE 29–6   Fan cycling for a condenser fan motor.

case the unit is installed outdoors. If yours is an indoor installation, the condenser fan cycling switch is really not crucial, and it may be bypassed. Indeed, many ice machines are supplied from the factory without a fan cycling switch.

Motor Speed Selector Switch

Some older furnaces and other applications have a manual switch that the owner can use to choose a low-speed or a high-speed operation for a fan motor (Figure 29–7). Suppose you find that the motor will

not run because the speed selector switch is not making contact. Inquire if the owner actually uses the speed selector switch (most owners will be unaware that it even exists). If the speed selector switch is never used by the owner, simply remove it, choosing one motor speed to use and capping off the unused speed.

Internal Overload

You have arrived at the job site and determined that the condenser fan run capacitor has failed, and it caused the condenser fan to fail. That, in turn, caused the compressor to go out on internal overload, and it will not reset.

The internal overload has externally accessible wiring (Figure 29–8). The internal overload switch cannot be replaced. You can replace the entire compressor, but that will be very expensive. What is the

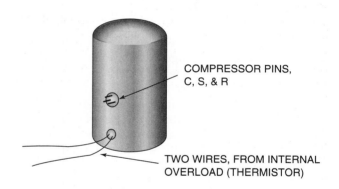

COMPRESSOR PINS,
C, S, & R

TWO WIRES, FROM INTERNAL
OVERLOAD (THERMISTOR)

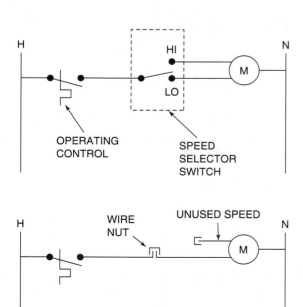

FIGURE 29–7   Removal speed selector switch.

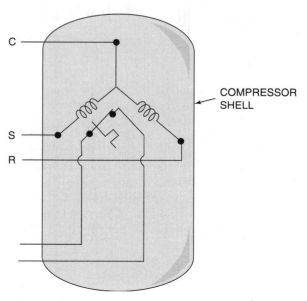

COMPRESSOR
SHELL

FIGURE 29–8   Internal pilot-duty overload.

COMPRESSOR

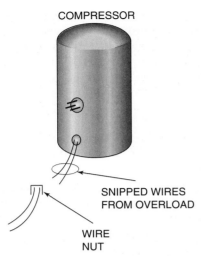

SNIPPED WIRES
FROM OVERLOAD

WIRE
NUT

FIGURE 29–9 Compressor with
bypassed overload.

risk if you bypass the internal overload by cutting
the wires leading to it and placing them together in
a wire nut (Figure 29–9)? Obviously, you will lose
the protection that is provided by the overload. If
the condenser fan fails again for any reason, it may
cause the compressor to fail also. But is that so bad?
The cost of replacing the compressor at that time will
be the same as replacing it today, so there is nothing
lost in delaying that repair. But the advantage in bypassing the overload is that maybe the condenser fan
will never fail, and the unit will be replaced entirely
after it has reached the end of its useful life. In that
case, the customer will have avoided the expense of
replacing the compressor. The risk that the customer
assumes in bypassing the overload is small in comparison to the potential benefit.

Retrofit/Upgrade?

Sometimes a needed repair presents an opportunity
for upgrading equipment that would otherwise not
be justified. For example, suppose that a customer
with a standing pilot furnace asks you if it should
be upgraded (modified) to an electronic ignition system (see Unit 22 on electronic ignition controls for
gas-fired furnaces). In truth, the savings that will
be realized by turning off the pilot flame when room
thermostat is satisfied will probably not be sufficient to justify the expense of retrofitting the furnace. However, suppose you have diagnosed that
this same standing pilot furnace has a failed gas
valve. If you have to replace the gas valve anyway, it
may now be worth the *incremental* cost to upgrade.
Only the difference in cost between the gas valve

replacement and the retrofit must be justified by
the savings.

UNIVERSAL REPLACEMENT PARTS

Once the service technician has completed troubleshooting, there is usually an electrical part that
needs to be replaced. If you are working for a company that provides service exclusively on one brand
of equipment, chances are that you will have most
of the exact replacement parts that you need. However, many technicians will need to be prepared to
perform repairs on any brand of equipment. There
are three strategies for having repair parts on hand:

1. Carry as many different parts as possible
   on your truck. You will need a very large
   truck, you will likely spend a lot of money
   on gas, you will have a lot of money invested
   in your inventory, and you will probably
   need an inventory system to help you find
   the exact replacement part that you need.

2. Whenever you need a part, go to the whole-
   sale parts house to purchase the exact replacement. You will spend a lot of time on
   the road.

3. Carry parts on your truck that can be used
   for many different applications. You will
   avoid many (not all) trips to the wholesaler
   while minimizing the number of parts that
   you need to stock.

 Following are some suggestions for parts
to carry:

1. A 40 VA transformer with a 120/208/230 V primary and a 24 V secondary. This will replace
   20-VA and 40-VA transformers on residential
   and small commercial air conditioners.

2. A 115/230 V reversible motor in 1/4 and 1/3
   HP sizes. These will replace many evaporator blower and condenser fan motors.

3. A cold control (box thermostat) set at 35°F
   and another set at 0°F, with a differential of
   less than 5°F. Many cold controls sense
   evaporator coil temperature, and the cut-in
   and cut-out temperatures are specific to the
   design of that particular box. However, the
   35°F and 0°F cold controls can be installed
   to sense box temperature instead of coil
   temperature and therefore can be used on
   all coolers and freezers.

SUMMARY

Many service calls require that a decision be made whether to repair or replace components. In many cases, the customer is involved in the decision-making process. As far as return call prospects are concerned, it is always better to replace with a new component. Your reputation might depend on the number of return service calls you have to make. As your experience increases, your ability to make repair/replace decisions will improve.

PRACTICAL EXPERIENCE

1. Practice connecting a variable-speed drill motor to a small motor and rotating the armature at low speed.
2. Locate compressors in the shop that have internal overloads. Check the overloads with an ohmmeter.

REVIEW QUESTIONS

1. When working on relatively inexpensive units such as a refrigerator or window air conditioner, it is often impractical to do repair work that requires opening the refrigerant system.

   T_____   F_____

2. On a home service call, you find a blower motor in a furnace that does not rotate freely. The only procedure is to replace the motor.

   T_____   F_____

3. Is it possible that a multispeed motor will operate at a higher speed even though it has failed at the low speed?

4. In cooling applications, will the humidity be higher or lower if the fan motor is operated at low speed? _____

5. In heating applications, what might the customer complaint be if the blower motor is operated at high speed? _____

6. Is it practical to bypass a failed fan-cycling switch of an icemaker?

7. A speed-selector switch of an older furnace has failed in the open position. You do not have a switch that will mechanically fit in the system. Is it practical to hardwire the motor to operate at a single speed?

8. Is it practical to bypass a compressor internal overload that has failed?

9. When carrying replacement transformers on your truck, it is better to carry a 20-VA transformer rather than a 40-VA transformer.

   T_____   F_____

10. In a heating system, it is dangerous to operate the blower at a high speed.

    T_____   F_____

UNIT 30

TESTING AND REPLACING COMMON DEVICES

OBJECTIVES

After completion and review of this unit, you should be familiar with test and repair methods involving:

- Manual switches.
- Automatic switches.
- Cutouts.
- Cycling switches.
- Low-voltage transformers.
- Overloads.
- Circuit breakers.
- Contactors.
- Black boxes.

After you have completed your troubleshooting using voltage, you will want to retest the individual device that you have identified as the problem. This serves two purposes.

1. It confirms your troubleshooting conclusion, eliminating the occasional "brain failure" experienced by all technicians at one time or another.

2. It identifies the specific mode of failure for the component. This is sometimes required for warranty replacement of failed parts.

MANUAL SWITCHES

Usually, you have determined that a manual switch has failed because you have read voltage across it, indicating that it is open. Remove the manual switch from the circuit. Using your ohmmeter across the terminal contacts, you should measure almost zero ohms when the switch is supposed to be closed and infinite ohms when the switch is supposed to be open. If you measure infinite ohms across the switch when it is in the closed position, the switch has failed.

When you replace a manual switch, check the "amp" rating. This is the maximum current that the switch contacts can safely carry. Also, check your actual circuit volts. The replacement switch *amp rating* must be the same or higher than the amp rating of the failed switch. The replacement switch *volt rating* must be at least as high as the actual circuit volts. There is no electrical problem if the replacement switch has a much higher amp or volt rating.

AUTOMATIC SWITCHES

Automatic switches are those that sense something (pressure, temperature, humidity, flow, level, etc.) and open or close a set of contacts when a certain "set point" value of the sensed variable is reached. A voltage reading across the switch terminals indicated that the switch is open. After the switch has been disconnected from the circuit, the switch contacts can be tested again with an ohmmeter to determine if they are open or closed. However, with an automatic switch, you also need to determine if the switch should be open or closed while you are testing it.

High Pressure Cut-Out

The high pressure cut-out HPC is an automatic switch. When it fails, it usually fails in the open position. If it failed in the closed position, and there were no other trouble, there would not have been a service call. With the power off, measure the resistance of the contacts. They should be closed (zero ohms). If contact resistance is high or infinite, the HPC has failed. Only after you are convinced that the HPC has failed do you remove the pressure sensing line from the system

Condenser Fan Cycling Switch

The condenser fan cycling switch closes when the high pressure rises high enough that the condenser fan is needed. When the compressor is not running, it should be open because the high-side pressure is too low to close it. If the condenser fan cycling switch fails, it will usually fail in the open position, causing the condenser fan to not run and maybe causing the compressor to cut out on the HPC. If the condenser fan switch fails in the closed position, there will probably be no service call.

As with manual switches, when replacing an automatic switch, the switch contacts must be rated for amps at least as high as the failed switch and for volts at least as high as the actual circuit voltage.

Transformer

When troubleshooting why a circuit doesn't work, you may eventually be led to the transformer, and you find that you are not getting 24 V from the transformer secondary. You must then measure voltage at the transformer primary. If the correct voltage (usually 115 V or 230 V) is present, then the transformer has failed. From these readings, many novices would mistakenly conclude that the secondary has failed.

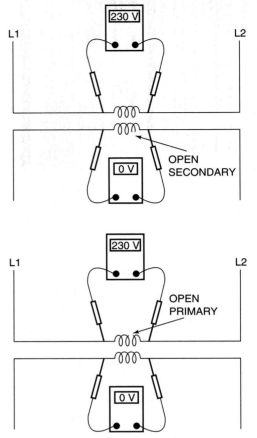

FIGURE 30–1   Open primary gives same readings as an open secondary.

But, in fact, either winding could have failed. As shown in Figure 30–1, with either winding open, you will get the same set of voltage readings.

Sometimes you need to know *which* winding has failed. On warranty replacements, you normally need to fill out a card describing the failure. The manufacturer who is receiving return of the failed part would much prefer a description that says "primary coil open" or "secondary coil open" than a simple "broken" or "no 24 V."

In order to determine which coil has failed, you need to remove the transformer from the circuit and measure the resistance of the secondary and the primary. It is possible for either winding to be either shorted or open, but you will usually find that the

ADVANCED CONCEPTS

Manufacturers request return of failed parts so that the failure mode can be analyzed by their engineers. If they detect a pattern of failures, they will take steps to modify the design to improve the reliability of the product line.

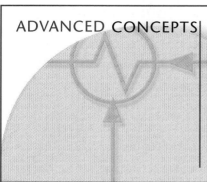

ADVANCED CONCEPTS

The VA rating is the product of the secondary voltage times the maximum number of amps that can be supplied at the secondary voltage. It can be expressed as:

$$VA = Volts_{sec} \times Amps_{sec}$$

The preceding 20-V transformer would be capable of supplying slightly less than one amp at 24 V, as follows:

$$20\,VA = 24\,V \times Amps$$
$$Amps = 20/24 = .83$$

failure is an open coil, and more often than not it is the primary rather than the secondary that has opened.

When replacing the transformer, you must check three numbers: the primary voltage, the secondary voltage, and the VA rating. The primary voltage and secondary voltage must match the failed transformer. The VA rating of the replacement transformer must equal or exceed the failed transformer.

Transformer VA Rating

The **VA (volt-amps) rating** is a measure of how much power the transformer can supply at the rated voltages. In other words, a transformer with a 115 V primary and a 24 V secondary and rated for 20 VA might be sufficient to supply enough power to supply two contactors. However, if the circuit required enough 24 V power to supply four contactors, a 40 VA transformer might be used.

OVERLOADS

An overload may be of the **line-duty** or the **pilot-duty** type. A line-duty overload (Figure 30–3) is used on compressors in small systems. It is wired in series with the common terminal of the compressor, and

therefore it carries the total current draw of the compressor. If the current draw of the compressor exceeds the rating of the overload, the switch opens. On larger systems, a pilot-duty overload is used (Figure 30–4). The sensing portion of the pilot-duty overload carries the compressor current, but the switch portion of the overload does not. The switch on the pilot-duty overload is wired in series with the compressor contactor coil. On some air-conditioning systems, a thermal pilot-duty overload is used (Figure 30–5). With this system, the overload sensing element is a thermostat that is imbedded into the stator windings of the compressor. If the windings draw too much current, the winding temperature will rise above the set point of the overload switch. When the overload switch opens, the compressor contactor is de-energized, and the compressor stops.

When you find an open overload switch (indicated by a voltage measured across it), there are two possible failures:

1. The overload has failed.
2. The compressor was drawing too much current, and that caused the overload to trip. When it reset, it tripped again, and the cycle continued until the overload gave up.

MORE TRANSFORMER ADVANCED CONCEPTS

Even though a transformer secondary is rated at 24 V, its actual output may be a few volts higher or lower. As the load on a transformer increases (more 24 V loads are added in parallel with existing loads), its output voltage decreases. The larger the load, the more a transformer's voltage output will drop below its open (no load) circuit rating. Most systems will work over a voltage range that varies some from the rated input voltage. However, an electronic control can burn out if the circuit voltage exceeds the maximum for that control. The standard for acceptable output voltage variation is set by the National Electrical Manufacturers Association (NEMA). It is shown in Figure 30–2. At rated load, the output voltage should be 23 to 25 V. At minimum load, it may be as high as 27 V.

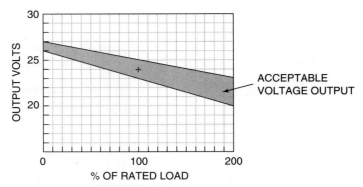

FIGURE 30–2 Voltage variation for a 24-V transformer.

If you don't check, you could wind up replacing the overload, only to find that the replacement overload also trips when the compressor runs. Before you replace the overload, you must check to see if the problem is simply a defective overload, or if there is also a defective compressor. You can do this by temporarily bypassing the overload with a jumper wire. Start the compressor and measure the current draw. *Do not run the compressor for more than a few seconds with the overload bypassed.* Compare the amp draw of the compressor with the rated full-load amps. If the compressor amp draw is below the FLA, you can simply replace the overload. If the compressor draws more than the rated FLA, further investigation is required to determine why. The compressor may be low on oil, the condenser may be dirty, there may be a refrigerant restriction, etc. If the compressor does not start at all, there is another problem besides the open overload.

Universal Replacement Overload

Although it is always best to replace overloads with an exact replacement, sometimes you are unable to determine the model of compressor or overload that you are working on. In this case, you can use a universal replacement based on the locked rotor amperage of the compressor. To find the locked rotor amps, do the following (the compressor motor windings must be at ambient temperature to use this method).

1. Disconnect the power supply and temporarily bypass the defective overload with a jumper.
2. Temporarily remove the wire from the start terminal of the compressor.
3. Place a clamp-on ampmeter around the wire to the run terminal or the common terminal.
4. Momentarily reconnect the electrical power and read the ampmeter.
5. If the unit does not have a start capacitor, multiply the amp reading by 1.33 to obtain the locked rotor amps.

FIGURE 30–3 Line-duty overload.

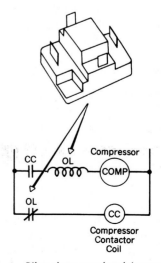

FIGURE 30–4 Pilot-duty overload (current sensing).

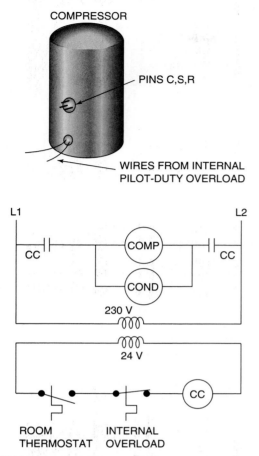

FIGURE 30–5 Pilot-duty overload (temperature sensing).

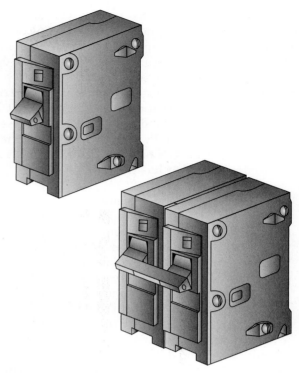

FIGURE 30–6 Single-pole and two-pole circuit breakers.

phase circuits, respectively. You must determine which circuit breaker is the one that supplies power to your unit. If, for example, you're working on a 115-V unit, you don't need to consider the circuit breakers that are ganged together in pairs or in threes.

6. If the unit does have a start capacitor, multiply the amp reading by 1.10 to obtain the locked rotor amps.

7. Select a new universal replacement overload from the locked rotor amps determined earlier.

CIRCUIT BREAKER

The circuit breaker (Figure 30–6) protects the wiring that supplies power to the air-conditioning or refrigeration unit you are troubleshooting. If you measure zero voltage at the unit, go to the circuit breaker panel (you might need some help from the customer to find it).

You may see individual circuit breakers, pairs of circuit breakers with their trip levers connected, or groups of three circuit breakers with their trip levers connected (Figure 30–7). These are for single-phase 115-V circuits, single-phase 230-V circuits, and three-

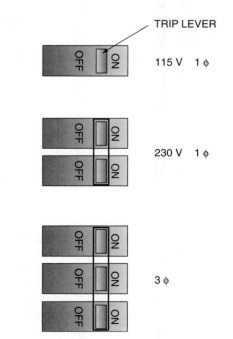

FIGURE 30–7 Circuit breaker arrangements.

Look at the circuit breakers. You will note that the trip levers are all aligned. However, if one of the breakers is tripped, its trip lever will be *slightly* out of alignment. Inspect the alignment carefully to determine which circuit breaker has tripped. If you cannot identify the tripped breaker, push each trip lever toward the ON position. Each breaker that is not tripped will feel solid. However, the tripped breaker will be "soft" and will allow you to push it a slight bit toward the ON position. That is your clue that the circuit breaker has tripped.

👉 Sometimes a technician will turn each circuit breaker off and then back on, resetting each one. In commercial applications, this can be a disaster. Many stores, for example, will have cash registers that are also computers, and they keep records of the day's transactions. By indiscriminately turning off circuit breakers, all the data could be destroyed. This is only one example of the problems you can cause by momentarily interrupting power to circuits when you don't know what's connected. You could cause significant grief to the customer and embarrassment to yourself. Don't do it.

To reset the circuit breaker, push the trip all the way to the OFF position, and then push it back to the ON position. If there is still no power to the unit, there are two potential problems:

1. There is a short in the unit that caused the circuit breaker to trip as soon as you reset it.
2. The circuit breaker is defective, and it will not reset.

Disconnect the power wiring from the unit (or trip the disconnect switch to Off if there is one). Then reset the breaker again. If it resets this time, then there is a short in the unit. If it does not reset, the circuit breaker is defective.

Testing the Circuit Breaker Electrically

All the preceding assumes that if the circuit breaker stays in the ON position, that it is supplying power to the unit. This is not necessarily so. If you remove the panel that covers the wiring connected to the circuit breakers, you will see that there is one wire connected to each circuit breaker (Figure 30–8). Using your voltmeter, place one probe on a ground

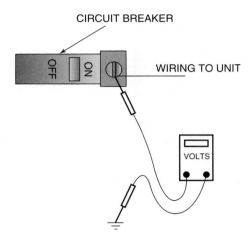

FIGURE 30–8 A voltage reading indicates that the circuit breaker is making contact.

connection, and move the other probe to each wire attached to a circuit breaker. If there is voltage present, the circuit breaker switch is closed. A zero reading between any circuit breaker and ground indicates that the switch inside the circuit breaker is not making contact, even if the trip lever indicates that it is ON.

👉 Sometimes the circuit breaker is functioning normally, but you must turn it off so that you can do some service work on a unit. How do you identify which circuit breaker to turn off? Usually, you will find that each circuit breaker is labelled on the door of the circuit breaker box, identifying what loads are served by each circuit breaker. But on some systems, the identifications have been lost or not kept up to date as the system was modified. There is a tool now available that consists of two electronic parts. One part is a transmitter. It is fastened to the power wiring of the unit that you wish to deenergize. The other part of the tool is a receiver. It is taken to the circuit breaker box, where it will provide an audible or visible signal when it is attached to the circuit breaker that matches the circuit where the transmitter is attached.

CONTACTORS

Contactors contain both a load (the contactor coil) and a switch or switches (contacts). Sometimes the contacts are visible, and you can tell from simply looking at the contactor if the switches are closed ("pulled in"). Other contactors have the switches concealed beneath a removable cover. You can either remove the cover to observe the contacts, or you can check for voltage across the contacts by measuring across L1–T1, L2–T2, and L3–T3. If you read voltage across the switch, it is open. If you measure proper voltage across the coil, and you also determine that the switches have remained open, the contactor has failed. It may be a failed coil, or the coil could be normal and the moving switch mechanism is mechanically stuck. Or it could be that the switches moved to the closed position, but because of badly pitted contacts, one or more of the switches has not actually made electrical contact.

Measure the resistance of the coil. If it has infinite resistance, it has failed. Otherwise, the failure is probably mechanical. On small contactors, you will simply replace the entire contactor. On larger contactors, the coil and the switches may be individually replaceable. However, in some cases, the extra time required to obtain and change the coil or contacts is more costly than it would be to simply replace the entire contactor with one that you stock on the truck.

When you replace a contactor, the coil voltage (24 V, 120 V, 230 V) must match. The number of poles for the replacement must equal or exceed the failed contactor. You can use a three-pole contactor to replace a two-pole contactor by simply leaving one of the poles unused. The amp carrying rating on the switches (20 a, 30 a, 40 a, etc.) must equal or exceed the amp rating on the failed contactor.

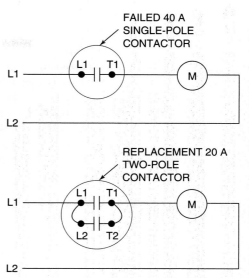

FIGURE 30–9   Using a two-pole contactor to carry twice its rated amps.

Suppose you have discovered a failed single-pole 40-a contactor. You don't have a 40-a contactor on the truck, but you do have a two-pole 20-a contactor. You can use the two individual switches in parallel to replace a single 40-a switch. Each of the two poles will carry half of the load (see Figure 30–9). This arrangement will also be found on some factory-wired units.

ADVANCED CONCEPTS ABOUT CONTACTOR CONTACTS

The current carrying capability of contacts is rated several ways, depending on the load being switched. *Inductive full-load amps* are the maximum amps a contact can switch when it is controlling an inductive load such as a motor. This rating is for continuous duty. The inductive contact rating may also be expressed in horsepower, being the approximate maximum size of electric motor that may by switched by the contact. *Inductive locked rotor amps* are the rating of maximum current for short periods of time when a motor is first started. The initial current drawn by a motor is 4 to 6 times higher at start up than the full load.

The *resistive rating* is the maximum amps a contact is capable of switching when it is controlling a resistive load such as an electric heater. The resistive load rating is normally higher than the inductive rating since the resistive load causes less arcing at the contacts during switching and inrush is very low. Sometimes the resistive load rating is also given in equivalent kilowatts of electric heating for various voltages.

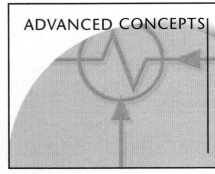

ADVANCED CONCEPTS

The principal of operation of the ohmmeter is that a battery inside the meter sends electrons out through one of the probes, through the device whose resistance is being measured, and then back through the other probe to the meter. The voltmeter converts the rate of electron flow to a resistance reading. When the ohmmeter is first attached to the terminals of the capacitor, the battery inside the ohmmeter is able to push electrons onto the capacitor plates relatively quickly. However, as the capacitor plate approaches its limit in its ability to store electrons, the electron flow out of the ohmmeter diminishes. This shows up as an increasing resistance.

CONTROL RELAYS

Control relays are, in theory, the same as contactors, with the following differences.

1. Contactors have switches usually rated for 20 a or more, whereas control relay contacts are usually rated for 18 a or less.
2. The switches on contactors are always normally open, but the switches on control relays come in many variations, including NO, NC, SPDT, DPDT, 3PDT.

The troubleshooting and replacement guidelines for control relays is the same as for contactors.

CAPACITORS

Capacitors can fail in one of four ways.

1. The plates are shorted together (shorted capacitor).
2. There is a discontinuity between the two capacitor terminals (open capacitor).
3. There is a path from one of the capacitor plates to the capacitor casing (grounded capacitor).
4. The capacitor plates do not hold the correct amount of charge.

Of course, it is not necessary to diagnose or test the capacitor if you find that it has blown itself apart, as is sometimes the case.

Many technicians use an ohmmeter (set to the highest resistance scale) to check capacitors. If the ohmmeter shows that either terminal is grounded to the casing or that there is zero or infinite resistance between the pins, the capacitor has failed. If your ohm reading between pins shows a resistance that initially falls toward zero and then reverses so that

the resistance is increasing, the capacitor is *probably* good. However, some capacitors may check out okay with an ohmmeter, but then fail when they are under load in a circuit. This resistance test on capacitors is best done with an analog ohmmeter. You can watch the needle swing to the right (toward zero ohms), and then turn around and swing to the left as the resistance increases. Early models of digital ohmmeters (and some modern models as well) are not suitable for using to test capacitor resistance. Because the actual resistance is continuously changing, some digital ohmmeters cannot "zero in" on one correct reading.

Capacitors are best checked with a capacitor tester, and they will almost always give you a correct determination of whether a capacitor is good or failed. It not only measures whether or not the capacitor stores electrons, but it also determines the microfarad capability of the capacitor. Some of the modern digital multimeters include a capacitor checking option. When set on this option, the multimeter will read out the microfarad capability of the capacitor being tested. You must still check to see if the capacitor has grounded to the casing.

When replacing a capacitor, you can replace it with one of an equal or higher voltage rating. However, the capacitance rating of the new capacitor should match (within 10%) the capacitance of the failed capacitor. Figure 30–10 shows what failures you can expect if you don't use the correct rated replacement capacitor.

Replacement Capacitor Compared to Original	Possible Results
mfd rating too low	Capacitor fails
mfd rating too high	Stalled motor, lower running amps, low motor rpm
Voltage rating too low	Capacitor fails
Voltage rating too high	No problem

FIGURE 30–10 Effect of not matching the ratings of a failed capacitor.

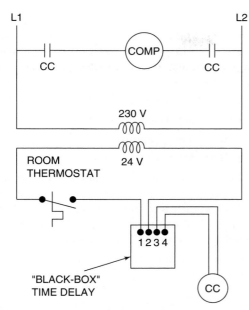

FIGURE 30–11 4-Wire electronic time delay.

ELECTRONIC "BLACK BOXES"

Each year, more and more control functions are being accomplished with electronic "black boxes." The term *black box* refers to an electronic box. We don't need to know how it works; we only need to know what it is supposed to do. For example, Figure 30–11 shows a circuit that uses an electronic time delay. The time delay is a "black box." All we know is that when voltage is applied to terminals 1–2, one minute later, the same voltage should appear at terminals 3–4. If you measure 24 V at terminals 1–2, and after one minute you measure zero volts at terminals 3–4, the black box time delay has failed, and must be replaced. Do not attempt any repairs on a black box device.

Other black box devices may have more than one input signal. The general strategy is to know what all the input signals must be and what the output signal should be. If you check all the inputs, and they are all proper, and if you are not getting the correct output, replace the box. There is usually no testing (i.e., no resistance checks) done on the box itself to determine if it has failed.

SUMMARY

A completed service call includes finding the problem, usually a failed component, and returning the system to normal operation. It is usually necessary to determine how/why the component failed to complete the proper repair. It is not sufficient to simply replace failed components. The newly installed components might fail immediately or within a short time, requiring a return service call. Failed components must be investigated to determine the cause of failure. This will lead to finding all the problems and a successful service call.

PRACTICAL EXPERIENCE

From a group of failed components, measure the resistance of the components coils, contacts, and so on to determine what the failure indication is.

REVIEW QUESTIONS

Choose all correct answers.

1. An operating air-conditioning system has a shorted high-pressure switch. During a routine service call to check the thermostat:
   a. The failed pressure switch would be found rapidly.
   b. The failed pressure switch would not be found.
   c. After the thermostat problem was fixed, the system would function.
   d. The thermostat problem could not be fixed until the high-pressure switch was repaired.

2. It has been determined that a manual switch has failed. The rating on the switch is 20 a, Volt 120. The switch may be replaced with one rated at:
   a. Amp 15, Volt 220.
   b. Amp 35, Volt 220.
   c. Amp 30, Volt 24.
   d. Amp 10, Volt 440.

3. The voltage measured across a low-voltage transformer are primary voltage (115 V); Secondary voltage (0 volts). This indicates:
   a. Normal reading, transformer okay.
   b. Transformer needs cleaning.
   c. Transformer failed; replacement needed.
   d. Transformer secondary open.

4. A condenser fan cycling switch in a normally operating system:
   a. Would have 115 volts across its contacts.
   b. Is connected in series with the compressor contactor coil.

c. Is connected in series with the condenser fan.

d. Would have zero volts across its contacts.

5. A compressor is cycling on and off. Voltage checks indicate that the overload is the electrical device causing the cycling. This indicates:

   a. The compressor has failed.

   b. The overload has failed.

   c. The compressor contactor has failed.

   d. The system requires troubleshooting.

6. In the circuit of Figure 30–5, the condenser fan is operating while the compressor motor is not operating. A possible cause is:

   a. A failed transformer.

   b. A failed compressor contactor.

   c. A failed compressor.

   d. A failed internal overload.

7. The compressor contactor in Figure 30–11 is not energizing. The voltage measured across the contactor coil is zero volts and remains zero volts for 10 minutes.

   a. The black box should be replaced.

   b. If the voltage measured between 1 and 2 is zero volts, the black box is not the problem.

   c. If the voltage measured between 1 and 2 of the black box is 24 V, the black box is the problem.

   d. The compressor contactor coil is open.

8. In the circuit of Figure 30–11, the voltage between L1 and L2 is 230 volts. The voltage between 1 and 2 of the black box is zero volts. The voltage across the thermostat is zero volts. The probable trouble is:

   a. Failed transformer.

   b. Failed compressor contactor coil.

   c. Failed room thermostat.

   d. Failed black box.

9. In the circuit of Figure 30–11, the compressor contactor is not energizing. The voltage measured between 1 and 2 of the block box is 24 volts. The voltage measured between 3 and 4 of the black box is 24 volts. The probable trouble is:

   a. Room thermostat.

   b. Compressor contactor.

   c. Black box.

   d. Compressor.

10. In the circuit of Figure 30–11, 230 volts are applied between L1 and L2. The room temperature is above the selected temperature. After a period of five to ten minutes the compressor contactor energizes, and the compressor operates.

    a. The compressor contactor contacts are stuck.

    b. The room thermostat is bad.

    c. This is normal operation.

    d. The black box is bad.

UNIT 31

DDC CONTROLS

OBJECTIVES

After completion and review of this unit, you should be familiar with some direct digital control circuit concepts such as:

- Computer-based building automation.
- Building controls strategies.
- Economizer control.
- Load shedding.
- Night purge.

SYSTEM BASICS

Direct digital controls (DDC) are a part of computer-based building automation systems. A computer is integrated into the control system as shown in Figure 31–1. A control valve is being used to control the supply air temperature form an air handler. Information from a sensor in the supply airstream is converted to a digital form and fed to the computer. Based on the information from the sensor (and possibly other sensors as well), the computer sends an output signal that is converted to an electric signal that will modulate the electric water valve. The relationship between the input signal from the sensor(s) and the output signal from the computer follows a program that is stored in the computer. From the computer terminal, an engineer or technician can reprogram the operation of the control system.

A variation of the DDC control is shown in Figure 31–2. Here, the output signal from the computer is eventually converted to an air pressure signal that varies from 3–15 psi and is used to operate pneumatic controls. Pneumatic controls have been used for many years. The operators (valve actuators or damper motors) can modulate, depending on how far the incoming air pressure can move a piston or a diaphragm (Figure 31–3). DDC controls grew in popularity for the following reasons:

1. They are more reliable and accurate than pneumatic controls, which require a well-maintained clean, dry, compressed air source.

2. The calibration of the sensors and the set point of the controller may each be adjusted from the computer terminal.

3. Very complex control logic schemes can be programmed into the computer software.

4. The controls can react very quickly, without "**hunting**." Hunting is a term used to describe a control signal that oscillates wildly around a set point before settling down. This happens with pneumatic controls if you try to set them up to respond very fast.

5. The cost over the lifetime of the control system is low, compared with pneumatics.

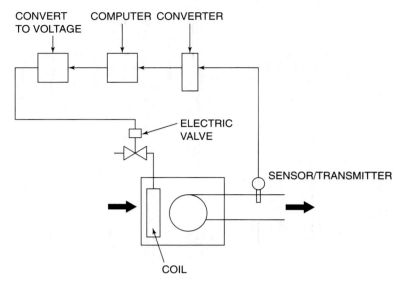

FIGURE 31–1 Computer-based control system.

6. The control scheme and logic is easily changed. It can be accomplished by merely reprogramming the software instead of changing hardware.

In addition to the normal functions of controlling temperature within a building, the DDC control system can be integrated into a total building automation system. It can sense the indoor and outdoor temperatures each morning and compare it with past building operating history in order to determine the latest possible starting time for the heating and air-conditioning equipment so that it achieves comfort by the time the occupants arrive.

At the end of the day, the building automation system can determine what is the most energy efficient time to shut down the heating and cooling systems. They can be turned off before quitting time, and the building will "coast" down to its nighttime conditions while still maintaining comfort

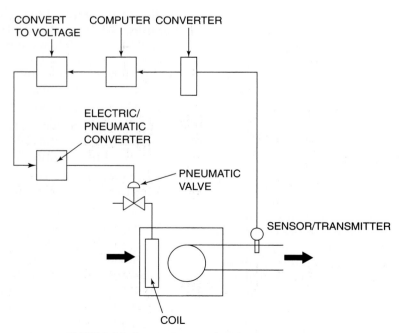

FIGURE 31–2 DDC control with pneumatic valve.

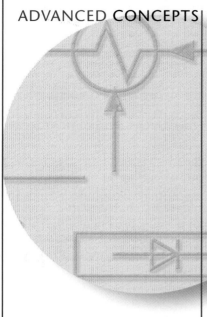

ADVANCED **CONCEPTS**

Modulating controls have several different modes of operation. In order to explain them, we will describe the liquid level control system shown in Figure 31–4. The level in the bucket (similar to temperature in a room) is being controlled at a relatively constant set point. There is a loss of water from the bucket (similar to heat loss from a room), and there is a source of make-up water (similar to heat input from a furnace) to keep the level constant. A modulating control valve gets its input signal from a control arm that moves up or down in response to the position of a float. The water flow out of the bucket is the load. The system as shown is in equilibrium. That is, the flow of water out of the bucket is 10 gpm, and the input from the water valve is also 10 gpm. Because the flow in exactly matches the flow out, the level is remaining constant. If the load changes to 8 gpm output (Figure 31–5) the level in the bucket will begin to rise. As it rises, it will move the control arm, closing off the control valve. The level will continue to rise until the input has been reduced to exactly 8 gpm, once again matching the load. In this case, the control mode is called **proportional**. That simply means that the corrective signal (movement of the control arm) is proportional to how different the actual level is from the desired level (set point). If the level is 1-in. higher than the desired level, the control arm moves a certain amount. If it is 2-in. higher than the desired level, the control arm moves twice as much. In other words, the corrective signal to the control valve is proportional to the error signal. Prior to DDC controls, pneumatic controls were simply proportional controls. But with DDC, more sophistication is available and is commonly used.

The problem with proportional control in the preceding example is that if the load decreased from 10 gpm to 8 gpm, we were stuck with a new equilibrium level that was slightly higher than our set point. This is called offset. DDC controls commonly add an additional mode to correct this offset. It is called **proportional plus integral**. After a new equilibrium has been reached, the integral mode looks at the level, compares it with the set point, and adjusts the output to eliminate the offset. This would be analogous to using a turnbuckle to adjust the length of the vertical part of the control linkage to restore the original level, even though the load is now 8 gpm instead of 10 gpm. Still another refinement in control modes available with DDC controls is called **proportional plus derivative plus integral** control, sometimes called **PID** control. With PID controls on our level control example, the turnbuckle adjustment to eliminate the offset would not be made at a constant speed. Rather, the turnbuckle would initially be adjusted quickly. But as the actual level approached the desired level, the rate of adjustment of the turnbuckle would slow down. This may seem like an absurd level of overkill, but with DDC controls, there is little or no additional cost because the control modes are determined by the software in the computer.

---

for the last occupants of the building. This is commonly referred to as providing the optimum start/stop times.

The building manager can get almost unlimited reports from the computer, including monthly energy use. These reports, coupled with other reports on where the energy is being used and other energy-saving measures described later, can save 50% or more of the energy usually consumed by a similar building without computerized controls.

BUILDING CONTROL STRATEGIES

Modulating Controls

All the controls described prior to this point in the text have been on-off controls. For example, a room thermostat has a set point of 74°F. If the room temperature rises to 75°F, the air-conditioning system turns on. When the room temperature falls to 73°F, the air-conditioning system turns off. Whenever the

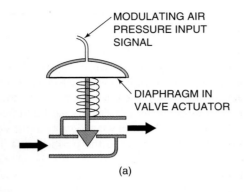

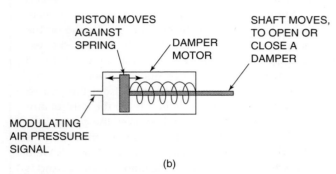

FIGURE 31–3 Pneumatic devices. (a) Normally open pneumatic valve with actuator. (b) Damper motor.

air-conditioning system operates, it operates at 100% of its capacity, and the control scheme maintains room temperature by only operating the cooling for as much time as necessary.

The first control scheme shown in this unit (Figure 31–1) would be difficult to do with on-off controls. Suppose we were trying to maintain a supply air temperature of 60°F from a cooling system. If we tried to do this with a conventional rooftop air conditioner, when the air conditioner was off, the supply air temperature would be the same temperature as the return air (i.e., 74°F). The thermostat would turn on the compressor, and the supply air temperature would fall to 54°F, turning the compressor off. The

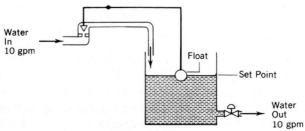

FIGURE 31–4 Level control.

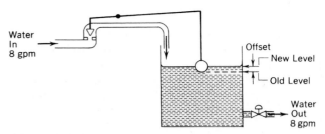

FIGURE 31–5 Level control-change in load.

temperature would then rise once again, very quickly, resulting in unacceptable cycling of the compressor. However, with the controls shown in Figure 31–1, the cooling is being provided by chilled water, and the amount of cooling that is being provided to the air can be modulated (adjusted) by modulating the amount of water that is allowed to flow through the coil.

Although most small commercial and all residential applications will use on-off control, most larger installations will use hot and chilled water for building heating and cooling and will control the building temperatures with modulating controls.

Economizer Control

Figure 31–6 shows an air-conditioning system with an **economizer**. In many large buildings, there are interior areas that have no outside walls. There is heat generated in these areas from lights, people, and equipment that must be removed, even when it is cold outside. Some air-conditioning systems must be available year-round to cool these interior zones. But systems with economizers can take advantage of the cold outside air for "free cooling."

When it is hot outside, the dampers are positioned so that the outside-air and exhaust-air dampers are mostly closed, while the return-air damper is wide open. This allows only the minimum required amount of outside air into the building (too much outside air would dramatically increase the air-conditioning load).

When outside-air temperature drops, at some point it reaches a condition where its **enthalpy** (heat content) is lower than the enthalpy of the return air. This is sensed by enthalpy sensors in both the outside-air duct and the return air duct, and their signals are sent to the computer. When the outside-air temperature is sufficiently low, the computer sends an output signal to the damper motors. The return-air damper will close, while the exhaust-air and outside-air dampers open wide. This allows the use of 100% outside air.

As the outside-air temperature continues to fall, the amount of cooling required from the chilled

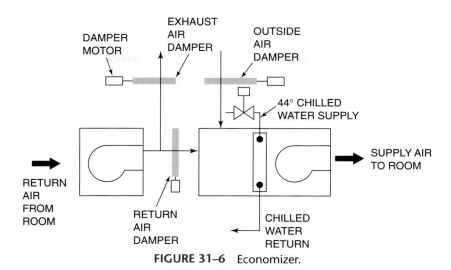

DAMPER
MOTOR

EXHAUST
AIR
DAMPER

OUTSIDE
AIR
DAMPER

44° CHILLED
WATER SUPPLY

RETURN
AIR
FROM
ROOM

RETURN
AIR
DAMPER

CHILLED
WATER
RETURN

SUPPLY AIR
TO ROOM

FIGURE 31–6 Economizer.

water system also is reduced, until there is no need for mechanical cooling (all the cooling is being provided by the outside air). If the temperature of the outside air continues to fall, the computer will signal the outside-air and return-air dampers to begin closing, while the return-air damper opens slightly. As the outside-air temperature continues to fall, the quantity of outside air continues to be reduced, using only as much outside air as is necessary to satisfy the room cooling demand.

Load Shedding

The largest single component of cost for an electric utility is not the cost of fuel to generate the electricity. In most cases, the largest single cost is the debt repayment on the loan that was taken by the utility company to finance the construction of the electrical generating station. Unfortunately for the electric company, they must build (and finance) enough generating capacity to satisfy the peak electrical demand, but most of the time (when it is not exceptionally hot outside), some of the electrical generating capacity sits idle because it is unneeded. Therefore, electric utilities have devised two methods of attempting to level their load.

1. **TOU rates** Time-of-use rates charge customers different rates per kilowatt of electricity consumed, based on the time of day when it was consumed. Some rate schedules provide electricity at 3:00 a.m. (when demand is typically very low) for 10% to 30% of the price of the same quantity of electricity consumed at 3:00 p.m.

2. **Demand charges** These are imposed on customers *in addition to* the electrical consumption charges. For example, the utility company might charge $5.00 per kilowatt based on the highest rate of kilowatt usage by the customer during the previous month.

Load shedding is a strategy whereby the building automation senses the total electrical consumption at each moment and compares it with the previous highest rate of consumption for the month. If it appears that the building is on its way toward setting a new peak, certain nonessential equipment will be turned off automatically to avoid setting a new peak demand. The computer in the building automation system has been previously programmed with all the decisions about which equipment will be shut off, when, and for how long. With the computer control, the load shedding strategy can actually become quite sophisticated. For example, instead of a simple "rolling shutdown" of preselected equipment, the computer can be programmed to change the room temperature set points for the period where there is risk of setting a new demand.

Night Purge

In the typical system without night purge, the air-conditioning equipment is started an hour or two before the occupants arrive. At the time when the occupants are expected to be gone, the night setback temperature begins. Typically, the temperature of the building will rise during the night, so it might be 80°F when the equipment is started again the next morning. This is because much of the heat that started into the building during the afternoon takes that long to actually make its way into the occupied space.

With a night purge system, outside air is introduced into the building during the evening. This has two beneficial effects.

1. The building will be cooler just before the cooling equipment must be started. This will allow for a later equipment start time and will require removal of less heat from the building before the occupants arrive (the "pull-down" period).
2. The quality of the indoor air is improved because the air contaminants that collected in the building during the day will be purged to outside.

Of course, the building automation system senses if the outside air is precooling the building too much. This would be undesirable, because then the building owner would incur additional heating costs. However, a properly set up building automation will allow the introduction of only just the right amount of air.

Chilled Water Reset

The typical air-conditioning system in a large building consists of a chiller that produces water at 44°F and air handlers that use that 44°F water to cool the room air (Figure 31–7). However, during many times of the year when peak cooling is not required by the air handlers, it would be desirable to reset the chilled water supply temperature to higher than 44°F. A chiller can produce a ton of cooling with 48°F leaving water temperature with much less electricity than a ton of cooling with 44°F leaving chilled water. Each degree of chilled water reset reduces the chiller energy costs by about 1 to $1\frac{1}{2}$%. This may not sound like much, but it represents a potentially major reduction in operating costs.

The building automation computer can be programmed to look at all the air handlers and select the one that, at that instant, requires the coolest chilled water to satisfy the room cooling demand.

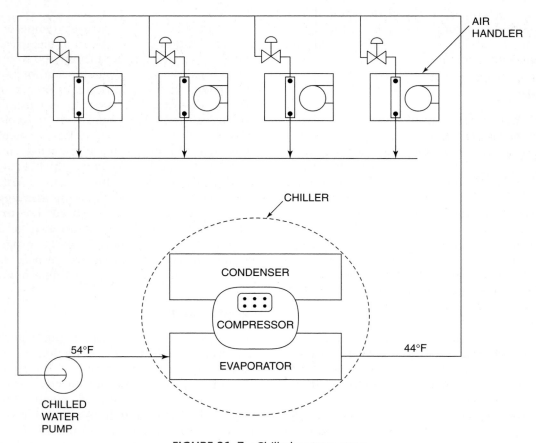

FIGURE 31–7 Chilled water system.

The computer will then send a signal to the chiller, resetting the chilled water supply temperature as high as possible, but not so high that it would create a problem for the zone that is demanding the lowest chilled water temperature.

Condenser Water Reset

At design conditions, condenser water will usually be supplied to a water chiller at about 85°F. Some systems maintain that temperature constant by either cycling the fan on the cooling tower or using a temperature control valve to bypass water around the cooling tower (Figure 31–8). Other systems will allow the condenser water temperature to drift down as it becomes cooler outside. This results in lower operating costs for the chiller due to lower head pressure. However, the actual savings due to lower condenser water temperatures are limited. Chiller efficiency does not continue to improve with each degree of reduction in condenser water temperature. At some temperature, as condenser water temperature (and head pressure) continues to fall, the chiller efficiency actually begins to deteriorate, and energy consumption *increases*.

With a DDC computerized control system, it can be programmed to continuously monitor the condenser water temperature and the electrical consumption of the chiller. The control system can then allow the condenser water temperature to drift only to the point where the best chiller efficiency is achieved.

Fan Cycling

Historically, chilled water/air handler systems have been controlled so that the air handlers operated continuously, while the temperature of the air being supplied to each zone was modulated by modulating the flow of hot or cold water through a heating or cooling coil. However, the amount of air supplied by the air handlers was determined by the design engineers to be the quantity of air needed to maintain comfort conditions when it is approximately 95°F outside. Whenever it is cooler than 95°F outside, there is actually more air being supplied than is necessary. This has led to the development of variable air volume (VAV) systems, in which terminal air units actually modulate the quantity of air being supplied to each zone. They are more energy efficient than constant volume systems. But some of the benefits of variable volume systems can be achieved with constant volume systems using fan cycling.

By continuously measuring room temperature and other variables, it is possible, at times, for the computer to turn off some of the air handler fans for short periods of time, without adversely affecting the comfort of the occupants. Fans may be cycled off for 5 to 15 minutes, at a frequency of every 20 minutes to every hour.

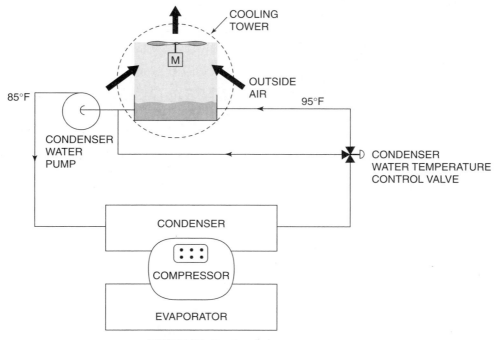

FIGURE 31–8 Condenser water reset.

Another use of fan cycling is to avoid setting a new electrical demand for the electrical billing period (see "Load Shedding"). Some applications only initiate fan cycling as part of the load shedding strategy.

Fire and Smoke Alarm Systems

The fire and smoke control systems in large buildings have two functions:

1. Detecting and reporting alarms
2. Taking action in response to alarms

The fire alarms can include smoke detectors, heat detectors, manual alarms, and sensors to indicate that water is being released from a fire sprinkler system.

Upon activation of any of the preceding alarms, a signal is sent to the computer console. It will respond, as programmed for each situation, to sound audible alarms, display graphically the location of the sensor that has detected a problem, and automatically take over the control of the air systems (to prevent the spread of fire), and even fire doors and elevators.

Lighting Control

Lighting control is usually not within the realm of duties of the heating/air-conditioning/refrigeration person in the maintenance department of a large building. But lighting comprises a significant portion of the heat load in a building. Any steps that can be taken to operate the lights for no longer than required by the occupants will yield energy savings for the cooling system. Lighting is easily controlled through the same building automation system that controls the building environmental systems.

The computer program that operates the light switches has several functions, in addition to turning the lights on and off at preprogrammed times:

1. The occupants (if there are any) must be warned before the lights automatically go out. This is an obvious safety issue. Commonly, the computer controls the lighting control panels to momentarily shut off the lights 10 minutes prior to actually shutting off the lights for the evening. This will give any remaining occupants a chance to leave the building while the lights are still on.
2. A means may be provided for occupant override of the programmed operation. If somebody is working later than usual and doesn't want to leave the building when the warning of shutdown comes, there must be a means for the occupant to have the lights left on. There are three ways that this is usually accomplished. The simplest is to merely provide switches that the occupant can operate to tell the computer not to turn off the lights for another hour. A second method is to have motion detectors that sense whether there are occupants remaining in the area where the lights are scheduled to be turned off. The computer can then automatically skip that area until the occupants have left. A third method of the occupant advising the computer that the lights must be left on is through the use of the occupant's telephone that is tied into the computer. The computer can also keep a record of which occupants are staying late (for security purposes) by requiring the input of a security code that is unique for each occupant.

SUMMARY

Direct-digital control for commercial buildings has been with us for some time. Advancements are being made in the use of electronics in the control of home air-conditioning systems. It is important that you stay abreast of these advances. Continually seek out information on new systems. Continue your education in HVAC/R systems. Many changes are taking place and they are changing rapidly.

REVIEW QUESTIONS

1. DDC in this textbook stands for:
   a. Double-direct current.
   b. Direct-digital control.
   c. Dual-digital concepts.
   d. Directing-double concepts.
2. In the circuit of Figure 31–1, a sensor converts supply air temperature to a voltage.
   T_____   F_____
3. In the circuit of Figure 31–1, the converter receives the voltage from the sensor along with other voltage inputs and converts them to a digital signal.
   T_____   F_____

4. In the circuit of Figure 31–1, the computer receives the digital signal from the converter. The computer uses the information from the converter to produce a digital output that represents what the program says the system should do at this time.

   T_____   F_____

5. In the circuit of Figure 31–1, the digital voltage is converted into a proper signal for the control of the electric valve.

   T_____   F_____

6. In the circuit of Figure 31–2, a converter has been added that converts the output voltage of the "Convert to voltage" to a pneumatic signal.

   T_____   F_____

7. Economizer control using free cooling refers to using outside air when it is cool enough to lower the temperature in given rooms.

   T_____   F_____

8. In economizer control, dampers are used to control the mixture of inside air with outside air providing the proper air mixture and room temperature.

   T_____   F_____

9. TOU rates usually provide for a higher rate at 3:00 a.m. than at 3:00 p.m.

   T_____   F_____

10. Give the benefits of a night-purge system.

GLOSSARY

A Abbreviation for ampere.

AC (ac) Abbreviation for alternating current. Also sometimes used as an abbreviation for air conditioning.

Air over (AO) motor A motor designed to drive a fan, and be cooled by the stream of moving air.

Alternating (AC) current An electric current that reverses its direction at a fixed frequency.

Alternating current An electric current that "alternates." Its flow is considered to change direction with each half cycle.

Alternator A device that produces alternating current by rotating magnets past stationary coils of wire.

Ammeter An instrument used to measure the rate of current flow in a circuit.

Amp An abbreviation for *ampere.*

Amp hour An abbreviation for ampere hour.

Ampacity Current-carrying capacity expressed in amperes.

Amperage A term referring to the total amount of current flowing in a circuit.

Ampere The unit of measure for the rate of flow of electrons. This value is obtained by dividing circuit voltage by its resistance in ohms.

Amplifier An electronic device used to control the amplitude of a signal, either voltage or current.

Analog meter A meter that uses a moving pointer instead of a digital display.

Anticipator A small resistor in a room thermostat that heats up the bimetal element slightly faster than it would be heated by the rising room temperature alone.

Anti-recycle timer A clock/switch combination that prevents a unit from restarting until after it has been off for a predetermined time.

Aquastat Honeywell brand name that is used generally to denote an immersion-type water-temperature-sensing thermostat.

Arc A flow of ions through the air between two electrodes or contact points. Material, usually metal, is transferred from one point to another.

Armature That part of a generator, starter, or motor that rotates inside the unit.

Armored cable Flexible metal conduit.

Atom The smallest particle of matter that has the characteristic chemical properties of an element.

Automatic reset An automatic safety control that will reclose after the abnormal condition that caused it to open returns to a normal value.

Autotransformer A means of reducing motor speed by using different transformer taps to supply different reduced voltages to the motor.

AWG American wire gauge. A system of designating wire size (diameters). The smaller the number, the larger the gauge of the wire.

Back EMF The voltage produced by an inductive load that opposes the applied voltage.

Battery   An electrochemical device for storing mechanical energy in chemical form so that it can be released as electricity when needed.

BHP   Brake horsepower. A measure of the output power of a motor.

Bimetal   A temperature-sensing element that bends when its temperature changes.

Bimetallic   Something made of two dissimilar metals, usually invar and copper, that have been fused together. Changes in temperature cause the bimetallic strip to change shape (bend).

Bin switch   Senses when the ice maker bin is full, and turns off the refrigeration.

Bin thermostat   A thermostatically controlled bin switch.

Bleed resistor   A resistor connected across the terminals of a capacitor to slowly dissipate the stored electrical charge.

Blower relay   A control relay in a furnace that controls the operation of the furnace fan.

Bonnet   The area of a heat exchanger in a furnace through which the room air passes.

Bonnet switch   A thermostat that senses bonnet temperature and closes when the temperature climbs to approximately

Branch   One of the paths for current flow in a parallel circuit.

Breaker   A circuit bleaker.

Brush   A conductor arranged so as to make contact with a rotating surface, such as with the armature of a motor or the slip ring of an alternator.

Brush holder   A device used in motors and generators to hold the brushes in position on the commutator.

Brushes   The stationary contact for bringing power to the rotating contact surface on the rotor of a brush-type motor.

BX   Type of flexible metal conduit. Also called armored cable.

Cable   Insulated wire, usually stranded, used for conducting electrical current.

Cad cell   A device whose resistance decreases when it senses light.

Calibrate   The adjustment of a device so that it actually operates at the temperature that is indicated on the device.

Capacitance   The electrical storage capacity of a capacitor.

Capacitor   A device that can store electrical energy and is used in motor circuits to increase starting torque or improve motor efficiency.

Capacity   The quantity of available electricity that can be delivered from a system or device, such as ampere-hours from a battery.

Carbon pile   A series of carbon discs held together loosely by an insulated holder; generally found in a battery load tester. The tighter the discs are pressed together, the less the resistance to current flow.

Casing ground   A ground wire attached to the casing of equipment to prevent the possibility of a shock resulting from a ground fault.

Cell   A combination of two dissimilar metals in an electrolyte. Two or more cells make a battery.

Celsius (C)   A temperature scale used in the metric system. The freezing point of water at sea level atmospheric pressure is 0°C and the boiling point 100°C.

Centigrade   A term often used for the Celsius scale of temperature measurement.

Centrifugal force   That force that tends to impel an object or parts of a object outward from a center of rotation.

Centrifugal short   Shorted windings of an armature or rotor that only occur when centrifugal pressure builds up while the unit is rotating.

Charging rate   The amount of amperage flowing from the generator or alternator to charge a battery.

Circuit ground   The portion of a circuit used as a common potential point.

Clockwise (cw)   A rotation in the same direction as the hands of an analog clock.

Closed   Not open. Generally refers to a set of contacts in a switch or relay that are connected.

Closed circuit   A completed circuit. A circuit in which there is no interruption of current flow.

Cold control   Jargon for the line-voltage thermostat in a small refrigeration system.

Color code   The system of color used for wires and some connectors in electrical circuits to aid and facilitate the tracing and troubleshooting of electrical system and subsystem problems. Also, the system of colored bands on electrical components that indicate each component's electrical value.

Combination gas valve   Includes main valve, pilot valve, and pressure regulator in one unit.

Common   (a) A terminal that is a part of two or more circuits; (b) The neutral side of a 115-V system; (c) Another term used for *ground*.

Commutator   The part of a generator or motor armature that consists of copper bars. Brushes "ride" on the commutator to collect or supply current to the armature windings.

Complete circuit A continuous path for electrons to flow from a higher potential to a lower potential.

Condenser A term often used, though not properly, for capacitor. A condenser is actually the part of the air-conditioning system that dissipates heat collected in the evaporator.

Conductor A path for electrical current; any material that allows electrons to flow.

Conduit Tubing that is used to enclose and protect electrical wiring.

Contactor Heavy-duty switch or switches closed by a coil in a different circuit.

Continuity A completed circuit or device that has no electrical interruption.

Control circuit The low-amperage circuit that includes the switches and coils for contactors and relays, but not the large motors that they may control.

Control relay Light-duty switch that is operated by a coil in a different circuit.

Cooling anticipator A resistor in a cooling thermostat that causes the compressor to be energized sooner.

Coulomb A quantity of electrons of 6,250,000,000,000,000,000 (6.25×10^{18}).

Counterclockwise (ccw) Rotation in a direction opposite the direction of the hands of an analog clock.

Current limit relay A relay that protects a circuit from overload by opening its contacts and breaking the circuit when the current flow reaches a predetermined high.

Current relay A start relay used on compressors that senses the motor speed by sensing the current through the run winding.

Current The flow of electrons in a circuit.

Cut-in The pressure or temperature at which the switch in an automatic pressure or temperature control will close.

Cut-out The pressure or temperature at which the switch in an automatic pressure or temperature control will open.

Cycle A complete event from start to finish.

Cycling A motor repeatedly turning on and off.

DC (dc) Abbreviation for *direct current*.

DDC (direct digital control) A computer-based means of programming the operation of operating controls.

De-energized A term given an electrical control device, such as a relay, when no power is applied to the coil.

Defrost termination switch A thermostat that senses evaporator temperature and opens when the temperature rises to approximately 45° F.

Defrost timer A clock/switch combination that initiates and terminates the defrost cycle in a freezer.

Delay relay See *Time-delay relay*.

Delta A method of connecting the three coils of a three-phase system. (The end of the first coil is connected to the beginning of the second coil; the end of the second coil is connected to the beginning of the third coil; the end of the third coil is connected to the beginning of the first coil.)

Demand charge A charge made by the electric utility based on the peak usage of electricity.

Density Quantity per unit of volume; compactness of matter.

Dielectric A nonconductor of electricity, such as an insulator.

Differential The difference between the cut-in and cut-out settings on a controller.

Diode A device that permits current to flow in one direction only; a kind of an electrical one-way check valve.

Diode, zener See *Zener diode*.

Direct current (DC) An electrical current as produced by a generator or battery that flows in one direction only.

Direction of electron flow Electrons flow in a circuit from negative to positive potentials.

Discharge To remove energy, such as to "discharge" a capacitor.

Distilled water Water that is free of impurities by distillation; it is first heated to a steam vapor and then cooled back to liquid.

Domestic hot water Hot water for use in sinks, baths, and showers.

Door switch A momentary switch that is operated by the opening or closing of a door.

Double-shaft motor A motor with a long shaft that extends out of both ends of the motor.

DPDT Double-pole double throw (switch).

DPST Double-pole single throw (switch).

Draw The amount of current required to operate an electrical device is often referred to as *current draw*.

Dripproof motor A motor constructed with ventilation openings at the bottom so that falling liquid at an angle of 15 degrees from vertical will not enter the motor.

Drop in voltage See *Voltage drop*.

Dry switch A switch that is operated by a device, but is not electrically connected to that device, such as an auxiliary contact on a contactor.

DTS Defrost termination switch.

Duct heater   An electric heating element mounted inside a duct.

Dynamic electricity   Electricity in motion.

Economizer   A mode of operation of an air-conditioning system where cool outside air is used to supplement the mechanical cooling.

Efficiency   The ratio of useful energy output, divided by the energy input.

Electric potential   Electrical pressure (voltage).

Electrical horsepower   A unit of measure of electrical power. One horsepower is equal to 746 *watts*.

Electrical interlock   A scheme where a control relay is energized at the same time as a load. The control relay contacts energize a second load.

Electricity   A form of energy produced by chemical change, induction, heat, light, or friction.

Electrode   One of the conductors of a cell.

Electrolyte   A liquid conducting medium in which the flow of current is accompanied by a movement of material in the form of ions.

Electromagnet   A coil of wire with a soft iron core capable of being magnetized only as long as current is applied to the coil.

Electromotive force (emf)   That voltage (force or pressure) that causes electron movement in an electrical circuit.

Electron   The negatively charged, lightweight, movable particle of an atom.

Electronics   The branch of physics that studies the behavior and applies the effects of the flow of electrons.

Element   A substance that cannot be separated into two or more substances.

EMF (electromotive force)   A scientific term for voltage.

EMT   Galvanized, lightweight conduit.

Energized   A term given an electrical control device, such as a relay, when power is applied to its coil.

Energy   Capacity for performing work. Electrical energy is measured in *watt-hours*.

Enthalpy   Heat content of air.

Evaporator   That part of an air-conditioning system that picks up unwanted heat.

Fahrenheit (F)   A temperature scale used in the English system. The freezing point of water at sea level atmospheric pressure is 32°F and the boiling point is 212°F.

Fan-limit switch   A fan switch and a limit switch in a common enclosure that use a common sensing element to sense bonnet temperature.

Farad   The unit of capacitance.

Field   A term referring to an area under the influence of electrically charged object. See also *Magnetic field*.

Field coil   That part of a motor or generator that is mounted inside and to the frame. It is positioned around the armature.

Field, magnetic   See *Magnetic field*.

Field wiring   Wiring that is done by the installing technician, as compared to factory wiring.

Filament   The resistive element of a bulb that glows to produce light when current is forced through it.

Finishing relay   A control relay that keeps an ice maker operating to the end of the cycle, even after the bin switch has opened.

Fish wire   Tape made of tempered spring steel used to pull wires through conduits and wall openings.

FLA   Full load amps.

Float switch   A switch that opens or closes in response to the liquid level in a tank.

Flow switch   An automatic switch that senses flow (usually of a liquid).

Flux   A substance used in soldering, brazing, and welding that promotes fusibility. See also *Magnetic field*.

Flux density   The quantity of force per unit area. See also *Magnetic field*.

Flux, magnetic   See *Magnetic field*.

Foot   A unit of measure. There are twelve inches to a foot.

Foot-pounds   A measure of torque or turning force; measured by a prony brake.

Force   Energy acting to produce power or acceleration.

Frame size   A set of physical dimensions of motors, as established by NEMA.

Freezestat   Jargon for a thermostat that opens when the sensed temperature drops below 32° F.

Frequency   In alternating current, the number of cycles per second. Frequence is measured in *hertz*.

Fuse   An electrical overload protective device for interrupting a circuit when the current flow exceeds safe limits for a particular circuit. The fuse will melt to open the circuit when overloaded.

Fused disconnect   A main disconnect switch plus fuses in each leg, all in one box.

Gage   A measuring device, such as a wire gage, used to check wire size.

Ganged switches   Two or more switches that are mechanically linked to operate together.

Gauge   A mechanical dial-type measuring device used to check pressure.

Generator   A device that is used to convert mechanical energy into electrical energy.

Germanium   A semiconductor material.

GFCI Ground fault current interrupter   An outlet that disconnects the power supply if it detects an unequal amperage flow in the two legs of a single-phase supply.

Ground   The common in an electric circuit. In power distribution systems, the earth or electrical connections to the earth.

Ground circuit   The wiring or body metal that makes up the ground of an electrical circuit.

Ground wire   A connection to a conducting material that is at zero electrical potential.

Grounded (ground fault)   A failure mode wherein a device has allowed unintended contact from current-carrying wiring to ground.

Grounded circuit   A circuit that is intentionally or un-intentionally shorted to ground.

Growler   An electrical device used to check an armature to determine if its windings are open or shorted.

Hall effect   The change in conduction of a special semiconductor device when the device is in the influence of a magnetic field.

Hard-start kit   Contains a start capacitor and solid-state start relay to give added starting torque to a PSC motor.

Harvest cycle   The part of an ice cuber operation where the cubes are melted out of the mold.

Head pressure control   A control scheme to reduce condenser capacity to prevent the high-side refrigerant pressure from dropping too low.

Heat anticipator   An adjustable resistance in a heating thermostat that causes the switch contacts to open sooner.

Heat sequencer   A time-delay relay used to sequence the operation of several electrical strip heaters.

Heat sink   A metal plate or bracket used to hold a heat-producing component, such as a diode, for the purpose of dissipating heat from the component.

Helix   A term used to describe a spiral and cylindrical shape, such as a wire that was formed around a round core.

Henry   The unit of measure of electrical inductance.

Hertz   The unit of measure for frequency. Standard alternating current frequency in the United States is 60 hertz.

High leg   The leg connected to the terminal of a delta transformer that is opposite the center-grounded leg.

High limit   A safety thermostat that opens in the event that a sensed temperature becomes abnormally high.

High pressure cut-out   A pressure switch that opens on a rise in pressure.

Holding contacts   See *Seal-in circuit.*

Horsepower   A unit of measure for energy. One horsepower is required to lift 550 pounds (249.5 kg) one foot (0.3 m) in one second.

Hot circuit   That portion of a circuit not at ground potential, electrically insulated, and usually at a potential (voltage).

Hot lead   A wire or conductor connected to a high side, not the ground or neutral terminal.

Hot   Jargon for the electrically pressurized leg of a 115-V power supply.

Hot-wire sensor   A device that senses air flow. A *hot wire* senses air flow by changing resistance when cooled by air movement.

HPC   High-pressure cut-out.

Humidistat   A switch that closes to turn on a humidifier when the sensed humidity becomes too low.

Hunting   A control problem characterized by repeated wide deviations above and below setpoint of a controlled device.

Hydrogen   See *Hydrogen gas.*

Hydrogen gas   The lightest and most explosive of all gases. Hydrogen (H) gas is emitted from a battery during charging and mandates that certain safety precautions be observed.

Hydrometer   A device used to measure the specific gravity of the electrolyte in a battery.

Identified terminal   The terminal on a capacitor that is marked to show that it is the terminal that is connected to the plate nearer to the outside casing.

Impedence relay   Lock-out relay.

Inch   An English standard measure. There are twelve inches in one foot. One inch (1 in.) is equal to 25.4 millimeters (25.4 mm) in metric measure.

Inducer fan   A small fan that helps pull the products of combustion through the heat exchanger in a furnace.

Induction   The current induced in a conductor when the conductor cuts across a magnetic field or the magnetic field moves across the conductor.

Induction motor   A motor in which a current is induced in the rotor.

Inertia   The tendency of a body at rest to remain at rest and a body in motion to remain in motion.

Insulate   To bar the flow of electrical current by the use of a nonconducting material.

Insulation   Nonconducting materials used to prevent the passage of electrical current, such as the plastic covering on wire.

Insulation tape   A plastic like tape used to insulate bare or exposed wire.

Insulator   A nonconducting material used to prevent the passage of electrical current.

Interlock   A scheme where when one device is energized, it energizes a second device.

Ion   A positively charged atom

J-box   Jargon for junction box. An enclosure where mechanical connections of wiring is made.

Junction   A connection of two or more wires.

Junction block   An accessible block for the connection of two or more wires for the convenience of checking electrical circuits.

Kilo (K)   A prefix used in the metric system to indicate 1000. For example, a kilogram is the same as 1000 grams.

Kilometer   A metric measure used for distance. One mile, for example, is equal to 1.609 kilometers (1.609 km).

Kilowatt-hour   A quantity of energy equal to a rate of one kW being consumed for one hour.

Klixon   A trademark brand name, commonly used as a generic term for a line-voltage overload.

kWh   Kilowatt-hour.

Lamination   Thin plates of metal placed next to each other to form the core of a coil or other electrical device.

Lead   A malleable metal, bluish gray in color. The chemical symbol for lead is Pb.

Lead peroxide   A combination of lead and oxygen, chocolate brown in color. As found in a storage battery, the chemical symbol is PbO_2.

LED   Light emitting diode.

Left-hand rule   A method used to find the direction of electron flow or the direction of the lines of force around a current-carrying wire. See also *Right-hand rule*.

Level switch   See *Float switch*.

Light-emitting diode   An electronic component that converts electrical energy to light energy.

Limit control   A limit switch or control is designed to open if the temperature at the sensing point is above a predetermined limit.

Limit switch   A safety switch (usually a thermostat) that will open if the sensed temperature or pressure rises above a maximum safe value, or drops below a minimum safe value.

Line drop   The voltage loss in conductors in a circuit due to their resistance.

Line-duty   A switch designed to carry the heavy load of a compressor or other load.

Lines of force   See *Magnetic field*.

Load   An electrical device connected into a circuit that draws current.

Loadstone   A natural magnetic iron ore that attracts and holds pieces of iron or steel.

Lock-out   A condition that occurs when an electronic ignition system fails to sense a flame, and the control box will not energize anything until it is reset.

Lock-out relay   A control scheme using a control relay that locks out the operation of a device until the circuit is de-energized to reset itself.

Low limit   A safety thermostat that opens if the sensed temperature becomes too cold.

Low pressure control   An automatic switch that opens on a decrease in pressure.

Low-voltage thermostat   A thermostat whose switch is designed to operate a 24-V load.

LPC   Low pressure control.

LRA   Locked rotor amps.

LSV   Liquid solenoid valve.

Magnet   A ferrous material, usually steel, with a north and south pole, having the property of attracting other ferrous materials.

Magnetic   A material having the properties of being attracted or magnetized by a magnet.

Magnetic field   The area around a magnet where magnetic lines of force occur.

Magnetic flux   See *Magnetic field*.

Magnetic lines of force   See *Magnetic field*.

Magnetic poles   The north (N) and south (S) ends of a magnet.

Main winding   Same as *Run winding*.

Make   A term used to indicate that an electrical control device, such as a switch or a set of points, is in the ON or CLOSED position.

Manual reset   An automatic safety control that will remain open even after the abnormal condition that caused it to open returns to a normal value.

Matter   That which occupies space and has weight. All things are made of matter.

Mechanical interlock   A scheme where a device is energized, and the operation of that device mechanically operates a switch that energizes a second device.

Mega (M) A prefix indicating one million (1,000,000).

Megohm meter An ohmmeter with a very high scale, used to sense the impending breakdown of insulation. One megohm equals one million ohms.

Meter A term applied to an electrical measuring device, such as a voltmeter or an ammeter.

Mica A nonmagnetic nonconducting mineral material. Mica is often found as an insulator between riser bars of a commutator.

Micro A prefix indicating one millionth (1/1,000,000).

Microfarad The commonly used measure of capacitance, equal to a millionth of a farad.

Microprocessor A small computer using an integrated circuit.

Microswitch A small momentary switch used to sense the physical position of a device.

Mile An English measure of distance. One mile is equal to 5280 feet (1610.4 meters) or 0.622 km.

Milli A prefix indicating one thousandth (1/1,000).

Millimeter A metric measure for length. One inch (1 in.), for example, is equal to 25.4 millimeters (25.4 mm).

Momentary switch A switch that opens (or closes) when it is pressed, but then returns to its normal position when it is released.

Motor A rotating electromagnetic device used to convert electrical energy into mechanical energy.

Multiplex Using the same wire to carry more than one piece of information.

NEC National Electric Code.

Negative The electrical potential of a object that has gained some extra electrons, said to be negatively charged.

Negative pole That end of a magnet that would naturally orient itself with the magnetic south pole.

NEMA National Electrical Manufacturers Association.

Neutral (a) An object with an equal number of positive and negative particles is neutral. (b) The part of a 115-V circuit that carries current, but is at zero electrical pressure.

Neutron A tightly bound, collapsed combination of an electron and a proton.

Nichrome A type of wire that is used as the element in a resistance heater.

Nichrome wire A wire that has a high resistance. When a current passes through it, it becomes hot.

Normally closed A coil-operated switch (or valve) that is closed when the coil is not energized.

Normally open A coil-operated switch (or valve) that is open when the coil is not energized.

Nucleus The central portion of an atom that contains protons and neutrons.

Off See *Open*.

Ohm A unit of electrical resistance that opposes the flow of current. The symbol for ohm is Ω.

Ohmmeter An electrical instrument used to measure resistance.

Ohms The measurement of resistance to the flow of electricity.

Ohm's law The law of electricity determining the relationship between voltage, current, and resistance. Simply stated, voltage equals current times resistance $E = (I \times R)$.

On See *Closed*.

OPC Oil pressure control.

Open A break, intentionally or unintentionally, in an electrical circuit. A circuit is intentionally opened with a switch or a set of contact points. No current will flow in an open circuit.

Open circuit See *Open*.

Open winding A break in the winding of a field, armature, or coil that prevents the flow of current.

Operating control The thermostat or pressurestat that normally turns a compressor or other device on and off to maintain the setpoint.

Oscillate To move or swing freely back and forth in a motion similar to that of a clock pendulum.

Oscillograph A device used to present a display of electrical signals graphically through mechanical means.

Oscilloscope A device used to present a display of electrical signals graphically on a cathode-ray tube.

Overload (a) A condition where a device carries more current than it is rated to carry; (b) An automatic switch that senses current and opens if the current exceeds a preset rating.

Parallel circuit Two or more circuits, electrically side by side, the positive and negative terminals of which are connected to their respective positive and negative sources.

Permanent magnet A magnet made of tempered steel that has the ability to hold its magnetism for long periods.

Permeability The ability of a material to conduct magnetic lines of force.

Phase (a) One of the legs of a power supply. (b) The alignment (in time) of the voltage and the amperage in a wire.

Photocell A semiconductor device that produces a voltage when exposed to light.

Photoresistor A semiconductor device that changes resistance when exposed to light.

PID Proportional plus integral plus derivative control.

Piezoelectric The property of crystal material to produce a voltage when subjected to pressure.

Pilot-duty switch A light-duty switch that operates a low-amp draw device such as a contactor.

Pilot generator A series of thermocouples that is heated by a pilot flame, and produces 250–1000 mV to operate the gas valve.

Pilot light A lightbulb that is wired in parallel with a load to indicate when that load is energized.

Pilot safety switch A switch that senses the high temperature produced by a pilot flame in a furnace. If there is no flame, the switch opens.

Polarity Refers to pole identification, such as N–S poles of a magnet or + and − poles of a battery.

Polarize To give polarity, temporary or permanent, to a conductor or ferrous material.

Pole Either end of a magnet or of an electrical potential. See *Polarity*.

Pole piece See *Pole shoe*.

Pole shoe That part of a rotating device, such as a motor, that is used to hold the field coils in their proper position around the armature.

Poles The number of moving electrical switches that are operated by a coil in a contactor or control relay.

Positive Any atom or atoms with less than the normal number of electrons.

Positive pole That part of a magnet that tends to orient itself with the magnetic north pole.

Potential Electrical force, measured in volts. Potential may be positive or negative.

Potential relay A start relay used on compressors that senses the motor speed by sensing the amount of back EMF being generated by the start winding.

Potentiometer A variable resistor that may be adjusted manually.

Power The rate of doing work, usually expressed in horsepower.

Power factor A measure of lag between the time when peak voltage is reached and the time when peak current flow occurs.

Powerpile See *Pilot generator*.

Pressure switch An electrical switch device that is actuated by pressure or a change in pressure.

Pressurestat An automatic switch that opens or closes in response to a change in pressure.

Primary control The main controller on an oil burner.

Primary winding The winding of a transformer that is connected to electric power.

Printed circuit A circuit on which the wiring has been etched into a board.

Prony brake A device used to measure "foot pounds" of "torque."

Proportional control The modulating control mode where the corrective output signal is proportional to how far the actual sensed temperature (or pressure) is different from the setpoint value.

Proton A positively charged particle found in the nucleus of all atoms.

PSC motor Permanent split-capacitor; a type of split-phase motor that uses a run capacitor, but no start relay.

PTC device Positive temperature coefficient device, whose resistance changes with its temperature. It is used in the starting circuits of PSC motors.

Pumpdown A refrigeration control scheme wherein all refrigerant is pumped out of the evaporator prior to allowing the compressor to shut down.

Purge blower A fan that blows clean air through the heat exchanger of a furnace.

Push button A momentary switch.

RAM Random access memory.

Range The bounds of the available setpoints of a controller.

Rectifier An electrical device used to convert alternating current to direct current. See also *Diode*.

Regulator An electrical device used to control the output of a generator or alternator by controlling the current and voltage.

Relay An electromagnetic switch; a low-current-operated control device used to open and close a high-current circuit.

Reset relay See *Lock-out relay*.

Resistance That property of an electrical circuit that tends to reduce the flow of current. Resistance is given in ohms; see *Ohm*.

Resistance heating Producing heat by passing a current through a resistance wire.

Resistor A resistance device that, when installed in an electrical circuit, provides a reduction in current flow. See also *Rheostat*.

Revolutions per minute (RPM) The number of time a rotating member turns in one minute.

Rheostat A variable resistance device used to regulate the flow of current in a circuit.

Right-hand rule   A method used to determine the polarity of a coil. See also *Left-hand rule*.

Rigid conduit   Steel conduit made to the same dimensions as standard pipe.

RLA   Running load amps.

Roll-out switch   A switch that opens if the flame in a furnace escapes from inside the heat exchanger.

ROM   Read-only memory.

Rotor   That part of an alternator that rotates. The rotor of a motor is also referred to as an *armature*.

RPM   Abbreviation for *revolutions per minute*.

Run winding   The winding in a split-phase motor that is energized continuously when the motor is running. Also called *Main winding*.

RVS   Reversing valve solenoid.

Safety control   Any switch that does not operate during normal operation of the system, but shuts the system down if some variable gets outside of acceptable limits.

Sail switch   A flow switch that senses the flow of air.

SCR (silicon-controlled rectifier)   An electronic device that opens to allow flow through an anode-cathode when a predetermined current flows through the gate and then shuts off when the flow through the anode-cathode falls.

Seal-in circuit   A circuit that continues to energize a relay coil, even after the switch that initially energized the coil opens.

SealTite   A brand of weatherproof flexible conduit.

Secondary circuit   The circuit that is connected to the secondary of a transformer.

Secondary winding   The output side of a transformer.

Selenium rectifier   A wafer-type rectifier, now seldom used, to convert alternating current to direct current.

Semiconductor   A material that falls midway between good conductors and poor conductors. (See *Germanium* and *Silicon*.) Also, the devices, diodes, transistors, control rectifiers, triacs, etc., that are made from semiconductor materials.

Series   See *Series circuit*.

Series circuit   Two or more components connected electrically end to end, so that the positive pole of one connects to the negative pole of the next.

Service factor   The multiple of full-load amps that a motor can draw for short periods of time without damaging the motor.

SF   Service factor.

SFA   Service factor amps.

Shaded pole   An extra pole on a stator winding that is wrapped with a copper ring, causing it to be out of phase with the stator winding.

Shaded pole motor   A motor type characterized by no start winding and used on very small loads.

Short   The grounding, intentionally or unintentionally, of a current-carrying conductor or device.

Short circuit   See *Short*.

Short cycling   Rapid repeated starting and stopping of a motor or other load.

Shorted winding   The grounding of the winding of a field, armature, or coil. The short may also be turn to turn instead of to ground. See also *Centrifugal short* and *Short*.

Shunt   A parallel circuit used to divide, usually in unequal percentages, the amount of current flowing from one point to the other through each circuit.

Silicon diode   See *Diode*.

Silicon   A semiconductor material.

Silicon-controlled rectifier   An electronic device used to control the flow of electricity to a high-amp load.

Slip   The difference between the synchronous speed and the actual speed of a motor.

Slip ring   That part of an alternator rotor that provides a "race" for the positive and negative brushes to make electrical contact with the rotor.

Slippage   The difference between synchronous speed and operating speed. For example, if a two-pole sync speed is 3600 rpm and the operating speed is 3400 rpm, the slippage is 200 rpm.

Slow blow fuse   A fuse that can carry four to six times alts rated current for a few seconds.

Sneak circuit   A circuit that provides for the operation of a component when operation is not wanted.

Solenoid   A coil around a movable magnetic core. The core becomes magnetized and moves when the coil is electrically energized.

Solenoid valve   A valve that is operated by a solenoid coil.

Solid state device   An electronic component that is built from semiconductor materials.

Solid state relay   A start relay used on compressors that allows the Start winding to remain in the circuit for a predetermined time.

Solid state   A term used to describe an electrical device that operates without mechanical function.

Spark   A flow of electricity (electrons) through the air between two electrodes or contact points.

SPDT   Single-pole double throw (switch).

Specific gravity The weight of a solution with reference to water (H_2O), which has an assigned value of 1.0.

Splice The joining of two or more conductors at single point by use of fasteners or other devices.

Split phase motor A motor that has a start winding and a run winding.

SPST Single pole single throw (switch).

Stack control A primary control that incorporates an integral stack switch.

Stack switch A switch that senses flue stack temperature in an oil-fired system to determine if flame has been established.

Start winding The winding in a split-phase motor that is energized for only a few seconds on start-up.

Starter A combination of a contactor and an overload for starting a motor.

Starting torque A measure of the ability of a motor to start up against a mechanical resistance to the rotation of the motor.

Static electricity A charge of electricity that remains at rest until a suitable path is provided for its discharge.

Stator That part of the alternator that is stationary around the rotor. Referred to as a field in motors.

Step-down transformer A transformer whose output voltage is lower than the input voltage.

Step-up transformer A transformer whose output voltage is higher than the input voltage.

Stinger leg See *High leg*.

Stratification The "layering" of air caused by buoyant warm air collecting near the top of a room.

Strip heater Jargon for an electric resistance heater.

Strobe light A term often used for *stroboscope*.

Stroboscope An electrical neon lamp used to precisely time the ignition of an automobile engine.

Subbase A wallplate that contains the terminals for thermostat wires and accepts the mounting of a thermostat.

Substitute To replace a part suspected of a defect with one of known quality.

Sulfating The formation of the inactive coating of lead (Pb) on the surface of battery plates due to inactivity or nonuse.

Sulfuric acid A heavy, corrosive, high-boiling, liquid acid that is colorless. Mixed with distilled water, it forms the electrolyte used in storage batteries.

Switch A mechanical device used to open or close a circuit.

Synchronize To cause two or more events to occur at the proper time with respect to each other.

Synchronous motor A motor that rotates at the same speed as the rotating magnetic field in a motor.

Synchronous speed The rotational speed of the magnetic field in a motor.

Tachometer An instrument used to measure the revolutions per minute of a rotating member, such as an engine.

Temperature switch An electrical switch device that is activated by temperature or a change in temperature.

Terminal A device attached to the end of a wire, cable, or device used to make an electric connection.

Thermal fuse A one-time thermostat that opens when it is subjected to a temperature higher than its temperature rating.

Thermal overload A thermostat that senses motor temperature and opens if it gets too hot.

Thermistor A resistor with the characteristic of changing its value, in ohms, with a change of temperature. See also *Resistor*.

Thermocouple A junction of dissimilar metals that produces a voltage that is proportional to temperature.

Thermopilot relay A brand name to describe a pilot safety switch that is held closed by a hot thermocouple.

Thermostat A heat-sensitive mechanical device found in the cooling system to regulate engine coolant flow; an electrical switch actuated by heat.

Thinwall conduit Galvanized, lightweight conduit. Also called electro-mechanical tubing, abbreviated EMT.

Three-phase The electrical power system in which voltages are generated 120 electrical degrees out of phase with each other.

Throws The number of electrical positions available to a pole on a switch.

Time-delay fan switch An automatic switch that closes (to turn on a furnace blower) motor one minute after a 24-V heater is energized.

Time-delay fuse A fuse that can carry four to six times its rated current for a few seconds.

Time-delay relay A relay with a timed circuit so that the points do not open (or close, depending on application) until a predetermined time interval after power is available.

Tolerance An allowable or permitted variation in dimensions.

Torque A twisting or turning effort measured in foot-pounds (ft-lb) or inch-pounds (in-lb).

Totally enclosed air over (TEAO) motor   A motor constructed with no ventilation openings that requires external airflow over the motor enclosure for cooling.

Totally enclosed fan cooled (TEFC) motor   A motor constructed with no ventilation openings that includes an external fan with a shroud to move cooling air over the motor.

Totally enclosed non-ventilated (TENV) motor   A motor constructed with no ventilation openings.

TOU rate   Time of use. Electrical consumption prices are based on the time of day when it is used.

Transducer   A device that changes one form of energy to another.

Transformer   Nonmechanical devices used to step up or step down AC voltages.

Transistor   A solid-state semiconductor that is a current amplifier or switch.

Triac   An electronic component that controls the AC voltage that is allowed to pass to a load.

Trickle heat circuit   A small current that is allowed to pass through compressor motor windings to act as a crankcase heater during the off-cycle.

Undercut   To cut away mica from between the commutator bars of an armature.

Undercutter   A device (tool) used to undercut armatures.

Unloader solenoid   A solenoid on a compressor that operates to disable one or more cylinders.

VA rating   For transformers, the product of the output voltage times the maximum amps that can be supplied by the secondary winding.

Vacuum   A pressure that is below that of atmospheric pressure.

VAV   Variable air volume.

Volt   A unit of measure of electromotive force.

Voltage applied   The actual voltage read on a voltmeter at a given point in a circuit.

Voltage available   The voltage delivered by a power supply.

Voltage drop   The net difference in electrical pressure when measured on both sides of resistance (load). The loss of voltage caused by circuit resistance.

Voltage   The electrical pressure that causes current to flow in a closed circuit.

Voltage regulator   See *Regulator*.

Voltmeter   An electrical instrument used to measure voltage.

VOM   Volt-ohmmeter.

Watt   A unit of measure for electrical power. Amperage times voltage equals wattage.

Watt-hour   A unit of measure for electrical energy calculated by multiplying watts by hours.

Wattmeter   An electrical instrument used to measure watts.

Waveform   A graphical presentation of an electrical signal. Waveforms are normally seen on an oscilloscope. They are sometimes presented on an oscillograph.

Wild leg   See *High leg*.

Wye   A method of connecting the three coils of a three-phase system. (The ends of the three coils are connected together and the beginnings are brought out.)

Zener diode   A diode that conducts current in the normal manner in the forward direction but breaks down at a predetermined voltage in the reverse direction. Its main use is as a voltage regulator.

Zone valve   A hot-water valve on a boiler system that controls the flow of water to a particular zone.

Zone   An area of a building that is controlled by its own thermostat.

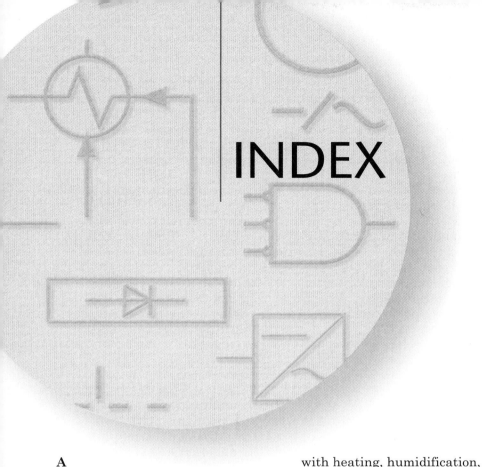

INDEX